# MANUFACTURING ENGINEERING

# MANUFACTURING ENGINEERING
## Principles for Optimization
### Second Edition

**Daniel T. Koenig**, P.E.

*Steinway & Sons*
*Long Island City, New York*

Taylor & Francis
*Publishers since 1798*

| USA | Publishing Office: | Taylor & Francis<br>1101 Vermont Avenue, N.W., Suite 200<br>Washington, DC 20005-3521<br>Tel: (202) 289-2174<br>Fax: (202) 289-3665 |
| --- | --- | --- |
| | Distribution Center: | Taylor & Francis<br>1900 Frost Road, Suite 101<br>Bristol, PA 19007-1598<br>Tel: (215) 785-5800<br>Fax: (215) 785-5515 |
| UK | | Taylor & Francis Ltd.<br>4 John St.<br>London WC1N 2ET<br>Tel: 071 405 2237<br>Fax: 071 831 2035 |

**MANUFACTURING ENGINEERING: Principles for Optimization, Second Edition**

1 2 3 4 5 6 7 8 9 0   B R B R   9 8 7 6 5 4

This book was set in Times Roman by Edwards Brothers, Inc. The editors were Ellen K. Grover and Mary Prescott; the production supervisor was Peggy M. Rote. Cover design by Michelle Fleitz.
Printing and binding by Braun-Brumfield, Inc.

A CIP catalog record for this book is available from the British Library.
⊛The paper in this publication meets the requirements of the ANSI Standard Z39.48-1984 (Permanence of Paper)

**Library of Congress Cataloging-in-Publication Data**

Koenig, Daniel T.
  Manufacturing engineering : principles for optimization / Daniel T. Koenig.—2nd ed.
    p. cm.
  Includes bibliographical references (p. 415) and index.

  1. Production engineering. I. Title.
TS176.K625   1994
671—dc20                                                                94-18098
                                                                              CIP

ISBN 1-56032-301-9

To Marilyn, Alan, and Michael
  for their patience and encouragement during
  the preparation of the manuscript.

# CONTENTS

# PREFACE TO THE SECOND EDITION

The manufacturing engineering field is one of the most dynamic of all the sub-specialties of engineering. Since I wrote the original manuscript for the first edition only eight years ago we have seen considerable change. Computer-Integrated Manufacturing (CIM) is now common practice, as is Manufacturing Resources Planning (MRP II) and Just In Time (JIT). All three of these topics were covered in their more or less primitive form in the first edition. They are developed considerably, with a manufacturing engineering viewpoint, in this edition. We also find industry striving for more teamwork between and within the functions. So we see apparent innovations such as total quality management (TQM) and concurrent engineering have become popular. They have changed management's approaches toward the uses of manufacturing engineering within the business continuum. These topics are also developed fully in this edition.

While manufacturing engineering always had a large cross-functional cooperative component, the past few years have brought this almost intuitive process to the forefront of industrial practice. It is more visible than ever before, and how we practice manufacturing engineering requires an understanding of the discipline and how it relates with this new paradigm. In this book I have expanded those interdisciplinary topics into chapters of their own to show how manufacturing engineering acts as either the core of or the catalyst of MRP II, JIT, TQM, and concurrent engineering. These topics are covered in considerable depth in the second edition in a manner that keeps proper perspective with the base theory of manufacturing engineering. Also, keeping with the format of the first edition, there are ample review questions to allow students to develop a full understanding of the topics.

As expressed in the preface of the first edition, creating a textbook requires willingly given support and advice from many people. I wish I had space to list all of my colleagues in my previous company, General Electric Co., and my current company, Steinway & Sons, who have given their advice and counsel for this project. I have had the opportunity to go from one great company to another and to continue my practice of manufacturing engineering. I continue to learn and apply new ways of improving productivity, and happily give my gratitude to all of those who have given their views. I have also had the privilege to discuss and learn manufacturing engineering with many colleagues through the auspices of the American Society of Mechanical Engineers and the Institute of Industrial Engineers. To those valued colleagues, I give special thanks for having performed the vital function of critical peer review for many of my ideas expressed in both editions of this book. I want to thank all of those who have agreed to express their opinions and help me form mine in the ever expanding field.

Having once written a book on a subject, and then agreeing to review it and add to it for a second edition, knowingly creates a temporary abandonment of family. Recalling the time it takes, and the patience required of loved ones during the process, makes it difficult to agree to do it again. My wife Marilyn's encouragement for me to do so I find to be a great act of personal sacrifice and love. She encouraged me to do so because she understood how important I think it is to have properly educated manufacturing engineers populating our factories. If we are all to enjoy the standard of living modern technology promises, well trained engineers are required; thus this second edition. She knows she has my love and gratitude for having put up with the process once more. As usual she has been my great confidant and counselor in seeing this project through.

*Daniel T. Koenig*
*Trumbull, Connecticut*

# PREFACE TO THE FIRST EDITION

Having practiced manufacturing engineering for roughly two decades, I have found that manufacturing engineers have not done an adequate job in documenting the management procedures and philosophies necessary for success. Thus, engineering students often enter this field with little understanding of how it works, except for the strictly technical or nuts and bolts activities. Unfortunately, this is not enough for a successful career in manufacturing engineering. Manufacturing engineers are constantly in the forefront of integrated corporate programs, so they must be well grounded in the management practices and philosophies of their profession. The purpose of this book is to instruct junior- or senior-level undergraduate and postgraduate engineering students in manufacturing engineering management strategies so that they may optimize future manufacturing processes and procedures.

Like so many engineers of my generation, I began my first manufacturing job with only a vague idea of what I was supposed to do. This book would never have been written had it not been for the many dedicated colleagues at the General Electric Company with whom I learned the trade, for those who had the patience to instruct young engineers and, later not so young, junior managers in the pragmatic as well as theoretical aspects of the job.

I wish I had space here to list all the General Electric engineers and managers who influenced me beneficially over the years: those who encouraged me to implement many of the techniques described in this book, those who counseled me on theory and practicality, and those who helped me develop an understanding of sound management practices. To all my colleagues and former managers I express my appreciation.

Having the idea of writing a book and actually doing it are two different things. The fact that this book has been written is due in large part to my wife

Marilyn, who constantly encouraged me to push on with it even though it meant many hours of solitude for her. I thank her for converting my script into a typed manuscript and for editing it. In addition, I thank my secretary, Doris Dyson, for typing the various revisions as I explored different ways of presenting the material.

*Daniel T. Koenig*

# MANUFACTURING ENGINEERING ORGANIZATION CONCEPTS

## A FABLE: THE COMPANY THAT COULD AND THE COMPANY THAT COULDN'T

Let me tell you a story about the Company That Could and the Company That Couldn't. Both of these companies were vying for a lucrative market for the same new product, which was different from anything that had been made before. The market for this product was enormous, and both companies were very eager to enter it.

The Company That Couldn't specialized in selling. They could sell refrigerators to Eskimos. They did not understand manufacturing well. They were content to let equipment vendors equip their factory and were miserly about providing manufacturing management support. They were convinced that sales were of primary importance and all other functions were secondary. Their philosophy was to emphasize sales, and whatever resources were left over, which were not much, were parceled out to finance, design, and manufacturing.

The Company That Could specialized in nothing. They believed in maintaining a balance of skills within all of their functions. Their salespeople could sell, but not refrigerators to Eskimos. They sold to serve the needs of their customers. Their engineers designed products that their salespeople could sell and their factory could produce. Most important, they tried to determine how best to make their products and did not depend on vendors to equip their plants.

The competition to introduce the new product to the marketplace was fierce. At first, the Company That Couldn't had a substantial lead over the Company that Could. Their order books were full and their factory was swamped with production requirements. In order to produce the new product to meet customer demands, they relied on their traditional method. They asked their equipment vendors to supply them with equipment to do the job. The vendors, exceedingly happy to do so, provided the Company That Couldn't with all sorts of equipment which, they claimed, "was the ultimate in making the new product." Perhaps it was, but the Company That Couldn't failed in making the product. The price to make it was higher than the selling price, and the output did not meet demand. Why? They did not have an understanding of the equipment. It was probably too complex. They also did not have the knowledge to manage the factory. Their management staff did not know how to train their workers to make the product on the vendor-specified and vendor-designed equipment. They also did not match their product design to the equipment they bought. All of this happened because the Company That Couldn't did not pay attention to supporting a balanced organization.

The Company That Could had a much different approach. They found out what the market wanted in the new product. Then they designed the product and presented it to their marketing, finance, and manufacturing departments. Marketing reviewed it for what the customers really wanted in the product. Manufacturing reviewed it for ease of manufacturing and how much it would cost to make. They determined whether existing equipment could do the job, or whether they had to develop something new. They thoroughly explored how they would train their operators to make the product. They determined what the critical design elements were and how many units of the new product they could produce in a given time period. They also estimated the costs for development and implementation. Finance reviewed these costs and stated what the company could afford.

After several iterations of this process, the Company That Could was satisfied that it could successfully enter the market for the new product. And as a result, the Company That Could took the market away from the Company That Couldn't. They were able to produce products that were of better design and higher quality, to produce them at a lower cost, and to deliver them on time. This was possible because the Company That Could had a balanced approach. It worked as a team with all facets of its operations participating. Perhaps even more important, before it tried to sell the new product, it learned how to make it so that it could be sold at a profit.

The moral of this story is that manufacturing engineering is a vital part of a company's success. Learning how to produce at the lowest costs and still meet the constraints of the design and the marketplace is essential for any company. This book is about the techniques and philosophies of manufacturing engineering that make industrial organizations "companies that could."

## THE INDUSTRIAL MATRIX

Let us first look at the makeup of a typical industrial organization to see how manufacturing engineering fits in this matrix. An industrial organization must, at a minimum, consist of marketing, design, finance, and manufacturing. These basic functions can be further refined to another level. For example, design engineering can be broken down into research and development, applications, and product service.

Now let us look at the basic responsibilities of each of the major functions.

Marketing: Define what is sellable within the charter of the organization and then sell it.

Design: Based on what is sellable, design the product in accordance with good scientific and engineering principles and produce the specifications for the product.

Finance: Raise the necessary funds for the organization and dispense them in accordance with recognized fiscal practices. Keep track of all funds to optimize their uses.

Manufacturing: Produce the product at the lowest possible cost, in the shortest possible time, and in such a way that it meets all design specifications.

The four functions are interrelated, of course. Marketing cannot obtain orders for products that manufacturing cannot produce or design cannot engineer. Finance must recognize that it is supporting a manufacturing entity; its policies, for example, cannot be based on principles developed for the successful operation of banks and insurance companies. Design cannot specify products that are beyond the scope of manufacturing to produce. Nor can it ask for nuances that improve the elegance of the design but do little to improve the salability of products. Manufacturing must live within the budgetary levels deemed prudent by finance and must make the product within the specifications required by design. Manufacturing must also deliver finished goods in accordance with the desires of the customers as defined by marketing. Marketing must take into account the effects on manufacturing of all delivery promises made to customers.

## THE MANUFACTURING MATRIX

How does manufacturing engineering fit into this organizational matrix? Manufacturing consists of two major categories: operations and support. It is necessary to understand the duties of these categories to properly understand the role of manufacturing engineering.

Operations is the producing arm of manufacturing. Here management directs the activities of people, machines, and processes in producing the product in accordance with an overall schedule. Operations receives the equipment, instructions, raw materials, and master schedule from the service groups, then applies labor to produce the product.

The service groups consist of materials, quality control (sometimes known as quality assurance), and manufacturing engineering. Their function is to provide direct support to operations in the form of raw materials and equipment to work with and information on how to use both.

## Materials

Materials has the responsibility for producing a master production schedule in accordance with orders received and anticipated by marketing. The materials subfunction, then, controls and monitors the production schedule of operations, the transmission of specification information from design to manufacturing, and the supply of manufacturing instructions from manufacturing engineering to operations. It has the other major function of purchasing raw materials and supplies and ensuring that they arrive on time to support the overall master production schedule while minimizing inventory costs. It normally does all this via management of the Manufacturing Resources Planning (MRP II) program.

## Manufacturing Engineering

Manufacturing engineering has the responsibility for instructing operations in how to make the product, the sequence, and the facilities to use. It also has the overall responsibility for planning the nature of the factory and its present and future equipment.

In addition, manufacturing engineering evaluates capacities per time frame for marketing to use in sales strategies; evaluates manufacturing capabilities for design engineering to use as constraints on product specifications; and evaluates current manufacturing performance for overall monitoring and for modifiers to capacity and capability evaluations.

Manufacturing engineering is responsible for the maintenance of current equipment and the evaluation and purchasing of new equipment. It also provides this service for nonproducing facilities such as buildings, offices, vehicles, and other miscellaneous items.

Another function frequently assigned to manufacturing engineering is that of process control. Measurements of quality are continuously made during a process, usually as part of the traditional system of evaluating productivity performance. This is a relatively recent amalgamation of duties and strives to combine the requirements for high quality with those for improved productivity necessary for corporate survival.

## Quality Control

Quality control has traditionally been the liaison between manufacturing and design. This function interprets design's specifications for manufacturing and develops the quality plan to be integrated into manufacturing engineering's methods and planning instructions to operations. Quality control is also responsible for recommending to management what level of manufacturing losses (cost of mistakes in producing the product) can be tolerated. This is based on the complexity of the product design, specifically the degree of preciseness necessary in tolerances. Quality control traditionally monitors manufacturing losses by setting a negative budget that is not to be exceeded, and establishes routines for measurement and corrective action.

Within the past decade or two, quality control has become increasingly involved with marketing and customers in establishing documentation systems to ensure guaranteed levels of product quality. This new role has led to the new title quality assurance, to differentiate it from traditional in-house quality control.

Quality assurance strives through documentation of performance and characteristics at each stage of manufacture to ensure that the product will perform at the intended level. Whereas quality control is involved directly with manufacturing operations, quality assurance is involved with the customer support responsibilities generally found within the marketing function. Many industrial organizations have chosen to establish an independent quality assurance subfunction within the manufacturing function and have placed the technical responsibilities of quality control, namely process control, within the manufacturing engineering organization.

## MANUFACTURING ENGINEERING RELATIONSHIPS WITH OTHER FUNCTIONS

Manufacturing engineering interacts with the major functions of the industrial organization as well as the subfunctions within manufacturing. Manufacturing engineers are essential in future business planning activities led by marketing, where factory capabilities and know-how on optimizing costs are paramount in any strategy. Their inputs are vital to finance for planning future allocation of funds. And their definition of what is manufacturable and what is not, and the various degrees in between, greatly influences the design function and the type of design specifications produced.

## MANUFACTURING ENGINEERING WITHIN THE MANUFACTURING FUNCTION

Figure 1.1 depicts the organization of the manufacturing function with a specific breakout for manufacturing engineering.

There are several alternative ways to depict the manufacturing engineering organization. For example, there could be a producibility engineering unit under Advanced Manufacturing Engineering (AME) if there is a sufficiently strong need in a particular company's operation. Likewise, some managers choose to have separate methods engineering units if their products demand continuous redesign of workstations and fixturing. Another breakout could be that of a Computer-Integrated Manufacturing (CIM) systems unit with resources derived from AME and from the methods, planning, and work measurement unit. This would be

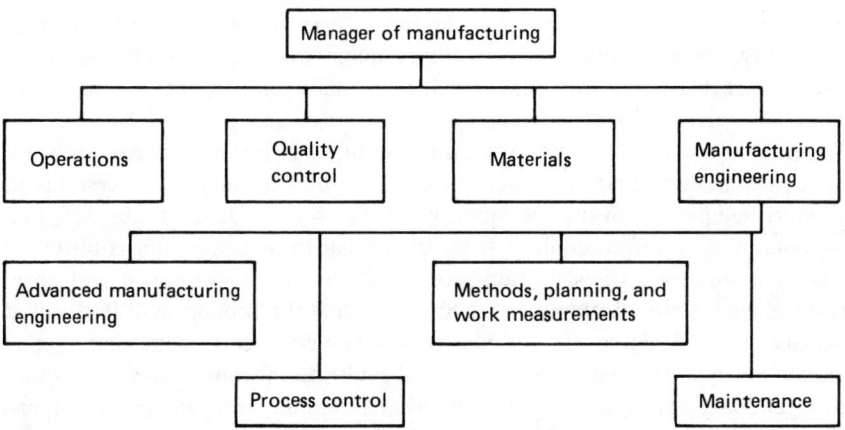

**Figure 1.1** Organization of the manufacturing function with a breakout for manufacturing engineering.

especially useful if the company was just beginning computerization. Process control could very easily be a quality control unit, and in many organizations it is. It is included here as a separate unit because of its inherently technical nature compared to the trend in quality control toward nontechnical activities supporting marketing.

The point here is that manufacturing engineering organizations are not cast in concrete, but can and should be modified to fit the specific needs of individual companies. The organization shown in Fig. 1.1 is the classical manufacturing engineering organization. These units cover all phases of manufacturing engineering's responsibilities. We will now define the charters of the four manufacturing engineering units.

## ADVANCED MANUFACTURING ENGINEERING

The major responsibilities of the advanced manufacturing unit are:

Area planning
Capacity analysis
Capability evaluations
New technology evaluations and needs
Producibility engineering
Computer-integrated manufacturing development
Investment project management
Long-range planning and forecasts

A separate engineer could have each or several of these responsibilities, or several engineers could share one area of responsibility. It matters little what the specific organizational structure looks like as long as all responsibilities are properly attended to. For simplicity, we will discuss each area of responsibility separately.

## Area Planning

This activity determines the present and near-future shape of the factory to meet all needs for facilities. Area planners use the master production schedule of the materials subfunction to determine what the factory must produce. They match the requirements for production with the present capabilities of the factory and then develop plans for additions and modifications to the factory's equipment so that production plans can be met. Area planners are the experts on factory floor layout, well versed in the capabilities of the many types of equipment and processes. They must also have intimate knowledge of the services available and needed in conjunction with these types of equipment and processes. For example, they must be aware of the electric power, voltage, amperage, phase, AC or DC power, transformer, and filtering requirements for control circuits; availability of potable and sanitary water; purity and pressure of compressed air; pressure and condensate control of heating and process steam; hook height above the floor and lifting capacity of cranes; number of cranes on a rail; and sundry other items involved in making machines work and getting materials to and from those machines.

Area planners work with materials schedulers, marketing analysts, design engineers, manufacturing engineers, and others to determine the needs of the factory for the near-term to medium-term future. They compare the requirements to current capacities and decide whether the factory has the necessary equipment to fulfill the future needs. If it does, the area planner simply verifies that fact for management. If there are shortfalls in capacity, the area planner will notify management and recommend a course of action to remedy the situation.

The area planner has another function, that of optimizing present capabilities so that the factory can produce products at the lowest possible cost. This responsibility is a corollary to that of determining the shape of the factory to meet future production goals.

The work tools of area planners are the various layouts (factory floor blueprints) and capacity evaluations for the equipment and processes. Area planners are systems engineers. Their task is to gather information from experts and determine whether a manufacturing flow exists. If it does, is the flow sufficient to meet production volume needs? If it does not, what must be done to rectify the situation? They obtain information on capacity from the capacity analysis and capability evaluation sections of advanced manufacturing engineering. Information on machine rates of speed comes from the methods, planning, and work measurements unit. The physical characteristics of the machine tool processes are obtained from the maintenance unit or the engineers responsible for new technology evaluation and needs or investment project management.

After assembling the necessary information, the area planner begins to place facilities on the layout to achieve the desired product flow. The layout covers either the entire factory floor or specific portions of it, depending on the scope of the task at hand. If the master production schedule calls for producing the same product, but more or less of it, the layout will probably cover only a portion of the factory and concentrate on optimization of current equipment, or perhaps an addition to the equipment inventory. If the plan calls for introducing a new product, a total layout will be required showing how to integrate the old and new products and how facilities are to be shared or dedicated. The flows will also have to be considered for interferences. Typically, the area planner will use templates derived from a computer software layout program to locate facilities on the monitor screen and will list the capacities and required services next to the templates. This procedure will reveal bottlenecks where there is too large a difference between facilities and capacities. After identifying bottlenecks, the area planner will determine the alternatives for solution, such as adding additional similar equipment or substituting improved or larger capacity equipment. Once a layout is produced, reviews with shop operations and other manufacturing engineering personnel are scheduled in order to obtain operational viewpoints and other technical inputs. Usually, valid points are made during these reviews which result in a need for further refinements and iterations of the layout.

As implied above, area planning is an art as well as a science. It involves many engineering disciplines, which must be artfully melded together to achieve an optimal factory plan. This melding is the task of the area planner, who is always a senior manufacturing engineer.

## Capacity Analysis

Capacity analysis is the detailed study of the amount of current product and product mixes that can be produced within a specified period of time. This information is vital for the makeup of the master production schedule and for marketing sales strategies.

Engineers in this discipline are very concerned with the product mix being imposed on the factory. They traditionally convert specific products to the basic elements of fabrication time, machine time, process time, and assembly time necessary to manufacture them. They create prototype workstation loads per product. Knowing the amount of product to be produced in a given time period and the work time per manufacturing element, it is possible to calculate capacity, which is normally expressed as capacity per workstation. The engineer can then specify which manufacturing areas have the capacity to produce the load and which areas do not. If a factory is required to make more than one product at a time, which is often the case, the job of determining the capacity becomes more complex, and as a practical matter capacity analysis is not usually done. Instead, the engineer will show limits; that is, 100% product A–0% product B, and 0% product A–100% product B. The engineer will also list the number of hours per workstation required for each product. With this information the factory limits are established. This is sufficient because the master scheduler, using the MRP II system, is mortgaging time against workstations in order to load the factory. Hence, the scheduler is interested in work elements as devised by the capacity analysis engineer.

The work of the capacity analyst is normally transmitted to the area planner for inclusion in the master plan for the specific area. In many manufacturing concerns, capacity analysis is considered part of area planning.

## Capability Evaluations

Capability evaluation is the analysis of what tolerance can be created at what incremental cost, what critical processes can and cannot be performed, and what maximum size and weight can be handled in the factory. This information is critical for the design function's ability to continuously evolve the product or product lines. Design engineers must understand the limitations of the factory in order to design the products for the factory. Engineers involved in capability evaluations have the responsibility for determining the limitations and codifying the results for use by others.

These engineers require very detailed evaluations of equipment performance and an understanding of the present capabilities of equipment. They must understand the effects of deterioration of equipment on its ability to produce to close tolerances. Their primary function is to be aware of the performance level of the equipment and to periodically update the data base.

Capability evaluation engineers perform another vital function, that of technically evaluating proposed products to determine whether they can be made. If it is feasible to make a proposed product, then capability evaluations engineers will develop the preliminary optimized sequence and workstation selection to be used in making the product. This is done in close consultation with the area planner for integration with the factory's master plan. Like capacity analysis, capability evaluation is often considered to be part of area planning.

## New Technology Evaluation and Needs

The competitive marketplace requires that manufacturers continuously evaluate costs of production. Failure to do this invites competitors to erode away hard-earned market share and makes it more difficult to sell one's product. Just as the design function must continuously evolve a better design, so the manufacturing function must continuously evolve better ways of producing the product. The goal is always to reduce the cost of production. This is done by knowing how the product is made and searching for new types of equipment to produce the product at lower cost.

Considerable effort is expended in keeping current on new ways to produce the firm's products. This is usually done by maintaining close liaison with vendors who sell equipment traditionally used by the company, attending trade shows, participating in technical societies' programming, and, of course, regularly reading pertinent trade and technical publications.

Engineers involved in new technology evaluation produce a series of reports on new technologies of interest to the firm. These reports become the research basis for long-range planning and product introductions.

## Producibility Engineering

Producibility engineers ensure that designs produced by the design function are workable, that is, optimally producible in the company's manufacturing facilities. Producibility engineers deal in tolerance evaluations and changes in design to facilitate manufacture on optimal machines and facilities. Another task is to document detailed factory limitations and procedures for the understanding of design engineers. Their main goal is to lower the costs of making the product by obtaining design optimization relative to manufacturing needs. The producibility engineer is often said to be the manufacturing ambassador to the design function. They interpret manufacturing capabilities for the designers and convey the designers' real needs for product functionality to manufacturing.

The producibility engineering discipline has become more important as the need for productivity improvement in industry has increased. This discipline, once one of many in the spectrum of manufacturing engineering, has been singled out as decisive in helping industries compete in the modern high-cost era. The subject of producibility engineering and its relationship to the philosophy of concurrent engineering will be covered in considerable detail in Chapter 6.

## Computer-Integrated Manufacturing Development

This activity investigates productivity gains that can be made by applying computers to all phases of manufacturing. Engineers involved in CIM development are active in all three areas of the Computer-Aided Design/Computer-Aided Manufacturing (CAD/CAM) triad: machine/process control, measurements control, and planning control. They are responsible for developing and enhancing all CIM systems, then transferring operating control to other units within the manufacturing section for actual use.

The CIM development engineers work closely with their counterparts in the design and finance functions to conceive and implement the company's CIM plan so that an integrated system can exist. The integrated system is usually based on a common data base shared by all functions. The CIM development unit is charged with making sure that the integrated data base will serve the manufacturing section's needs while not being detrimental to the needs of other functions. The development and management of CIM systems will be considered at length from the viewpoint of manufacturing in later chapters.

## Investment Project Management

This segment of advanced manufacturing engineering works very closely with the finance function. It is responsible for "keeping book" on all funds allocated to the capital investment program. These engineers produce the investment plan and manage the projects from inception through approval by appropriate levels of management. When the plan has been approved, the investment project management engineers place orders for the equipment, monitor the progress of vendors in producing the equipment, and finally arrange for the equipment to be installed and debugged. Once the equipment is deemed operational by the intended users, control is released to the users.

Investment project management depends heavily on inputs from the area planners. Area planners, being responsible for determining the immediate and near-future needs of manufacturing to produce the product, are the natural initiators of capital investment projects. They provide the reasons for requests and preliminary information regarding cash flow and payback periods. The investment project management engineers, using this information, prepare the detailed documentation used to obtain the appropriate levels of management approval. They also use the services of the finance function to provide support documentation for the capital expenditure request.

Having received approval to spend money for capital equipment, the investment project management engineer will place an order for the equipment with an approved vendor. Vendor appraisal and selection normally take place during the preparation stages of the request for funds so that the investment project management engineer will know how much money to request. In consultation with the new technology evaluation/needs engineer and the responsible area planner, a list of appropriate vendors is drawn up. Vendors may be asked to visit to discuss the project, or visits may be made to vendor facilities to see similar equipment. The overall purpose is to determine exactly what piece of equipment will be purchased. If many vendors are capable of supplying the necessary equipment, requests for quotes are issued to the vendors inviting them to state prices and delivery times and to specify what their product is capable of doing. A request for quote is often required even when there is only one vendor.

This entire process tends to define the project for all involved. For the financial people, it quantifies near-term expenditures of money and allows comparison to budget plans. For shop operations, knowledge of incoming equipment allows plans to be made concerning staffing logistics and product output levels. For manufacturing engineering, this information is vital for factory layouts, support services planning, obsolescence evaluation, and productivity/producibility issues.

Once an order is placed, the investment project management engineer plans for receipt of the equipment, its installation, and its final turnover to shop operations. The engineer works with the area planner on matters concerning where the facility will be located, and oversees the maintenance unit for the actual installation. Debugging of the unit is done in conjunction with the methods, planning, and work measurements unit. Various elements are involved in installing capital equipment. However, the investment project management engineer develops the schedules, determines the project's critical path, and is responsible for its successful completion.

## Long-Range Planning/Forecasts

This activity develops the company's plans for technical improvement, capacity improvement, and capability improvement in accordance with its overall strategic and operating plans. Responsibilities include translating overall objectives and goals into technical and hardware plans, usually over a five-year or longer period. These specific plans allow the finance function to develop criteria for profits or other means of funding. They allow for an evaluation of the stated objectives and goals. This is done by an iterative process between the manufacturing and finance functions. Depending on the circumstances of the company, the long-range planning/forecast engineers will start the process either from a known dollar value for investments per year, or from a known or targeted production level over the forecast period. In either case, the iterative process applies.

The process is started by gathering inputs from the marketing and design functions on quantities and types of products to be sold over the forecast period. Accurate forecasting being very difficult, marketing is normally reluctant to make specific forecasts, and whatever forecasts it makes are usually vague and full of restrictions and disclaimers. Nevertheless, manufacturing must have a forecast and must be able to believe it. Hence, a very important assignment for the long-range planning/forecast engineers is obtaining a forecast in sufficient detail and verifying it. This may be done informally by asking marketing planning to explain how they arrived at their forecast, or formally by asking marketing to have another source, usually an external or internal consulting service, make a forecast. Whatever the source, manufacturing engineering must have confidence in the forecast before proceeding.

Given a forecast, the long-range planning/forecast engineers develop scenarios of manufacture in consultation with the area planners, capacity analysis engineers, capability evaluation engineers, and new technology evaluation/needs engineers. This will result in a facility capital investment plan for the forecast period. The plan is then reviewed with other functions and the iterative procedure begins.

## METHODS, PLANNING, AND WORK MEASUREMENT

Like the advanced manufacturing engineering unit, the Methods, Planning, and Work Measurement (MP&WM) unit may have one engineer for each major activity, several for one activity, or one for several activities. Again, the personnel allocation depends on the staffing needs of the particular organization.

MP&WM is the core of the industrial engineering discipline conducted by manufacturing engineering. This unit performs the traditional industrial engineering work carried out for the factory. It develops and implements nonfacility or personnel productivity improvement programs. Hence the MP&WM unit is the branch of manufacturing engineering that is closest to shop operations on a daily basis. It instructs the shop in how to do a specific job, trains the operators, determines how much time it should take to do each operation, and measures the shop's performance.

The major phases of the MP&WM unit's work are listed below and discussed in greater detail in the following sections. The activities are interrelated, but for the purpose of definition we can examine them independently.

Workstation methods
Workstation planning
Workstation and tool design
Time standards
Measurements of productivity
Variance control programs
Job rate evaluations
Operator training

## Workstation Methods

Methods engineers devise the basic techniques for producing products as required by the design specifications. They mate a workable design with the specific capabilities of the factory to produce the part or assembly at an optimal cost. Methods engineers use the best combination of machine feeds and speeds or process controls to achieve the results called for in the design specifications. They specify tools to be used for cutting; proper voltages, amperages, and wire feed rates for welding processes; the best suited jigs and fixtures for assembly; and so forth for the particular manufacturing operations. They outline the chronological sequence of events necessary to do the required work, but do not arrange in sequence anything that does not have an operational prerequisite. The specific sequencing is left to workstation planning personnel, who will optimize the entire sequence in time.

Methods engineers also spend considerable time and effort determining the positioning of work for proper presentation to the machine or process equipment—for example, how the workpiece should be secured and loaded to minimize tool chatter, droop, or process-induced stresses. Proper positioning helps to ensure that the finished part will have all the characteristics required by the design specification.

The specification of tooling falls within the domain of the methods engineer, who outlines what is necessary and what is to be accomplished, and then turns the task over to the tool designer for specific development. The methods engineer defines the constraints within which the tool designer operates.

Methods engineers normally are the manufacturing engineers most closely associated with factory floor operations. They have many opportunities to observe ongoing operations and evaluate how they are progressing. They are often called upon to participate in productivity improvement projects, which may range from minor tooling improvements to complete realignments of production techniques.

Methods engineers are frequently people who have had long experience in a particular factory or with a particular product line.

## Workstation Planning

Planners create the detailed step-by-step sequencing instructions for shop operations to follow. A properly sequenced set of workstation plans will always result in the optimum way of producing a part or assembly with repeatable results. The planners use the design specifications, engineering drawings, and methods instructions to develop a "road map" for the operators to follow. Their information output covers the exact detailed chronological sequence of tools to be used, the number of passes to make for a weld or the number of cuts to take on a specified machine tool, and the exact setup on the machine tool to be followed. Each step also includes the time allowed for the step to be completed. Good planning leaves nothing to chance. Every aspect of the job to be performed must be accounted for, leaving no room for supposition or inference.

The final function of the planner is to put the plans into the manufacturing organization's production control system. This may be a manual or computerized system. The only overall requirement is that the operator or the intelligent machine understand what is required and be able to respond effectively.

The planning function is the storehouse of the body of knowledge on how its products are manufactured. One ongoing activity is cataloging and codifying this information for future use. Many companies use Computer-Assisted Process Planning (CAPP) for this purpose. This technology allows easier repeatability of planning for like or similar parts and assemblies by applying a group technology code— a classification of parts by shape, manufacturing sequence, or materials from which they are made. In this way families of plans can be developed, grouped, and easily recalled, and only the superior plans will be used over and over again. Thus, CAPP encourages standardization and optimization and leads to improved productivity. By using the computer to search for similar plans already on file using their group technology code number, the number of new plans required per time period is reduced and the quality of planning is improved. In essence, all planning is being done by the senior or more qualified planners. Still another advantage of CAPP is the reduction of cycle time required to go from engineering drawing to completed planning. The computer can search for similar planning in a fraction of the time required to do it manually.

The computer is changing the nature of how workstation planning organizations are run. Drawings are becoming data inputs for Computer Numerical Control (CNC) machines, step-by-step sequences are generated from data bases rather than handwritten instructions for a specific part, and timing sequences are calculated by previously programmed systems rather than one at a time. All these abilities have increased the output of the planning function, but the output is still the same and still performs the very necessary function of telling the operator what to do and how to do it.

## Workstation and Tool Design

The design of tools is an exclusive responsibility of manufacturing engineering. Tool design engineers create tooling on the basis of constraints spelled out in the specifications prepared by methods engineers. Tool designers must understand not only the classical concepts of stress analysis, but also how tools are used and what they will produce. They must have a good background in metallurgy, welding, forming, cutting, and other associated technologies. A good background in processes is also required. While the tool design engineers specialize in the ultimate contact device between the material to be worked on and the workstation, the workstation engineers are more systems-oriented than tool designers. They are charged with designing a workstation that optimizes production time and achieves the most effective and productive utilization of space and time.

Tool design engineers work similarly to the product design engineers. They use the same techniques that go into the design of a product to be produced in the factory. The difference is that their output usually calls for a short production run and its manufacture is almost always assigned to a tool specialty shop. However, tool design engineers need considerably more than technical competence. They must understand the use of tools in manufacturing. For this reason, a lengthy on-the-job apprenticeship is usually needed to convert a theoretically trained engineer into a tool design engineer.

Workstation engineers deal with principles of motion economy, the efficiency of body movements. They must know how material will arrive at the workstation, how it should be handled there, and how the finished part should leave. They design workstations to minimize both human movement and machine movement and ensure optimal efficiency of the human-machine combination. The workstation engineer is as close to the stereotypical efficiency expert as anyone likely to be found in manufacturing.

Workstation design and tool design involve an amalgamation of unlikely skills. The broad systems thinking skills of the workstation engineer and the specific detailed thinking skills of the tool design engineer come together in the workstation. Here the workstation engineer must know what the contact tool can do and how it must fit into the scheme of the workstation. The tool design engineer must know the limitations of the workstation so that the tool design will flow within the human-machine system. The systems-oriented and detail-oriented engineers complement each other and act as consultants to each other.

## Time Standards

Plans show the amount of time it takes to complete each particular segment of work at the workstation. These times are developed by time standards engineers. The classical body of knowledge known as motion and time study, or efficiency measurements, deals with the most efficient ways of using the human body; for example, a wrist motion is more efficient than a full arm motion because it uses less energy and is faster. Motion and time study also deals with the motions of machines and how much time they take to complete a cycle; it is concerned, for example, with feeds and speeds for a lathe as specified by the methods engineer. Combining times for human motions with times for machine motions makes it possible to calculate the time it should take for any manufacturing operation. The time to perform a set of operations at a workstation is called a time standard.

Time standards engineers codify and tabulate time standards of operations for use by planners. They create computer programs that the planners may use to determine how much time any particular operation should take and to assign planned times to plans issued to shop operations.

The system for writing time standards is quite specific and borders on being tedious. Fortunately, computer systems are being devised that assist in accomplishing this task, but it still remains very detailed work. Engineers must first understand the workstation: what it contains, the parts likely to be processed there, how it is loaded and unloaded, the feeds and speeds, and the number of operators working. They must understand the workstation design, the methods to be applied, and the tools to be used. They then list in chronologic order all actions to be taken by human or machine or both together. For each item on the list they research the principles of motion economy for human actions and mechanical speeds for machine actions. Ultimately, they record the least amount of time it should take to do each item on the list. For items that are variable, they construct an equation that defines the spectrum of events. The total time of all items on the list is the time standard. Although time standard development is laborious, it is the only way of knowing what the ultimate optimal time for an operation should be. Any other method of setting planned time is usually a comparison to previous performances, good or bad, and the potential for improvement is not known.

## Measurements of Productivity

Using plans and time standards, methods engineers can determine how much time, in theory, it should take for a work element. Combining all the work elements of a job reveals the time necessary to complete that job. Knowing how long it takes to do a job makes it possible mathematically to compare actual work performance to theoretical or standard performance. This mathematical ratio becomes a measurement of efficiency. If enough measures of efficiency are taken, it is possible to discern trends. If the ratio approaches unity as time progresses, then productivity is improving. If the reverse happens, productivity is decreasing.

There are other measures of productivity such as dollars of product value output per direct worker—the so-called output over input measurement. These types of measurements tend to be too large-scale, and therefore of little use in managing shop operations. Efficiency trends tend to be the most direct and easiest to measure, and a considerable body of knowledge on this measurement system has been accumulated and classified.

Manufacturing engineers involved in productivity measurement are concerned with the validity of the data, hence the accuracy of the measurement trends. They spend their professional time developing data collection systems, which typically evolve about the way work is dispatched to the workstation and how progress is recorded. In addition, most computer manufacturers now offer industrial management systems, the heart of which is usually a data collection system, which often work in conjunction with MRP II systems. The goal of these systems is to capture completions of processing steps as soon as they occur so that elapsed time can be compared to planned or theoretical time, generating an efficiency measurement on a real-time basis. These systems employ user-friendly terminals at the workstation that instruct the operator in how to make entries and have a mechanism for allowing the operator to record start and stop times. With accurate information of this type, factory managers can tell the state of the operations at any time and take corrective actions immediately. They can accurately gauge their production rates and make judgments for future orders. Engineers involved in measurements of productivity play an important role in devising these systems and thereby considerably influence the information gathering and analysis necessary for success.

## Variance Control Programs

Variance control is very closely associated with measurements of productivity. Variance measurement is another way of saying efficiency measurement. The variance is the amount of change or mathematical distance from the norm, which in this case is 100% efficiency, that is, doing the work in exactly the planned amount of time. Thus the variance is the difference between the planned time and the actual time.

Variance control programs are any techniques that can be applied to reduce the mathematical distance between actual and planned times for accomplishing an assigned task. For example, if the time to perform a job was 129% of the standard, then the variance would be 29%. The variance control program would be aimed at bringing the 29% variance as close to zero as possible. To do this, engineers would look for all the factors that hindered the completion of the operation within the planned time and develop plans to eliminate them or, as is usually the case, reduce them. Strategies can range broadly, from ensuring that properly sharpened tools are being used to asking producibility engineers to review the manufacturability of the product.

Variance control programs tend to become complex because many unknowns must be evaluated and simultaneously accounted for. Hence engineers working in this area tend to be senior in manufacturing experience and thoroughly understand the nature of the products being produced. Many of the projects used in variance control programs require inputs from most of the organizations of the manufacturing function. For this reason, the programs require good project management skills. Such projects normally deal with underperforming workstations; hence they are highly visible and require mature judgment for their successful conclusion.

## Job Rate Evaluations

To function properly, an organization must have the proper people in the proper jobs. One of the tasks of MP&WM is setting the standard requirements for successfully carrying out specific jobs. This is job rate evaluation, or determining the value on a comparison scale of each job that must be performed for the factory to function. The engineers who do this determine the order of job importance so that management knows how to pay for the various positions.

Engineers usually do this by designing several key jobs that the particular factory needs in order to produce its products. Skilled lathe operator, welder, and assembler would be key jobs for a factory that fabricates, machines, and assembles to make its products. Once these key jobs are designed, the responsibilities assigned, and the availability of these skills on the local market determined, the jobs are ranked in order of importance. All other jobs are then ranked relative to these key jobs.

In this process it is essential that the skills associated with all jobs be properly evaluated. Major mistakes here would seriously jeopardize the company's ability to attract qualified persons to the respective positions. Factors that are taken into account are mental and physical skills necessary to perform the function, availability of such skills in the surrounding geographic area, comfort or discomfort levels associated with the job, hazards in performing the job, and cost of operator mistakes associated with the job or degree of criticality of not making mistakes. Each of these factors is given an evaluation on a point system so that a basically subjective system can be made as objective as possible.

The job rate evaluation system can set the tone of personnel relations. If it is done correctly, the factory will have established the basis for high employee morale. If not, a factory with continuous labor strife may be created. Engineers engaged in this activity thus have a great influence on the morale and the productivity potential of the factory.

## Operator Training

This phase of MP&WM responsibilities is directly associated with the shop operations work force. Methods engineers are responsible for instructing the operators in how to operate the manufacturing equipment in accordance with the prescribed methods, and are therefore usually themselves experienced in operating the factory equipment.

In training the work force, engineers use all the traditional methods: classroom, training aids, on-the-job training, pairing trainees with experienced operators, and so forth. In addition, the MP&WM unit may operate an apprentice training program that gives recipients both classical technical training and specific skills training.

Regardless of the size of the organization, all factories conduct some form of operator training. The modern industrial concern cannot exist without such programs.

## MAINTENANCE

The responsibilities of the maintenance unit are:

Machine tool and equipment maintenance and upgrade
Buildings and grounds maintenance

Stationary engineering
Disposal of excess and obsolete equipment
Toolroom

The maintenance unit is one of the few organizations within manufacturing engineering that has skilled operators reporting to it. It is similar to shop operations in administration of personnel, so that the operation of maintenance consists of more than directing the activity of professional engineers. It also includes several foremen and their respective skilled craftsmen. In many ways this unit resembles a multipurpose construction company.

## Machine Tool and Equipment Maintenance and Upgrade

Maintenance is responsible for keeping the factory's equipment operational so that production can continue. A related responsibility is that of continually upgrading or improving equipment to allow constantly better performance. Maintenance interrelates closely with shop operations to keep the equipment functioning, and with advanced manufacturing engineering to devise better ways to use the equipment to meet continually evolving design requirements.

Engineers work with a wide span of organizations: vendors selling parts and retrofit equipment, shop operations management requiring coordination of maintenance efforts and production requirements, advanced manufacturing engineers requiring information on whether and how equipment can be modified to meet design modifications, and MP&WM for tooling efforts and tooling design and impacts of methods on equipment.

The relationships are even more pronounced for the equipment upgrade activities with AME. Here maintenance engineers are involved in projects that border on capital equipment acquisition, requiring skills closely akin to those of advanced manufacturing engineers. The dollar values are usually less—a reason upgrades are usually attractive to management—but the procedure is as intense as in major acquisition projects.

The maintenance unit employs skilled operators such as electricians, carpenters, pipe fitters, steelworkers, welders, machine repairmen, and electronics technicians. These trades are supervised by foremen knowledgeable in such skills and well versed in job estimating techniques. A major function is responding to requests to fix malfunctioning equipment. Typically, the maintenance foreman will receive a request to repair a malfunctioning piece of equipment. The foreman's task is determining the cause of the malfunction and then getting it fixed quickly. Since in a producing operation quick repair is a vital necessity, manufacturing concerns pay premiums for expert skills in this area.

In diagnosing malfunctions, the foreman can consult with the maintenance engineers. Normally, if it is a serious matter, the engineer takes over and provides the required technical expertise, while the foreman directs the repair activities based on the technical decisions of the engineer. It should be noted that this occurs for very difficult problems and is not the norm. Typically, the engineers spend their time planning and monitoring upgrade projects, not tending to breakdowns.

Engineers are also responsible for maintaining the machine tool capability and replacement log, where the history of all equipment is kept. It includes the frequency of repair, types of repairs required, cost, deterioration in ability to hold as-new tolerances, and other information critical to the equipment performance. Although engineers are the keepers of the log, information inputs are received from all manufacturing engineering functions. For example, the information on new equipment is often provided by the AME unit, while that on deterioration in performance may be provided by maintenance foremen or methods engineers. These records are the basis for calibrating current production rate capabilities as well as substantiating the purchase of replacement equipment.

## Buildings and Grounds Maintenance

The engineers here are responsible for ensuring that building systems such as air-conditioning, heating, sanitary systems, and lighting work adequately. A major part of their work is preventive maintenance so that disruptions in services do not adversely affect the primary operations of the company. They become involved with lesser-dollar-value capital equipment acquisitions commonly called hotel load services.

In addition to maintenance, they are usually responsible for cleaning and grooming services within offices and factory. In this activity not only less skilled personnel but also many of the highly technical trades such as electricians, master mechanics, carpenters, and electronic technicians are involved. Air-conditioning, heating, sanitary systems, and energy and lighting systems in modern structures have considerable technical complexity and require sophisticated engineering and technical support.

## Stationary Engineering

If the manufacturing operation is large enough, a separate stationary engineering group is established to operate the power plant, air-conditioning system, and other hotel load services. Like the other parts of the maintenance unit, it is staffed with a mixture of engineers, foremen, and skilled tradesmen.

## Disposal of Excess or Obsolete Equipment

This task involves selling equipment that is no longer needed or no longer capable of performing to required specifications. It is traditionally assigned to the maintenance group because of their close working relations with used equipment dealers, which come about because maintenance engineers often look for repair parts and frequently obtain them from dealers in used or renewed or rebuilt parts.

Disposal of unneeded equipment is not a haphazard operation. Each year the manufacturing engineering organization will prepare a budget for disposal of equipment, containing the cost of and the value to be received for disposal. The net effect should be a net gain for the company to offset the cost of operations. Cost of disposal would include actual removal costs, advertisement costs, and carrying charges of inventory after the equipment is declared obsolete. In well-run companies the funds to be received by selling obsolete equipment are not simply a desirable goal, but are actually planned for. The budgets are developed by evaluating the physical plant inventory records to determine the useful life of equipment. This information, coupled with the identification of replacement equipment on the capital investment budget, yields a realistic value for the used equipment sales budget.

It is the responsibility of maintenance to develop sales plans for disposing of the equipment and maximizing the company's profitability. In a sense, maintenance engineers must develop skills of successful marketing personnel. They must research the market and contact and develop customers for their particular products.

## Toolroom

The toolroom is responsible for making repair parts, jigs, and fixtures for all of manufacturing. It is also responsible for redressing and grinding all tools used in manufacturing operations. It is a service shop operated by manufacturing engineering and administered by the maintenance unit. Engineers in this segment are engaged in designing repair parts to support maintenance activities only. Tooling, jigs, and fixtures are designed by MP&WM engineers and manufactured for them in the toolroom.

The toolroom is usually equipped with general-purpose machines of a smaller nature and stocked with raw materials used in making repairs. A small stock of materials suitable for frequent MP&WM methods tooling work is also kept on hand. Materials for special requests may be drawn from production stocks or ordered as required.

Operators assigned to the toolroom are usually senior personnel, having master craftsmen skills and experience. Toolroom operations are similar to those of shop operations units. Work is received in, scheduled, and dispatched to the operators. The main difference between shop operations and the toolroom is that detailed method instructions are not required. These operators, having the highest skills, do their own planning to produce the required products.

## PROCESS CONTROL

The last unit within the manufacturing engineering subfunction is process control. This is a fairly recent addition to manufacturing engineering. Indeed, many classical analyses of quality control/assurance still consider process control to be a part of that subfunction. However, with the ascendancy of quality assurance over traditional quality control, there has been a trend toward placing process control within manufacturing engineering. This is due to a desire on the part of manufacturing management to concentrate technical support services within the strongest and traditional manufacturing technical support function, coupled with the fact that quality assurance is more of a certification activity with a significant part of its attention focused on other than manufacturing activities.

The responsibilities of process control are:

Quality planning
Nondestructive test
In-process inspection
Incoming material inspection
Product test
Total quality management administration

## Quality Planning

This is the planning and strategy activity of process control, and is sometimes referred to as process planning or inspection planning. The engineers involved develop the plans for checking the adequacy of performance of shop operations to ensure that the final product performs as designed. Using plans and methods produced by MP&WM as a guide, quality planning engineers determine where inspections and nondestructive tests will be specified during the manufacturing process. They also specify the type of inspection or test to be conducted and, based on design engineering requirements, determine what will constitute acceptance or rejection. These quality plans are then integrated into the overall operational plans by which shop operations produces the company's products.

Normal manufacturing activities produce a certain percentage of deviations from drawing. Some are important, some of little consequence. It is the quality planning engineers' responsibility to evaluate these deviations and determine what the proper corrective action will be. They then ensure through MP&WM that the corrective actions are factored into manufacturing planning for rework. As the arbiter of quality via the deviation from drawing procedure, quality planning has the data base to evaluate performance of the various shop and support functions. A score-keeping function is possible and desirable; in this way quality planning can report to management whether quality levels are improving or declining.

Quality status can be reported by statistically evaluating the numbers of deviations and their seriousness. This leads naturally to an evaluation of the cost of doing the repair work caused by the deviation. Repair work, which constitutes manufacturing losses, is an important measurement of organizational quality levels. Manufacturing losses are a significant measure of the adequacy of attention to detail of the operators and their foremen. High losses indicate a poorly managed operation. Quality planning engineers are responsible for setting the manufacturing losses, budgets, and measurements policy.

## Nondestructive Test

A good deal of the evaluation of products during manufacture is done by using nondestructive testing techniques. Engineers assigned this responsibility have extensive technical knowledge of x-rays, magnetic particle testing, ultrasonic testing, and dye penetrant testing and how such tests can verify product integrity. They spend about half their time designing tests in accordance with the quality planning engineers' specifications, and the other half supervising or performing such tests.

Nondestructive Test (NDT) engineers must have considerable knowledge of the structure of materials so that they can properly interpret test results. Such test results are often difficult to interpret, and judgments are often made on a probabilistic basis. Thus the NDT engineer must understand the design of the product, the rigor of the intended service requirements, and how the product or component is manufactured. With this knowledge, the NDT engineer evaluates the components, making educated judgments on the adequacy of the part for its intended use.

This is not to say that NDT engineers evaluate products in a void. The interpretation of test results is often guided by extensive codes and codified interpretations of those codes. In fact, many product specifications require testing in accordance with clearly defined codes. NDT engineers must be knowledgeable about the various codes and the interpretations given to these codes by technical societies and worldwide government agencies.

NDT is by far the most abstract function of manufacturing engineering. Many books, articles, and dissertations for advanced degrees have been written on the subject, and it is probably the most thoroughly publicized manufacturing engineering technology. Still, NDT is simply one tool that good manufacturing managers use along with other methods to produce high-quality products.

## In-Process Inspection

This part of process control is commonly referred to as the "policeman on the beat." The common and mostly incorrect impression is that in-process inspection inspects quality into the product or component. Inspection is done to verify parameters considered critical for proper product performance. It is not done to tell an operator whether he or she performed the function correctly, but to double check what is presumed to be proper performance by the operator. This is a subtle difference and is often misunderstood. The inspector's function is to check system conformance; operators are supposed to know that a part is correct before submitting it for inspection. Operations where 100% correctness is essential and 99.9% correctness equals failure are double-checked and sometimes triple-checked through the inspection system. Where 100% correctness is not necessary, inspections are called out in the planning to provide other important information, for example, whether the human-machine system is still performing within the expected tolerance range, and whether the process is producing as expected or designed. Quite often raw material specifications can vary, resulting in insufficient results. It is important to understand that it is not the function of in-process inspection to check the knowledge or adequacy of training of the operator; this is an MP&WM function.

Inspectors receive their instructions from the quality planning engineers. These instructions, which are contained in the quality plan, specify what the inspector will inspect, how often, and in what sequence of the operation. The inspection may be a physical check with measuring tools such as verniers, micrometers, or electronic or optical instruments, or a verification of documentation prepared initially by an operator making the part. Often an inspection is a combination of both.

Usually inspections are done when a product or component passes from one major operation to another, such as from fabrication to machining. Therefore inspectors perform another very useful activity for manufacturing management. When their work is complete, the status of the part is recorded and management then knows that the part is correct per engineering design through that specific operation.

Staffing of this segment usually consists of inspectors and foremen. Inspectors may be drawn from the ranks of the operators or may be trained initially upon hiring for the inspection job. Inspection foremen almost invariably will be promoted from the rank of inspector.

## Incoming Material Inspection

Process control is responsible for verifying that raw material received for use in manufacturing is of adequate quality. This is accomplished by measuring or verifying the material in accordance with the design specification.

Incoming materials inspection engineers work with design engineers to develop material specifications that can be met by vendors at reasonable costs. Design engineers specify materials for use in the product. Incoming material inspection engineers review the tolerance band for the specified materials in terms of what can be purchased within technical, cost, and delivery time constraints. Once agreement is reached with design engineering, they specify the types of acceptance tests to be performed on materials before release for manufacturing.

Work is also done in close collaboration with purchasing to qualify vendors. Purchasing agents will find vendors who claim they can supply or produce the needed materials. It is the incoming material inspection engineers's responsibility to determine whether the vendor can actually do so and can sustain the required level of performance over the life of the contract. Ensuring that adequate materials are received is a vital task; faulty materials lead to faulty products that, at best, require high repair costs.

## Product Test

Product test can be thought of as the culmination of all process control work. It can also be thought of as a quality check of the inspection process itself. If the quality plan is adequate and carried out properly, then the product's performance should have been verified and a total test is redundant. For this reason, a test of the completed product is often nothing more than a contractual requirement that must be performed before the customer accepts the product. But product test is also more than proving that the whole is equal to the sum of the parts. It allows for gathering data that support the design theory of the product, for interpretations to be made for further improvements in design so that future products will be better than present ones, and for evaluation of design evolution toward better performance costs. In addition, it is a means of verifying design, since not all design parameters can be fully calculated or predicted.

Product test engineers work closely with design engineers to plan tests to provide useful data. They must also work in close harmony with all other phases of manufacturing engineering. Not infrequently, product testing will turn up deficiencies in design that require major revisions in manufacturing processes. This is particularly true if the company produces many prototypes and has short production runs. Therefore, manufacturing engineers are as interested in product test's results as are design engineers.

For complex products, product test becomes a very important part of the total process control function. It gives the company a high degree of confidence that the product will perform as the customer expects it to, and this is a valuable marketing tool as it helps to establish the proper reputation with the customers.

## Total Quality Management Administration

The final responsibility given to process control is that of administering the quality awareness program. As I have stated previously, quality cannot be inspected into a product or process. Acceptable levels of conformance to specifications and instruction need to be achieved during the creation process itself. It is impossible for the inspection process to change nonconforming work to conforming work just by virtue of the fact that the work is undergoing an inspection. The best we can hope for during inspection is that deficiencies are discovered and corrected. It is far better that performers of value-added work are cognizant of the need to do acceptable work in the first place and not rely on the inspection function to find their deficiencies. Thus the need to correct deficiencies should be minimized to the greatest extent possible. The mechanisms of fostering this awareness and the importance of doing acceptable work always are the essence of Total Quality Management (TQM).

The members of the process control staff given the task of administering the TQM process are both proponents and planners. In many aspects they are adjuncts to the quality planning engineers in that they are devising plans to assure that tasks are performed properly within the process flow. However, the main task they perform is to demonstrate to the entire company structure that everyone is responsible for achieving acceptable quality levels. Moreover, they are responsible for creating a mental set within the company that customer satisfaction can be achieved only through creating products and services that the customer wants. Inevitably, that means thorough attention to detail and doing defect-free work.

The task of creating this necessary awareness for TQM falls within process control. Along with creating awareness is the assignment of assuring that TQM principles are being applied. This is the audit function. Also, in order for an audit function to be of use, first techniques of how to comply have to be codified and taught. These, too, are functions given to process control. The techniques involved will be explained and demonstrated in the chapter devoted to process control.

## SUMMARY

The reader has now been introduced to the complexities of the manufacturing organization. In particular, the manufacturing engineering organization has been defined to demonstrate its vast area of responsibility within the manufacturing matrix. It should now be evident how the various units of manufacturing engineering work and how they are related to each other. Manufacturing engineering is the primary technical resource of the manufacturing function, but its influence goes beyond the limits of manufacturing and extends to the activities of the design, marketing, and finance functions of the industrial organization.

## REVIEW QUESTIONS

1. Manufacturing engineering is considered to be the core technical discipline within the manufacturing organization. Discuss the interrelationships of manufacturing engineering with the other functions; in particular, state how and in what circumstances manufacturing engineering would be the lead organization for manufacturing in these relationships.

2. Referring to Fig. 1.1 and to the list of responsibilities of process control, explain why process control is considered part of the manufacturing engineering organization rather than part of quality assurance.

3. An area planner is required to develop a manufacturing facilities plan to produce a new product. List and describe the various steps the planner would take to accomplish this task.

4. Describe the difference between workstation capacity analysis and capability evaluation.

5. The area planner in the role of investments project management engineer is responsible for conceiving, purchasing, and implementing new equipment projects. Develop a generic flowchart showing in chronological sequence the steps the area planner must take to perform this task.

6. Methods engineers are responsible for defining the sequence of operations to be followed in producing parts in a factory. Show in a generic sense the chro-

nological order to be used for the operations, and explain why the order of the sequence is significant.

7. Explain the difference between planning and time standards.

8. What is a variance control program, and why is it important for the success of a manufacturing concern?

9. List and define the types of engineers and trades likely to be found in a maintenance unit of a typical machined parts job shop.

# TWO

## MANUFACTURING ENGINEERING MANAGEMENT TECHNIQUES

The manufacturing function cannot just exist; it must have a stated reason for being. A manufacturing organization exists to make its company's products. But just making the product is not enough; it must make the product so that the company can make a profit. To do less would result in a situation where the firm itself would eventually cease to exist. Therefore stated reasons or justifications for existence must be both generic and specific. It is the definitions and explanations of these generic and specific reasons for existence that lead to the management techniques employed in all manufacturing subfunctions. In this chapter we will thoroughly explore these reasons for existence to understand how they translate into specific management techniques applicable to manufacturing engineering.

## OBJECTIVES AND GOALS

Objectives and goals define the generic and specific reasons for any organization to exist. They are general policy statements and also statements of specific intent. A system of objectives and goals consists of three basic steps:

Objective: A broad-based generalized statement of intent.
Goal: A measurable statement of specific intent bounded by a specific period of time.

Project: A specific plan with measurable steps that lead to the accomplishment of
a goal. One or a set of projects may be required for the accomplishment
of a single goal.

Clearly, this system progresses from very broad-based to specific action steps
that define singular items to be accomplished. Figure 2.1 illustrates this concept;
but notice that the arrows shown go in both directions, from broad to specific and
from specific to broad.

Figure 2.1 implies that direction or policy is proclaimed, then actions are
taken and the results are reported back to the policy authority. The concept of an
objectives and goals system requires feedback of the results to allow management
to evaluate the worth of the basic objectives and to determine whether or not goals
were accomplished. This feedback also shows whether accomplishment of the
goals supports the stated objectives.

An example of an objective would be: "Maximize profits by legal and ethical
means." This is a generalized statement of intent. It tells the world that the company's policy is to maximize profits, but only by legal and ethical means. This
statement sets the tone for operational procedures. It tells all personnel that they
must abide by general precepts of legal and ethical conduct in their pursuit of
maximized profits. There is nothing measurable about this statement. It does not
give specific instructions to management, nor does it define legal or ethical.
Therefore, this is a classical objective; it is a broad, general statement of overall
policy. Implementation of that policy is left to others.

Another example of an objective is: "Improve productivity." This is an objective that would be internal to a specific function of a company and not necessarily issued by the highest level of management. Such a statement would be
issued by the manufacturing function. It meets the definition of an objective. It
is broad-based and generic. It simply states that the function will improve its level
of productivity, but does not state how or how it is to be measured.

Objectives can be thought of as statements made by a politician campaigning
for office. They sound good and are definitely the right thing to espouse. But
once the campaign is over, there is no way to measure the officeholder's performance against the stated objectives. They are too broad and open to too many
interpretations. In the business world such intangibles are useless for performance
measurement. Therefore all policy statements, or objectives, must be quantified
by supporting statements of goals.

Let us look at an example of a goal statement: "Improve profits by 15% for

Broad-based

**Figure 2.1** The objectives and goals concept.

Specific

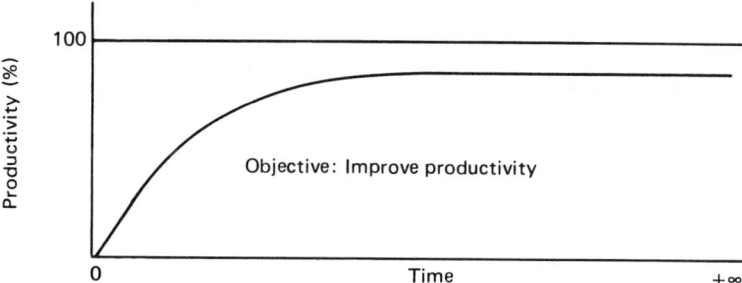

**Figure 2.2** The infinity analogy of objectives.

the next budget year compared to the performance of the current budge year." This is very specific: it refers to an amount that can be converted to dollars, and is bounded by a specific time period. It is measurable and allows virtually no room for interpretation. A goal sets a task to be accomplished and states it in such a way that it is easy to determine whether it has been accomplished.

An important characteristic of a goal is that it exists for only a certain clearly defined period of time. Once the deadline is reached, regardless of success or failure, the goal no longer exists and a new one must be proclaimed that fits the objective. The objective, of course, is the same as the one supported by the expired goal.

Goals are periodic, while objectives are ongoing and asymptotic. Figure 2.2 shows the infinity analogy of objectives. If productivity is the ratio of output of goods and services produced to input of capital, material, and labor, 100% productivity corresponds to the limit of unity for this ratio. Since this is tantamount to zero losses in translating all inputs to outputs, it can occur only in a time equal to infinity. That is, an objective can never be totally achieved in practice; it can only be approached asymptotically. Who is to say that the ultimate in productivity improvement has been achieved? It is always possible that a new process or a new machine will be invented that will advance the productivity improvement horizon.

Goals are not asymptotic. They are defined parts of the total objective that allow for measurable change in a defined time period. This is illustrated by the goal accomplishment line in Fig. 2.3. All values to be achieved are real numbers capable of full definition and readily plotted against the goal. The actual efficiency achieved can be determined for the quarterly time periods and entered on the graph. If the actual plot coincides with or is above the goal plot, then the goal has been achieved. If it is below the goal plot at the end of the year, then the goal has not been achieved.

## PROJECTS—THE MECHANISM FOR ACHIEVING GOALS

Let us look at some examples of projects that allow achievement of a goal. The goal represented in Fig. 2.3 is to raise the efficiency of the manufacturing function

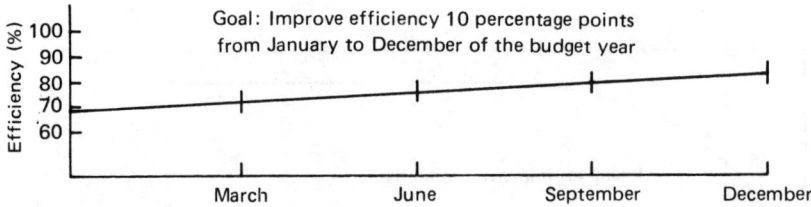

**Figure 2.3** The finite analogy of a goal.

from 70 to 80% by December 31 of the budget year. This is a directive, an order given to the appropriate subfunction of the organization to devise plans for achieving the targeted results. Such plans in the objectives and goals system are called projects. For the efficiency goal a project could be developed to improve the attention time of operators at the workstations. Another project could involve the development of workstation tooling to allow faster feeds and speeds to be employed. A third project might involve the retraining of operators in the specified method. A fourth project could be purchasing more efficient capital equipment. The point here is that many projects may be undertaken to achieve one prescribed goal.

How does the responsible individual know whether a specific project supports a given goal? Conversely, how does the manager know what projects to select to achieve the goal? There are no absolute rule answers to these questions. However, there is an important evaluation technique or logic that can be followed, which I call the rule of project selection pragmatism. The rule is as follows: *The project that costs the least in time, money, and people resources should be selected, provided that it allows achievement of the goal.*

Often no single project activity can be launched that will allow achievement of the goal. When this happens, I recommend implementation of the series corollary to the project selection pragmatism rule. The corollary is: *Select those projects that, when executed in series or parallel, will achieve the stated goal at the lowest possible cost in terms of time, money, and people resources.*

No matter how cost-efficient a project may be, if it does not allow achievement of the goal by itself or in conjunction with other projects, then it is an unnecessary tangent and should not be pursued. Each project must enhance the cause, that is, move the organization closer to ultimate achievement of the goal. Therefore, the following questions should be asked of each project:

1. If the project is implemented, what are the likely results?
2. What is the probability of success?
3. What are the consequences of failure?
4. Will failure deter or cancel other projects?
5. What are the costs of implementing the project?
6. Are there simpler ways of achieving the same results?
7. Will the planned results allow achievement of the specified goal, or partial achievement? If partial, how much?

In practical terms, a project is a series of sequential and perhaps parallel action steps that, when completed, achieve all or part of the requirements for accomplishing the goal. Each step is a single entity with a scheduled start time and a scheduled completion time. For each step there is a simple phrase or sentence that states what is to be done during the time period assigned. Each step also has one person assigned responsibility for its completion. When all steps are completed, the project is completed. Like a goal, a project must be evaluated on completion to determine whether the planned for results have been achieved. This is vitally necessary if the results of one project lead into the start of another.

Figure 2.4 is an example of a project. Note that the elements are listed. The steps are defined in simple phrases. The time to accomplish each step is clearly shown, and the individual responsible for accomplishing each step is identified. There are also other control features such as a number for identification and a method for recording actual starts and completions for each step. What is not included on a project sheet is an explanation of the state of progress. Since a project sheet is a working schedule, it would be cumbersome to include explanations for deviations from schedule. Therefore, it is common practice to reserve deviation reports for periodic progress reports, which may be either written or oral.

Projects are very important to the success of the manufacturing engineering organization. Therefore a few thoughts on how projects can go astray are in order. Failure of projects to produce the desired results is usually caused by one or more of the following.

1. The steps are not specifically defined; they are too general. Therefore it is difficult to know what must actually be done and when the step is complete. For example, the statement "Examine the QA records for number of failures at the workstation" is very open-ended. Quality assurance records for what type of failure at the workstation? A proper statement for this step could be, "Examine the QA records for the number of failures at the workstation caused by operator error." This is much more specific and focuses on the problem to be evaluated. Examining all failures for all causes may not be relevant to the problem at hand. Another constraint on the element description would be placing a chronological bound on the search—for example, restricting it to a period of six months or ten weeks. This would be helpful if it is suspected that the problem to be solved can be bounded.

2. The allowed time period for accomplishing each step is not observed. The time frame is established to do two things: allow completion of the project on time to support the time frame of the goal, and force conclusions to be drawn for the completion of each step. Frequently, manufacturing engineers will continue an investigation when the return for the effort is not worth the time and expense. Placing a time limit on each step forces evaluations to be made. It is acceptable in this type of work to draw conclusions once two-thirds of the data has been gathered and evaluated. In practical problems involving manufacturing, it is usually impractical to define 100% of the parameters and data. Strict observance of time periods is essential for efficient utilization of manpower. It is most important

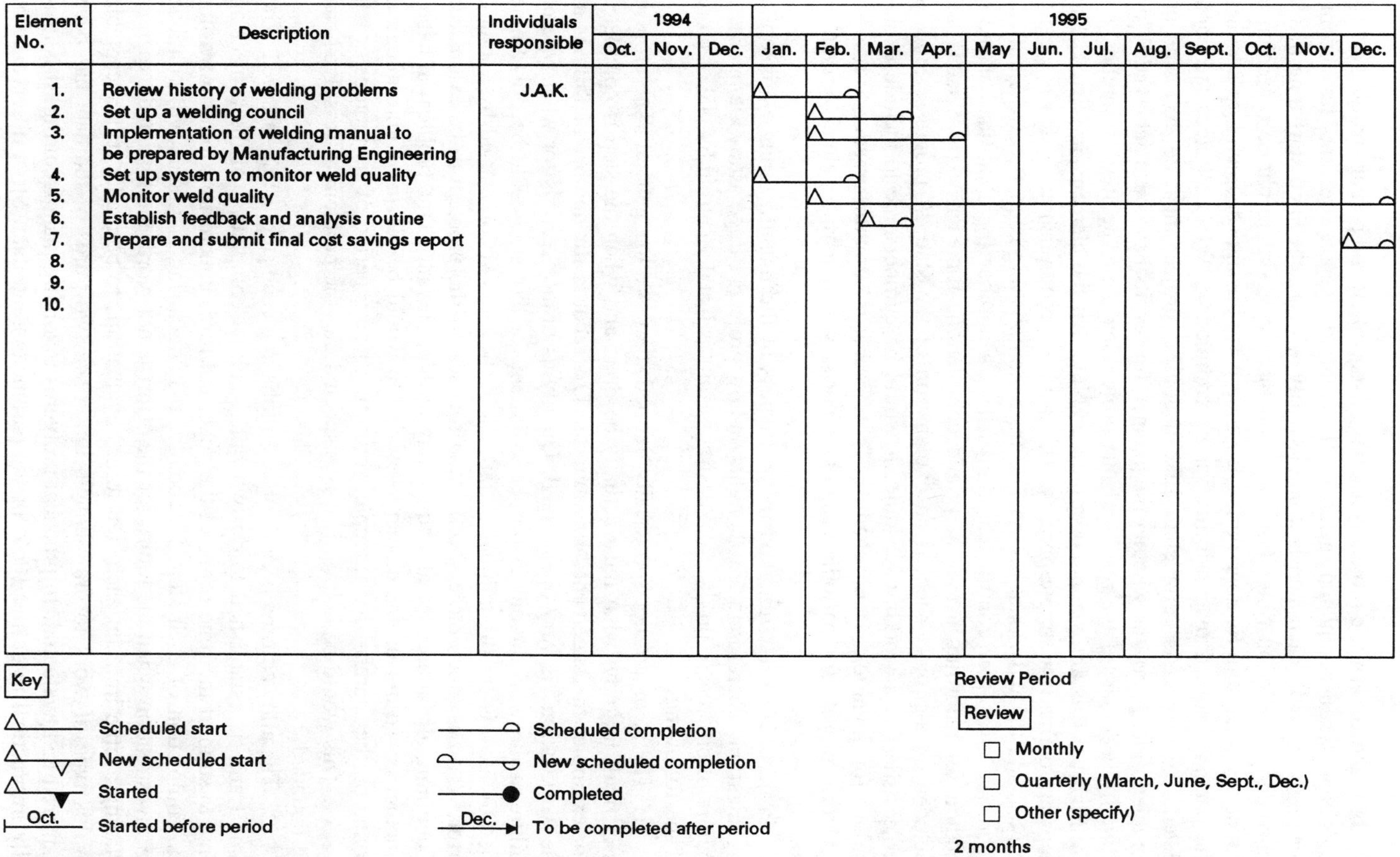

| Element No. | Description | Individuals responsible | 1994 | | | 1995 | | | | | | | | | | | |
|---|---|---|---|---|---|---|---|---|---|---|---|---|---|---|---|---|---|
| | | | Oct. | Nov. | Dec. | Jan. | Feb. | Mar. | Apr. | May | Jun. | Jul. | Aug. | Sept. | Oct. | Nov. | Dec. |
| 1. | Review history of welding problems | J.A.K. | | | | | | | | | | | | | | | |
| 2. | Set up a welding council | | | | | | | | | | | | | | | | |
| 3. | Implementation of welding manual to be prepared by Manufacturing Engineering | | | | | | | | | | | | | | | | |
| 4. | Set up system to monitor weld quality | | | | | | | | | | | | | | | | |
| 5. | Monitor weld quality | | | | | | | | | | | | | | | | |
| 6. | Establish feedback and analysis routine | | | | | | | | | | | | | | | | |
| 7. | Prepare and submit final cost savings report | | | | | | | | | | | | | | | | |
| 8. | | | | | | | | | | | | | | | | | |
| 9. | | | | | | | | | | | | | | | | | |
| 10. | | | | | | | | | | | | | | | | | |

Key

◁ Scheduled start
▽ New scheduled start
▼ Started
Oct. Started before period

◁ Scheduled completion
▷ New scheduled completion
● Completed
Dec. To be completed after period

Review Period

Review

☐ Monthly
☐ Quarterly (March, June, Sept., Dec.)
☐ Other (specify)

2 months

Figure 2.4 Project sheet format.

for those responsible for project completions to be very jealous of conserving time.

3. The project is incompletely sequenced. Either there are not enough steps to complete the project, or each step is too broad, requiring more than one item to be accomplished. For the example in item 1 above, we would not want to find all operator error failures and also evaluate the causes of these errors in one step. A better sequence would be to categorize the causes of the operator errors after the total numbers have been found and placed in the data base. This is the preferable procedure because the causes in the different categories would be quantified, and the major effort could be directed to the categories with the largest number of occurrences.

Combining steps usually results in losing the sharp focus of a project. Items may be overlooked, to the future dismay of the engineer. Of course, care must be taken not to be excessively precise. Moot elements must be eliminated. People are capable of interpretation; therefore computerlike sequencing is unnecessary. If the key steps are included, as understood by the engineer and the manager, then the fault of step combination will be avoided.

## SETTING UP THE OBJECTIVES AND GOALS SYSTEM

The objectives and goals system is a tool for proper management of an organization. However, the total cannot be utilized until the specific situation of the organization is analyzed. I cannot overemphasize the need to analyze the organization's reason for being before setting out to implement an objectives and goals system.

Many manufacturing engineering organizations tend to drift with the tide, doing what is asked of them by management, but never seeming able to get ahead of the situation. They are excellent problem solvers, using techniques learned after many years of reacting to business crises. Their planning is shallow; it is characterized by Long-Range Forecasts (LRFs) that are nothing more than shopping lists of equipment they would like to have or hunches about what would help them produce their products. There are no true goals, but projects abound to solve crises and to purchase items from their LRFs. This is the situation that exists in the many manufacturing engineering organizations that take the term "service organization" too literally.

In order to truly serve its company, a manufacturing engineering organization must be an initiator instead of a reaction. In order to initiate, it must plan ahead how to do it. Implementation of an objectives and goals system allows for continuous planning ahead, but only after identifying the objectives of the company. For manufacturing engineering that means discovering the basic objectives of the manufacturing function and then modifying them to be meaningful for manufacturing engineering activities.

## Determining the Organization's Objectives

For a newly launched organization, it is relatively easy to determine what the objectives are. They may be couched in other phrases, but they are there. For example, let us say that a computer system function is established within a company. The reason given may be: "To establish our company in the forefront of the computer revolution so that we can contribute to this exciting new technology to better serve our customers." The hidden objectives are to get ahead of the competition in implementing computers and, by doing so, to reduce operating costs. Thus, the objectives that this organization would use as its overall operating policy are:

1. Implement the utilization of computers in company operations.
2. Reduce the operating costs of company operations.

Objective 1 would lead to goals such as: "Implement the Honeywell DPS computer system in the purchasing unit by April of the next budget year." Objective 2 would lead to goals such as: "Reduce the operating expense of the Honeywell DPS computer system 25% by batch processing, effective in the third quarter of the budget year."

For existing organizations, the task of finding objectives is a little more difficult. There is no preamble stating why the organization was formed. Therefore, an evaluation of what the organization does and what it should do to support overall company policy is in order. Let us look at the function of which manufacturing engineering is a part, that is, manufacturing. Understanding the core objectives of this function will certainly lead to objectives for manufacturing engineering.

Manufacturing produces the product. Therefore, an objective of manufacturing would be to maintain the capability to produce its products. However, maintaining a capability is obviously not enough; it must be usable in a cost-effective or competitive way. Therefore, another objective could involve an attempt to reduce operator costs, that is, improve productivity. An associated objective would be to minimize all types of mistakes due to misunderstandings or breakdowns in communication, which would be commonly stated as, "Improve interfaces."

Summarizing, then, the objectives of the manufacturing function would be:

1. Maintain a capability to produce the company's products.
2. Improve productivity relative to producing the products.
3. Improve the quality of the products.
4. Improve interfaces with internal and external organizations.

The same objectives could be expressed differently (and often are), but the results would be the same. For example, "Improve productivity" could just as easily have been "Reduce operating costs," because improving productivity in materials utilization, personnel performance, machine tool operations, and so forth in actuality reduces operating costs. The words used for the objectives are chosen to emphasize the needs of the organization and the importance placed on these factors of the operation. Regardless of the choice of words, these are the four basic objectives of a manufacturing function.

Knowing the objectives of the manufacturing function, it is logical to conclude that the objectives of the manufacturing engineers subfunction must be in support. They could be the same—and, in fact, they usually are—but they could also be more specific or narrowly defined. This is illustrated in Figs. 2.5 and 2.6. Figure 2.5 illustrates the objective of improving productivity, a broad objective that is shared with all the manufacturing subfunctions. Notice, however, that the supporting goal is a manufacturing engineering goal. Figure 2.6 illustrates a narrower objective, conserving energy. This is a subset of the broader objective, of improving productivity, and more readily defines the manufacturing engineering organization's contributions to the broader objective.

There is no rule about how many objectives an organization can have, or whether the broader objectives should be mixed with the narrower ones. The only important rule is that objectives should be identified and that they should be true objectives—ongoing, and not changing over time.

---

Objective: Improve productivity

Goals:  Write appropriation requests for the 1996 facility plan based on the 1994 long range forecast. All ARs for major projects will be completed by the end of the fourth quarter of 1995.

| Projects: | Completion date | Savings | Priority |
|---|---|---|---|
| N/C chucker med. motor | 550 | $100 | Y |
| Test data collection equipment | 548 | 50 | X |
| Upgrade test equipment | 548 | 20 | X |
| Rebuild radial drill press | 540 | 15 | Z |
| M bar taping machine | 550 | 20 | Y |
| AC work center | 539 | 150 | X |
| DNC for frame machine center | 551 | 45 | Y |
| Keyway broach | 545 | 20 | Y |
| Welding robot | 542 | 60 | Y |
| Punch press robots | 548 | 120 | X |
| Bar insertion system fab rotor | 539 | 40 | X |
| Bar shaving fab rotor | 545 | 20 | X |
| Building repairs | 550 | | X |
| Rebuild punch presses | 536 | 40 | X |
| Rebuild gear cutter | 539 | 20 | X |
| N/C mach center MM | 551 | 100 | Y |
| Semiauto sub-arc—Bldg. 60 | 550 | 40 | X |
| Flux core weld station | 548 | 40 | Y |

**Figure 2.5** The objective of improving productivity.

```
Objective:  Conserve energy

Goals:    Reduce annual steam usage, $25,000, using sonic and thermal detectors to
          find leaking steam traps and problem areas on steam processes. Projects will
          be completed in October 1995.

                                                   Completion
Projects:                                          date          Savings
Audit steam traps and processes for problems       2/95
Analyze all steam applications for replacement
or improvement                                     4/95
Write appropriation requests to Advanced
Manufacturing Engineering as required              5/95
Order equipment/implement projects
after appropriation request approval               9/95
Write cost improvement                             10/95         $25,000
```

Figure 2.6 The objective of conserving energy.

## Establishing Goals

Once the objectives have been determined, it is time to determine what the goals for a particular period should be. Note that in Figs. 2.5 and 2.6 a goal is specified to support the objective. For simplicity, only one goal has been shown for each objective, although in actuality many goals may be required to support a single objective. The goals are not haphazard selections; rather they are the results of studies of needs done by manufacturing engineering.

The first step is to understand the overall business plan of the company, which specifies profitability levels to be reached for particular time periods. This, in turn, leads to allocations for operations, advertising, capital investment, and so on. Once the business plan is understood, its various components must be categorized under the ongoing objectives. Here there must be a 100% fit; the business plan cannot be in conflict with the objectives. Once the business plan is laid out in accordance with the objectives, the plan provides the starting point or prototype goals. Usually, these prototype goals are not identifiable with a specific function or subfunction. For example, a goal to make an overall $1 million profit for a product line does not automatically break down to a specific manufacturing goal. But by knowing the dollar amounts reserved for advertising, engineering, and all other nonmanufacturing categories, it is possible to determine the amount of costs allocated to production. Knowing the profit level required, we can then deduce the level of productivity needed. Hence, a goal can be set to reduce the cost of manufacture to produce the desired profit. This becomes the manufacturing goal. We must now reduce it to the manufacturing engineering contribution to the overall manufacturing goal.

Manufacturing engineering now asks, what can be done to support the manufacturing goal under the given objective? This must be determined on the basis of the mission of the manufacturing engineering organization. If the manufacturing goal is to reduce operating costs by a specified amount, then manufacturing engineering must review the potential within all of its units to reduce costs. These reductions can be through more efficient internal operations, such as introducing a computerized planning system to improve the efficiency of each engineer. Or the reductions can be a result of a product of the engineers' work—for example, improvements in shop operation cycle time due to computerized planning giving more standardized results than manual planning. Looking for opportunities that exist within the realm of assigned responsibilities will lead to goals that contribute to the overall manufacturing goal.

In this manner a catalog of goals can be developed in support of the overall manufacturing goal. It is the responsibility of the various levels of management to evaluate the potentials against the organization's capabilities to select goals to pursue. Some selections are quite evident, others are not.

**Example:** Under a goal to improve productivity by 10% for the measurement period, the Advanced Manufacturing Engineering (AME) unit may conclude that a new machine tool to replace an older one would more than achieve the goal. Hence, justifying and purchasing the new equipment may become one of the AME unit's goals to support the overall goal of manufacturing, which in turn supports the business plan and the ongoing objective.

A not so obvious selection of a goal is illustrated in the following.

**Example:** Suppose the manufacturing goal is to reduce operating expenses by 30% during the measurement period. The maintenance unit cannot as its share arbitrarily cut costs 30% to comply by reducing the number of skilled employees assigned, or cutting back on expenditures for spare parts by the dictated amount. If the maintenance unit did so, its costs would obviously be reduced substantially, but what would the effect on shop operations be? Operating expenses for shop operations would go up considerably due to delays in obtaining repair service. When service did arrive, it would be less likely that the repair part would be available. Therefore, for the maintenance unit to comply with the goal of reducing operating expenses by 30%, it must do something else. Perhaps a suitable goal in support of the overall goal would be: "Develop by the end of the fiscal period an operator dispatch and measurement system to improve the response time by 20% and the time for return of equipment to service by 10%." This goal does not mention reducing operating costs, but that certainly would be the indirect result. The machines would be running for a higher percentage of the time, which would mean more product output during a time period and less work in process inventory time, all of which would mean lower operating costs.

Selections of goals must support the overall goals of the organization. These overall goals (business plans), in turn, must support the ongoing objectives of the company. Let us now summarize the method of setting the period goals of the manufacturing engineering organization.

Receiving the period goals of the manufacturing function, the manager determines which goals pertain to manufacturing engineering. (For example, inventory reduction goals seldom concern manufacturing engineering and would not be selected.) The manager then negotiates with the manager of manufacturing to determine what percentage of the total applicable manufacturing goals will be assigned to manufacturing engineering. With this target set, the manager of manufacturing engineering further subdivides the target for each of the units within the subfunction. The unit managers then ask their engineers, project leaders, and foremen to develop specific goals and supporting projects.

What I have described above is a delegation of responsibility for planning down to the lowest possible operating level. The people who will perform the work are required to plan the work they will do to achieve the goal. Once the specific goals and supporting projects have been proposed, the procedure is reversed. At each reporting level, management evaluations are made to determine whether the specific goals and projects selected support the overall goals and can be completed within the given time period and with the available resources. When they are approved by the manager of manufacturing, the goals and projects selected become the operating plan.

The reviewing and selecting of goals is a year-round activity. It is an iterative process of evaluating current projects and making adjustments as outside conditions change. Traditionally, goals are set in the fourth quarter of the operating year, but prudent managers are evaluating and adjusting continuously so that at planning time the results need only be collated and sent out for approval.

## WORK PLANNING

Setting up the objectives and goals system does not guarantee the accomplishment of goals. Work planning does. Work planning is the meticulous attention to detailed steps that makes attainment of all the assigned goals possible. The basic elements of work planning, which are valid for any organization assigned a task to perform in a specified time frame, are:

1. Schedules: Lists of things to accomplish, including responsibilities assigned and due dates.
2. Action plans: Very specific plans for items to be accomplished by a specified individual, by a certain date.
3. One-on-one meetings: Periodic meetings between a higher-level person and a subordinate to discuss items pertaining to all phases of assigned work.
4. Communications: Documentation, both formal and informal of the status of scheduled items.

All work planning schemes must have these four elements. The titles of the elements may vary from organization to organization and sequences may be different, but all four elements must be present for a true work planning system. Without such a system, there is no management; there are only reactions to perceived crises. Let us now examine the elements of a work planning system individually, but keep in mind that they are all under way simultaneously and interact continuously.

## Schedules

If anything can be called a starting point of work planning, it would be the activity of making schedules. There is always a multitude of goals assigned to each unit of manufacturing engineering. Each goal may involve many projects, which in turn have many steps. All of these individual steps for all the projects and goals will have start and due dates. In addition to projects and goals, there may be other assigned activities that are not documented in the statements of objectives and goals but have definite starting and completion dates; an example would be financial report submittals. The scheduling step of work planning must compile all these requirements into an overall schedule of accomplishment for a particular unit or parts of a unit.

Manufacturing engineering organizations usually have computer programs to compile the start and due dates of all activities in one control chart or report. Whether or not a computer program is used, a master control chart or report is necessary for each unit. This is the only way a manager has to view the work in its entirety and see the work load distribution. Figure 2.7 illustrates a work planning schedule. For brevity, only a few items are shown; a typical work planning schedule could run for several pages. Also note that this example is in chronological order of completion dates. Sometimes a companion chart in chronological

**Schedule Control Chart; Manufacturing Engineering 1995**

| Item | Man-hours | Jan. | Feb. | Mar. | Apr. | May | Jun. | | Dec. | Project no. |
|---|---|---|---|---|---|---|---|---|---|---|
| 1. Audit steam traps | 28 | | | | | | | | | 16 |
| 2. Review history of welding problems | 46 | | | | | | | | | 22 |
| 3. Set up system to monitor welding quality | 8 | | | | | | | | | 22 |
| 4. Set up welding council | 18 | | | | | | | | | 22 |
| 5. Establish feedback and analysis routine | 9 | | | | | | | | | 16 |
| 6. Analyze all steam use for replacement and improvement | 48 | | | | | | | | | 16 |
| 7. Implement welding manual | 56 | | | | | | | | | 22 |
| 8. Write appropriations request for new steam traps | 9 | | | | | | | | | 16 |
| 9. Order and receive new steam traps | 22 | | | | | | | | | 16 |
| 10. Install new steam traps | 88 | | | | | | | | | 16 |
| 11. Write Cost Improvement | 6 | | | | | | | | | 22 |
| 12. Write Cost Improvement | 4 | | | | | | | | | 16 |

**Figure 2.7** A work planning schedule.

order of starting dates is also produced. Another item that is frequently on the schedule control chart is an estimate of man-hours to achieve individual line items. This is a useful device in distributing work and determining what type of effort will be required to complete the activity according to the schedule.

Once the control chart has been constructed, it is used as a reference schedule for reviewing progress. The manager uses it to make sure activities are started on time and completed on time, or to understand why deviations in the plan have occurred. It becomes the control document for virtually all the work going on in the specific area of responsibility. If the schedule control chart is constructed properly, all items, including those that are not part of the objectives and goals, appear on the chart. One should not infer, however, that a schedule control chart made up during the fourth quarter of the previous year would be continuously adequate through the present year. Usually this is not the case. Company plans are always sensitive to the marketplace, availability of labor may vary, other activities that must happen as a precursors to the unit's plans may not occur on time, and so forth. Therefore, the schedule control chart usually goes through several modifications during the year. A manager must determine the risks in accomplishing the schedule and be ready to replan activities on the basis of well thought out contingencies.

## Action Plans

The schedule shows the work to be accomplished by the entire unit or a larger entity. The action plan shows the work to be done by the individual. Depending on the size of the unit and the number of specific items to be accomplished, an action plan may be for several individuals or for only one person. If it is for more than one individual, then a project leader is normally assigned and is responsible for the work of all contributors.

If most of the unit's work is described in the objectives and goals statement, then the project sheets can be used as the action plan. Usually, however, an engineer will have several action plan sheets, some of which are project sheets from the specific objectives and goals statement, while the rest are action plan sheets covering miscellaneous activities.

The action plan is the individual engineer's work plan. It is what the engineer expects to be measured and evaluated against. Therefore, in drafting it, both the engineer and the manager should agree on what constitutes a satisfactory conclusion for each specific item of the plan. It must also be understood that it is the engineer or project leader's responsibility to report to the manager any obstacles that cannot be overcome in time to maintain the schedules. The traditional advice—to report back to a superior completion or reason for failure to complete—is critically important for a successfully run organization. Yet this simple rule is often neglected, leading to larger sums of money wasted in recovery efforts necessitated by lack of communication. For action plans to be dynamic, useful management tools, managers must require timely reports from their subordinates. Managers must educate their engineers to report any deviations from the plan together with proposals for corrective actions to bring the activity back on schedule. Through this commonsense approach, they can prevent salvageable projects from becoming unsalvageable.

## One-on-One Meetings

The key method for reaching understanding between a manager and an engineer relative to assigned work is a regularly scheduled (usually weekly) meeting between the two. The fact that there are no other attendees is a critical part of this element of work planning. When only two people are present, they give each other undivided attention. The purposes of the meeting are to review problems with ongoing work in a general and specific sense, to make sure the engineer understands what the manager wants, and to ensure that the manager understands the problems the engineer may be encountering.

This type of interaction, which is commonly referred to as a one-on-one meeting, is not structured. Its content depends on the current and near-term activities of the organization. The superior must come to the meeting prepared to discuss any aspect of the subordinate's work assignments, to point out good and bad performance trends, and possibly to suggest different approaches to particular problems. But the manager must be careful not to monopolize the meeting and must keep in mind that he or she should contribute a maximum of 50% of its content.

The engineer comes to the meeting with the same type of preparation, but with two additional targets: to give the manager a flavor for the work as it is progressing, and to obtain the manager's approval for any changes to current action plans that appear necessary.

The one-on-one meeting is the dynamic part of the work planning system— the part where the static plan becomes active and where policy decisions and changes are discussed and agreed to. It is the only way, except by actually doing the work, for the manager to measure the pulse of the entire organization.

It can also be thought of as a tutorial session where the superior can teach practical aspects of management to the subordinates. One of the roles of a manager is to develop skills within the employees. At the one-on-one meeting the manager can demonstrate the correct role model and instill good management principles.

## Communications

Lack of adequate communications has been and will continue to be the largest single cause of failure in any organization. Subordinates always have the duty of keeping their superiors informed about work progress. Managers must always give their subordinates enough information to adequately perform their assigned tasks. The work planning system requires the establishment of such up and down communications.

This element requires formal communications in the nature of periodic written reports and oral presentations. Two types of reports that have proved to be useful are the key results report and the quarterly review.

The key results report consists of a summary of the status of all key elements of work assigned to the unit. The manager will report in graphical and/or descriptive form the status of all activities compared to the plan. Data for the report come from updating the unit's work schedule and the individual's action plans and from the notes taken by the manager during the various one-on-one meetings with the subordinates. The key results report also contains financial data related to manufacturing engineering.

I have always found it useful to report financial deviations from budget by using descriptive methods for explanations and corrective action statements. A typical key results report for manufacturing engineering would contain the following:

1. A list of projects by title under their respective goals, showing actual and planned percentages of completion. Declarative phrases explaining deviations and planned corrective action should also be included where needed.
2. A section showing current measurements versus targets, such as manufacturing losses charged to manufacturing engineering; cost improvement status; productivity measurements; actual versus target data on critical financial accounts, such as funds spent for capital equipment purchases; and operator safety statistics.
3. One or two descriptive paragraphs outlining specific, significant accomplishments, for example, completion of an important project.
4. One or two paragraphs outlining specific concerns or directions the organization is taking on certain matters.
5. A section on personnel matters, covering morale, open positions, promotions, and so forth.

This report is sent to the manager's direct superior with copies to the subordinates and to other managers on the same reporting level within the manufacturing function. This is an easy way for a manager to keep all key associates informed and to make sure they all understand what the manager thinks is important.

The key results report is usually a monthly report, which makes it a document that can be used as a daily management tool. Since the report is due at relatively short intervals, it serves as a reminder to the manager to keep close watch on progress related to work planning schedules and action plans.

The quarterly review is an oral presentation, usually with visual aids, coordinated by the manager of manufacturing engineering but presented to the manager of manufacturing by the unit managers and key individual contributors. In other words, it is a presentation to a manager one reporting level higher than their immediate superior. The quarterly review does several things. First, it allows individuals further down the chain of command to report directly to senior management. Second, it allows the manager of manufacturing to ask questions directly of those managers and individual contributors with whom he would not normally have day-to-day contact. Third, it provides business exposure for junior personnel, who can get a flavor for the entire business picture by listening to the types of questions asked and other remarks made by the manufacturing manager. Fourth, it gives lower-level managers and individual contributors an opportunity to display their knowledge and ability to express themselves. Finally, it allows the manager of manufacturing to compare actual happenings with documented plans.

The quarterly review is an excellent communications technique. Since it is an oral review, the proceedings take on some aspects of a planning session. Invariably, the highest-level manager present will suggest changes, albeit small, that will affect one or more of the other three elements of work planning. Hence, this step also helps to convert a static written plan into a dynamic plan.

## FINANCIAL CONTROLS

We have discussed objectives and goals formulation and work planning. These activities lead to tasks that manufacturing engineering must perform. Now it is necessary to bring in the controlling factor that determines how much manufacturing engineering can do to develop an optimally productive factory. Financial controls put the restraint of reality on all programs. They make managers assess all activities in pragmatic terms, deciding which projects should be carried out and how to husband the always limited resources to achieve the best results.

Financial controls means staying within the budgetary levels imposed by the appropriate level of management. In making a long-range forecast, manufacturing engineering has the opportunity to propose that certain programs and projects should be started or continued and that others should be terminated. This almost always requires an assessment of what the costs and paybacks will be. The long-range forecast proposals are then weighed against available funds and proposals made by the other functions. Eventually a dividing up of the available funds takes place, resulting in an operating budget for the upcoming fiscal year and an official plan for the succeeding four years—a five-year plan in total. Usually the plan for the year after the new budget year is relatively firm and becomes less firm with each succeeding year. It is the budget year that is of primary importance for financial controls.

There are three main activities related to financial controls that must be undertaken by manufacturing engineering: (1) short-range forecasting, (2) periodic review of operations, and (3) daily accounts control. Short-range forecasting means predicting the amount of money to be spent in the next period (usually a month) and the succeeding two periods. The periodic review of operations is an analytical appraisal of how the forecast compared with the actual occurrences and determination of why there were differences. Finally, daily accounts control is a method of ensuring the specific accounts are not overspent and that the short-range forecast is attained. Let's look at these three activities in closer detail.

## Short-Range Forecasting

Figure 2.8 is a typical Short-Range Forecast (SRF) form. It lists the accounts for which the manufacturing engineering function is responsible and has space to predict what the related expenditures will be for the forecast periods. It also has space to show how the forecast differs, if it does, from the operating budget and to give reasons for the variations. The very fact that the SRF has places to show variations from budget implies that individual accounts may vary due to unforeseen circumstances. This leads us to the major challenge for management: how to allow variations in individual accounts and still maintain the overall budget level for expenditures.

The SRF is a monthly exercise conducted by all manufacturing subfunctions. Usually the first week of the current month is established as the due date for the SRF. This discipline causes all operations to look ahead over a period of 7 to 8 weeks for immediate concerns and up to 17 weeks (one quarter of the year plus the current month) for near-term concerns. It tends to force management to be concerned about the effect of current decisions in the near future, and thus tends to prevent expedient decisions that would cause problems later on.

> **Example:** Process control makes a concession to allow a part to proceed that is out of tolerance; that is, it takes a chance that the part will still work. This is done to make the quota for the immediate production goal. A month or two later the substandard part is used and the product fails in test. As a result, the manufacturing losses as shown in the overtime premium account are excessively high and shipment of the product to the customer is delayed. The actual expenditures are severely over the forecast, because of a decision that had only immediate short-term goals.

The SRF system tends to prevent problems such as that described above, because the manager would look at such a decision in terms of both what it might accomplish now and what the costs could be in the future.

The manufacturing engineering SRF is made up by the manager upon receipt of the SRFs from the various units. It is a system that should reach down to the project leaders, individual engineers, and foremen for inputs on their specific activities. Each manager must then critically review the inputs for practicality based on present circumstances.

> **Example:** A forecast requiring capital expenditures at three times the level of the previous forecast must be based on a change in equipment delivery rate or some other valid reason.

Every line on the forecast must be critically appraised before it is passed up the management chain. Managers are aware that their capabilities are being constantly evaluated, and one of the most visible signs of their worth is the validity of their SRFs.

FORECAST OF OPERATIONS AND EMPLOYEES

Subsection: _____     For the Period: _____

| Account Title | Account Nos. | Forecast | | | Deviation from Budget | | | Remarks |
|---|---|---|---|---|---|---|---|---|
| | | Next month | 2nd month | 3rd month | Next month | 2nd month | 3rd month | |
| Wages and Salaries | All 1 . . . | | | | | | | |
| Overtime Premiums | 210, 11, 12 | | | | | | | |
| Idle Time | 230 | | | | | | | |
| Other Premiums and Allowances | 220, 31, 40, 50, 51, 82, 86 | | | | | | | |
| Vacation and Statutory Holidays | 284, 85 | | | | | | | |
| Employee Benefits | All 3 . . . | | | | | | | |
| Total Compensation | | | | | | | | |
| Rentals | 460, 62 | | | | | | | |
| Stationery and Office Supplies | 504, 25, 26, 27, 28 | | | | | | | |
| Outside Services | 514, 16 | | | | | | | |
| Shop Tooling | 517, 18, 19 | | | | | | | |
| Welding Rod | 520 | | | | | | | |
| Shop Supplies | 521 | | | | | | | |
| Computer Charges | 530, 31, 862, 64 | | | | | | | |
| Transportation, Duty, Education, and Traveling | 542, 62, 54 | | | | | | | |
| Utilities | 571, 72, 83 | | | | | | | |
| Telephone | 576, 77, 78, 79 | | | | | | | |
| Rearrangements and other Building Maintenance | 584, 99, 587 | | | | | | | |
| Machinery Maintenance | 588, 89 | | | | | | | |
| Total Other Controllable Expenses | | | | | | | | |
| Assessments | All 8 . . . | | | | | | | |
| Expense Credits | All 9 . . . | | | | | | | |
| Total Applied Costs | | | | | | | | |
| Liquidations | 911 | | | | | | | |
| O/U* Liquidations | | | | | | | | |
| Employees | | | | | | | | |
| Hourly–Direct | | | | | | | | |
| Hourly–Indirect | | | | | | | | |
| Managerial and Professional | | | | | | | | |
| General Salaried | | | | | | | | |
| Total | | | | | | | | |

**Figure 2.8** Typical short-range forecast form.

The SRF is the manager's work plan spelled out in financial terms. If it turns out to be very different from what actually occurs, then it indicates that the manager's plan was not very good. For this reason, managers must review their plans in depth to be sure that the funds are available and are in agreement with the various unit action plans. Any good management plan requires comparison of actual with predicted expenditures. Therefore, those making predictions must expect critical and timely reviews of their performance in this area.

Figure 2.9 shows one line of an SRF filled out. The forecast shows a budgeted weekly expenditure of $3923.00, which is rounded to $4000.00 and expressed as $4.0 per week because forecasts are always in thousands of dollars. The fact that

| Account Title | Account no. | Forecast | | | Deviation from Budget | | | Remarks |
|---|---|---|---|---|---|---|---|---|
| | | Next month | Second month | Third month | Next month | Second month | Third month | |
| Machinery Maintenance | 588, 89 | 17.5 | 16.0 | 20.0 | 2.5 | 0.0 | 0.0 | Repair to CNC lathe, breakdown |

**Figure 2.9** One line of a short-range forecast form filled out.

the budgeted weekly expense is $4.0 is determined by looking at the "deviations" columns and knowing that a quarter is always 13 weeks, with one of the months having 5 weeks. Adding the three monthly forecasts together, subtracting the expected deviation, and then dividing by 13 gives the budget expenditure per week. The five-week months are traditionally the first month of each quarter, that is, January, April, July, and October. Therefore, for budget purposes, we have five-week periods and four-week periods. Note the deviation shown in the "next month" column. This shows that management has decided to allow an overexpenditure by $2.5 for the next period. The reason for this variance is explained in the "remarks" column. If not enough space is available, a notation would be made in that column signifying an explanation made somewhere else, usually at the bottom of the form. An analysis is made for each line of the SRF, but only significant variances are explained. If the variance of the example had been $0.5 or less, no explanation would have been required. The exact amount of variance requiring explanation will depend on the magnitude of the account and the judgment of the manager. For example, an account expending $250.0 per month would not require an explanation for a $2.5 variance. After each line has been evaluated and explanations entered where necessary, all the vertical columns are totaled and the forecasts are evaluated against budget to see how the unit of subfunction is faring. Here, the responsible manager must judge whether the forecast is acceptable. If the forecast is acceptable, it is forwarded. If not, it is sent back to the manager's subordinates for revision. Any revisions made to the SRF will of necessity result in changes in various action plans.

The last section of the SRF is an employee forecast. Here the manager is comparing the people on the staff, plus those considered necessary to conduct the plan, against the numbers that make up the wages, salaries, and benefits budgets. There are many reasons for employee numbers to be different from budget numbers: differences from expected productivity levels that were used to predict personnel needs when long-range forecast and budgets were prepared, changes in the scope of projects, and other changes in plans are a few of the more common reasons. Personnel forecasting is required to remind management that personnel allocations are just as much a budgetary item as stationery and other accounts, and that the expected levels of performance must be reached by all employees for the organization to achieve its goals.

## Periodic Review of Operations

The periodic review of operations is the measurement of the plan. It is carried out monthly by all those producing the SRF. The review starts with a comparison of the actual financial results with the month's forecast. Then, for all significant deviations, an investigation is made to determine the causes of variance. This investigation of causes yields the greatest benefit.

An SRF is an earnest attempt to predict the outcome of putting a plan into action. A simple plan leads to accurate predictions because all items or activities involved in the outcome can be anticipated and controlled. Unfortunately, most business plans are not simple, and the outcomes are not quite so predictable because a significant portion of the activity is out of the control of the project leader.

**Example:** A goal to expend 1000 hours of maintenance time to fix a series of machine tools by a specified date requires a knowledge of exactly what must be done and how quickly and consistently the repair team can work. The problem of the machine tools may be precisely determined, but the rate of work can only be approximated. A key individual may not come to work for a good reason, or another project may become more critical at the time, thus drawing away the key individual. It might then be necessary to have the job done by a less experienced person, who would spend more hours on the repairs than originally forecast. The result is a missed forecast and a deviation from the actual that must be evaluated.

We evaluate results so we may learn from previous errors and not repeat those errors. A deviation from forecast properly explained may lead to beneficial modifications in plans and make it easier for the organization to achieve its goals.

Figure 2.10 is an example of a form used in a typical periodic review of operations. Note that there are two sets of columns for comparing actual with forecast: current month and year to date. It is very important to be able to view one month's results in the perspective of the longer period of time signified by the year-to-date columns. This helps to keep responsible people from taking unnecessary actions when, in fact, the variances of the monthly results may not be indicators of true trends.

Figure 2.11 is an example of a completed review of operations. Note that a variance amount is shown for each account and that the year-to-date SRF column is a summation of all the previous months. Some managers may prefer to use the previous year to date plus the current month's forecast, especially if the unit is experiencing significant variance in many accounts and forecasting errors are compounding. Note also that not all variations are explained in Fig. 2.11. The same reasoning as for the SRF deviations is applied here. If the manager fully understands the causes of variance and can take necessary corrective action, complete explanations would be a waste of time. However, explanations require some sort of investigation, and this leads to an understanding of the cause of the problem. Therefore, wherever practical, all variances from the forecast greater than 10% should be investigated.

## Daily Accounts Control

Daily accounts control is tantamount to ledger management. This can be tedious and should be used only for accounts that are difficult to forecast or in which large variances would be very damaging. It is practically impossible for engineers to keep track of each entry into each account; to do so would mean the virtual abandonment of real engineering work. It is also not practical to assign a financial

REVIEW OF OPERATIONS

Subsection: _____     For the Period: _____

| Account Title | Account Nos. | Current Month | | | Year to Date | | | Remarks |
|---|---|---|---|---|---|---|---|---|
| | | Actual | SRF | Var. | Actual | SRF | Var. | |
| Wages and Salaries | All 1 . . . | | | | | | | |
| Overtime Premiums | 210, 11, 12 | | | | | | | |
| Idle Time | 230 | | | | | | | |
| Other Premiums and Allowances | 220, 31, 40, 50, 51, 82, 86 | | | | | | | |
| Vacation and Statutory Holidays | 284, 85 | | | | | | | |
| Employee Benefits | All 3 . . . | | | | | | | |
| Total Compensation | | | | | | | | |
| Rentals | 460, 62 | | | | | | | |
| Stationery and Office Supplies | 504, 25, 26, 27, 28 | | | | | | | |
| Outside Services | 514, 16 | | | | | | | |
| Shop Tooling | 517, 18, 19 | | | | | | | |
| Welding Rod | 520 | | | | | | | |
| Shop Supplies | 521 | | | | | | | |
| Computer Charges | 530, 31, 862, 64 | | | | | | | |
| Transportation, Duty, Education, and Traveling | 542, 62, 54 | | | | | | | |
| Utilities | 571, 72, 83 | | | | | | | |
| Telephone | 576, 77, 78, 79 | | | | | | | |
| Rearrangements and other Building Maintenance | 584, 99, 587 | | | | | | | |
| Machinery Maintenance | 588, 89 | | | | | | | |
| Total Other Controllable Expenses | | | | | | | | |
| Assessments | All 8 . . . | | | | | | | |
| Expense Credits | All 9 . . . | | | | | | | |
| Total Applied Costs | | | | | | | | |
| Liquidations | 911 | | | | | | | |
| O/U* Liquidations | | | | | | | | |
| Employees | | | | | | | | |
| Hourly–Direct | | | | | | | | |
| Hourly–Indirect | | | | | | | | |
| Managerial and Professional | | | | | | | | |
| General Salaried | | | | | | | | |
| Total | | | | | | | | |

**Figure 2.10** Periodic review of operations form.

analyst to each person who is required to manage portions of a company's money. Therefore, a middle-ground approach is used.

First, accounts that are out of control must be determined. This does not mean that accounts with consistent variances between forecast and actual should be classified as out of control. An account such as "electric utilities" may be totaling consistently higher than the budget and forecast every month, and still be well controlled. The reason may be energy prices are rising faster than predicted, and forecasts are consistently underestimating them. Daily accounts control is not intended to correct this type of situation. A correction here would typically involve senior members of the financial function who develop economic guidelines to be used and would probably result in a request that the manufacturing function develop more vigorous energy conservation measures. This, in turn, would result in major changes in the objectives and goals statement.

Once the accounts that are truly out of control have been determined, an effective plan must be devised to bring them into control. This is the purpose of daily accounts control.

Figure 2.12 is a form used to keep track of all forecast and actual amounts during a budget year for specific accounts. Note that there is space for budget,

REVIEW OF OPERATIONS

Subsection: __Mfg. Engineering__               For the Period: __May 1995__

| Account Title | Account Nos. | Current Month | | | Year to Date | | | Remarks |
|---|---|---|---|---|---|---|---|---|
| | | Actual | SRF | Var. | Actual | SRF | Var. | |
| Wages and Salaries | All 1 . . . | 31.4 | 30.1 | 1.3 | 141.7 | 131.4 | 10.3 | |
| Overtime Premiums | 210, 11, 12 | 0.3 | 1.0 | (0.7) | 3.6 | 5.0 | (1.4) | |
| Idle Time | 230 | 0.1 | 0.4 | (0.3) | 0.5 | 1.6 | (1.1) | |
| Other Premiums and Allowances | 220, 31, 40, 50, 51, 82, 86 | 0.4 | 0.9 | (0.5) | 6.1 | 13.6 | (7.5) | |
| Vacation and Statutory Holidays | 284, 85 | 4.9 | 6.4 | (1.5) | 31.9 | 25.6 | 6.3 | |
| Employee Benefits | All 3 . . . | 11.3 | 13.0 | (1.7) | 62.4 | 72.0 | (9.6) | |
| Total Compensation | | 48.4 | 51.8 | (3.4) | 246.2 | 249.2 | (3.0) | |
| Rentals | 460, 62 | 9.8 | 1.5 | 8.3 | 13.7 | 5.1 | 8.6 | $8,204 wrongly |
| Stationery and Office Supplies | 504, 25, 26, 27, 28 | 4.3 | 4.5 | (0.2) | 13.7 | 14.9 | (1.2) | allocated to coil |
| Outside Services | 514, 16 | 14.1 | 6.0 | 8.1 | 27.4 | 20.0 | 7.4 | plant |
| Shop Tooling | 517, 18, 19 | 0.3 | 0.8 | (0.5) | 2.8 | 3.2 | (0.4) | |
| Welding Rod | 520 | 0.0 | 0.6 | (0.6) | 1.1 | 2.4 | (1.3) | |
| Shop Supplies | 521 | 0.3 | 0.5 | (0.2) | 0.9 | 1.8 | (0.9) | |
| Computer Charges | 530, 31, 862, 64 | — | — | — | — | — | — | |
| Transportation, Duty, Education, and Traveling | 542, 62, 54 | 12.0 | 8.1 | 3.9 | 23.4 | 32.4 | (9.0) | |
| Utilities | 571, 72, 83 | 0.1 | 0.7 | (0.6) | 1.7 | 2.8 | (1.0) | |
| Telephone | 576, 77, 78, 79 | 1.0 | 3.5 | (2.6) | 6.3 | 14.2 | (7.9) | |
| Rearrangements and other Building Maintenance | 584, 99, 587 | 5.5 | 7.0 | (1.5) | 23.8 | 28.0 | (4.2) | |
| | | 0.0 | 0.1 | (0.1) | 0.0 | 0.4 | (0.4) | |
| | | 0.0 | 9.3 | (9.3) | 0.2 | 19.0 | (18.8) | coil facility |
| Machinery Maintenance | 588, 89 | 0.0 | 0.7 | (0.7) | 0.2 | 2.8 | (2.6) | delayed |
| Total Other Controllable Expenses | | 9.7 | 13.0 | (3.3) | 26.6 | 52.0 | (125.4) | |
| Assessments | All 8 . . . | 43.0 | 50.4 | (6.4) | 113.3 | 179.0 | (65.7) | |
| Expense Credits | All 9 . . . | 24.5 | 24.4 | 0.1 | 78.0 | 79.2 | (1.2) | |
| Total Applied Costs | | (7.8) | (1.0) | (6.8) | (7.8) | (4.0) | (3.8) | |
| Liquidations | 911 | 108.1 | 125.6 | (17.5) | 429.9 | 503.5 | (73.6) | |
| O/U* Liquidations | | 53.9 | 65.7 | (11.8) | 328.8 | 452.8 | (124.0) | |
| Employees | | *54.2 | *59.9 | (5.7) | *101.1 | *69.2 | *31.9 | |
| Hourly–Direct | | 24 | 24 | 0 | | | | |
| Hourly–Indirect | | 6 | 6 | 0 | | | | |
| Managerial and Professional | | 7 | 7 | 0 | | | | |
| General Salaried | | 3 | 3 | 0 | | | | |
| Total | | 40 | 40 | 0 | | | | |

**Figure 2.11** Completed review of operations form.

forecast, and actual amounts. This form allows the engineer to track forecasts for the previous three months and then compare them to actual and budget. We can see the entire picture or history of the account for the year on one page, and see whether there is any relationship between forecast and actual. Let us look at the month of April. Note that telephone and telegraph charges have been forecast three times in three different short-range forecasts: January, $2.0; February, $2.0; and March, $2.1. The actual turned out to be: April, $2.05. In this case, the actual and forecast amounts are very close. Also, the actual and budget amounts are relatively close. With the actual, the three forecasts, and the budget virtually the same, the account is clearly under control. No other measures need be instituted to track it.

Incidentally, note that in Fig. 2.12 the entries are not typed, but entered in longhand. This is acceptable because this is a control document for a specific individual, and would not normally be submitted up the chain of command. This is the type of document that an individual engineer would take along to a one-on-one meeting.

Now let us look at an account that does not pass this three-way check and is not under control, and see how we can get the forecasts closer to actuality and

Subsection: __Mfg. Eng.__  Account Title __Telephone and Telegraph__  Account No. __567__

Budget __20.0 K__
Target __10.0 K__

| | | Jan. | Feb. | Mar. | Apr. | May | Jun. | Jul. | Aug. | Sept. | Oct. | Nov. | Dec. |
|---|---|---|---|---|---|---|---|---|---|---|---|---|---|
| Budget | Monthly | 1.5 | 1.5 | 1.5 | 1.9 | 1.5 | 1.5 | 1.9 | 1.5 | 1.5 | 1.9 | 1.5 | 2.3 |
| Forecast | Cum.YTD | 1.5 | 3.0 | 4.5 | 6.4 | 7.9 | 9.4 | 11.3 | 12.8 | 14.3 | 16.2 | 17.7 | 20.0 |
| Jan. | | 1.5 | | | | | | | | | | | |
| Feb. | | 1.8 | 2.0 | | | | | | | | | | |
| Mar. | | 2.0 | 2.0 | 2.1 | | | | | | | | | |
| Apr. | | | | 1.6 | 1.5 | | | | | | | | |
| May | | | | 1.6 | 1.5 | 1.5 | | | | | | | |
| Jun. | | | | | 1.6 | 1.6 | | | | | | | |
| Jul. | | | | | | 1.5 | | | | | | | |
| Aug. | | | | | | | | | | | | | |
| Sept. | | | | | | | | | | | | | |
| Oct. | | | | | | | | | | | | | |
| Nov. | | | | | | | | | | | | | |
| Dec. | | | | | | | | | | | | | |
| Monthly Actual | | 1.72 | 1.44 | 1.93 | 2.05 | 1.55 | | | | | | | |
| Cum. YTD Actual | | 1.72 | 3.16 | 5.09 | 7.14 | 8.69 | | | | | | | |

**Figure 2.12** Form used to follow forecast versus actual amounts during a budget year.

budget. Figure 2.13 shows an account that is out of control. The forecast amounts are either significantly over or under actual amounts. In this case the forecast is quite useless for planning expenditures. It is evident that expenses are following a pattern quite different from that of the forecasts. This means that the individual responsible for this account must do considerably more to understand what is happening and to bring it under control. This example is typical of response accounts, that is, accounts where the manufacturing engineering organization must respond quickly to current situations. However, this does not excuse manufacturing engineering from performing on budget, since that would put the manufacturing function at the mercy of the lowest level within the organization, with policy being proclaimed by the actions of the most junior personnel.

Maintenance accounts are examples of response accounts. When equipment is malfunctioning, it must be fixed as soon as possible. Even with a chance-happening activity, it is possible to control the account and still be able to respond in an emergency. The key to successful control is to spend only the allocated funds for each month, and to spend each month of the forecast's funds simultaneously. This may sound like double-talk, but it is not. It simply involves knowing each month's budget (target) and anticipating when (in what month) the bill will be received and have to be paid. Studying Fig. 2.14 will show how this is done. Note that the budget is shown for each month. This is the amount the unit wants to have charged each month against this account. In May, $1.6 less than anticipated was charged to this account. That means that the remaining $1.6 will have to be paid at a later time. This then becomes the rollover or anticipated maximum charge against the next month's allocated funds. The column headed "special shop orders" is a list of all building maintenance activities that have occurred through June 21, the date of this report. This would include labor and material charges through June 21st. Additional charges on succeeding days would be entered in later editions of this daily report. Labor will invariably be charged in the month when it occurs because it is usually manufacturing engineering's own people doing the work, hence the billing should have minimal bookkeeping delays. Material is booked and paid for only after it arrives and the bill is received. One of the main purposes of the format is to determine the month when material will have to be paid for.

It is evident in Fig. 2.14 that Purchase Orders (POs) 3, 4, 6, 7, 11, and 12 are for material purchases; and they are placed in the month when the bill is expected to be paid. Usually the date promised by the vendor is used to place the future bill in the correct month. The total through June 21 is shown for each month. Figure 2.14 shows that June has reached its allocated fund level, hence no more work should be scheduled to be performed in June. Also note that the next three months already have charges booked against them. For July, $2.4 of the $6.0 budget has already been expended and the month has not yet started. A similar situation exists for August and September.

This system requires meticulously recording what has been spent and managing the operation to make sure that the results will be within the budget. This

Subsection: __Mfg. Eng.__ Account Title __Building Maintenance__ Account No. __587__

Budget __64.0K__
Target __55.0K__

| Forecast | | Jan. | Feb. | Mar. | Apr. | May | Jun. | Jul. | Aug. | Sept. | Oct. | Nov. | Dec. |
|---|---|---|---|---|---|---|---|---|---|---|---|---|---|
| Budget | Monthly | 6.0 | 5.0 | 5.0 | 6.0 | 5.0 | 5.0 | 6.0 | 5.0 | 5.0 | 6.0 | 5.0 | 5.0 |
| | Cum. YTD | 6.0 | 11.0 | 16.0 | 22.0 | 27.0 | 32.0 | 38.0 | 43.0 | 48.0 | 54.0 | 59.0 | 64.0 |
| Jan. | | 5.0 | | | | | | | | | | | |
| Feb. | | 5.0 | 5.0 | | | | | | | | | | |
| Mar. | | 6.0 | 6.0 | 7.5 | | | | | | | | | |
| Apr. | | | 5.0 | 5.5 | 6.0 | | | | | | | | |
| May | | | | 5.5 | 6.0 | 5.0 | | | | | | | |
| Jun. | | | | | 8.0 | 7.5 | 7.5 | | | | | | |
| Jul. | | | | | | 5.0 | 5.0 | 6.0 | | | | | |
| Aug. | | | | | | | 5.0 | 5.5 | 6.0 | | | | |
| Sept. | | | | | | | | 8.0 | 8.5 | 8.0 | | | |
| Oct. | | | | | | | | | 6.0 | 5.5 | | | |
| Nov. | | | | | | | | | | 5.5 | | | |
| Dec. | | | | | | | | | | | | | |
| Monthly Actual | | 2.6 | 9.8 | 8.2 | 5.9 | 3.4 | 7.7 | 8.3 | 2.4 | | | | |
| Cum. YTD Actual | | 12.4 | 22.2 | 30.4 | 36.3 | 39.7 | 47.4 | 55.7 | 58.1 | | | | |

**Figure 2.13** Forecast versus actual amounts for an account that is out of control.

54

Account No. 587    Account Title: Building Maintenance

Date: June 21

|  | Current month | Next month | Second next month | Third next month |
|---|---|---|---|---|
|  | June | July | Aug. | Sept. |
| Budget | 5.0 | 6.0 | 5.0 | 5.0 |
| Rollover from previous month | 1.6 | | | |
| Special shop orders | | | | |
| PO no. 1 | 0.5 | | | |
| PO no. 2 | 0.2 | | | |
| PO no. 3 | | 0.3 | | |
| PO no. 4 | | 0.5 | | |
| PO no. 5 | 0.4 | | | |
| PO no. 6 | | | 2.2 | |
| PO no. 7 | | | | 1.6 |
| PO no. 8 | 2.0 | | | |
| PO no. 9 | 0.1 | | | |
| PO no. 10 | 0.3 | | | |
| PO no. 11 | | 0.4 | | |
| PO no. 12 | | 1.2 | | |
| Total | 5.1 | 2.4 | 2.2 | 1.6 |

**Figure 2.14** Purchase order payment predictions.

means that the project's actions should be keyed to the budget or, which is more correct, the budget should be set to match the project schedules.

Of course, Fig. 2.14 as explained above represents an ideal case. Note that no contingency is shown. A maintenance account should always have a reserve fund to take care of emergency items, since all work cannot stop when the allocated money runs out. If the roof leaks on June 22, no prudent manager will wait until July to fix it. Therefore, in addition to the rollover, the manager will hold back a percentage of the funds for emergencies. This is done either by using a lower amount for the budget, or by entering a dummy purchase order in the contingency amount. In Fig. 2.14, PO 1 for $0.5 could be considered the dummy purchase order—a typical 10% contingency amount held in reserve. Also, in practice, the entire rollover never comes due for payment the next month. Therefore the contingency is somewhat greater than $0.5.

The format used in Fig. 2.14 is similar to that of the SRF. Hence, the figures for this specific account and all others tracked in this way can be used directly as the basis of the forecast. This technique, if followed daily, will result in response activities being controlled within 10%, which is excellent control for this type of account and acceptable for modern manufacturing management.

## SUMMARY

All manufacturing engineering activities must be controlled for results to be achieved as planned and not as a result of chance. In this chapter we have discussed the basic techniques used to plan for activities to occur and to track progress. We have also introduced the necessary financial controls.

To control manufacturing engineering activities, each activity must be scheduled—preferably before the fact, but in all cases scheduled. In this way a data base is created showing the time and resources required to handle a variety of assignments. With such a data base management can establish productivity targets based on real needs as constrained by time and funds, and can use the manufacturing engineering organization to study and implement projects in a tightly focused effort.

By controlling all operations with a plan, personnel needs and funds are accurately forecast and the company's business plan and targets achieved. If manufacturing engineering activities are permitted to exist in a reaction mode, there are never enough people to handle all the problems that occur, nor is there enough money to cover all expenses. By planning ahead, the cost of manufacturing engineering is reduced and its achievements increased.

## REVIEW QUESTIONS

1. An organization wishes to upgrade the level of education of its staff by in-house training. It intends to have 65% of its 250 people take a blueprint-reading course and 30% take a machine tool familiarization course. Finally it is considered necessary for at least 15% of the staff to complete both courses. The manager hopes to have the education plan completed within 12 months. Write an objective and supporting goals for this program.

2. Define the philosophical differences between an objective and a goal as related to manufacturing engineering.

3. Explain why adherence to the rule of project selection pragmatism and the series corollary are important to a profit-making organization.

4. The following is an excerpt from a press release issued by a new company. Determine what the objectives of the company are. "We have combined resources so that we can better serve the public in discerning the practicality and feasibility of mass transit systems."

5. A company decides that it must reduce its costs of manufacturing by 5% over the course of the next year. Determine what the objective is; then establish suitable goals statements for manufacturing engineering to support that objective. Include at least one goal each for advanced manufacturing engineering, methods planning and work measurements, and process control. A sum of $1,250,000 is available for investment; shop operations employs 250 people.

6. Refer to the schedule control chart in Fig. 2.7. Because of changes in personnel availability, the following items have had revisions to the completion dates. Construct a new schedule control chart showing the changes caused by these revisions: (a) Audit steam traps—completion delayed 6 weeks; (b) Set up system to monitor weld quality—completion advanced 3 weeks.

7. Define an action plan and explain how it would be used by an engineer and how the same action plan would be used by a manager.

8. A manager of manufacturing finds that a project to purchase a horizontal boring mill and a weld positioner are $12,000 over budget and $9000 under budget, respectively. Outline the items of information that should be included in the manager's key results report in addition to the basic over-budget and under-budget facts.

9. In reference to a short-range forecast, discuss why it is important to investigate all over-budget and under-budget forecast line items, even though the total forecast is under or equal to budget.

10. Explain how changes in the schedule control chart (question 6) will affect the short-range forecast and periodic review of operations.

11. List three typical account categories that would probably require daily accounts control. Give your reasons for each one.

# THREE

## FACTORY CAPACITY
## AND LOADING TECHNIQUES

Factory capacity and loading evaluations are critical for determining exactly what the factory can produce in its current alignment. This knowledge is essential for the company to be able to decide what products it can sell and when it will be able to deliver them. The task of determining the factory's capacity is assigned to the advanced manufacturing engineering unit. The task of loading the shop is assigned to the materials function's master scheduler for compliance with overall business plans and to production control, for the specific or short-interval scheduling. Manufacturing engineering plays a very important role in verifying that the shop loading is in accordance with the capacity evaluations. Therefore, capacity and loading are studied together, since the former leads directly to the latter.

## TECHNICAL CAPACITY

The ability to accurately determine the technical capacity of the factory is worth a considerable amount of money to a company. Consider what happens if a company bids and wins a contract to produce a complex product, and then finds out that it does not have the technical capacity to produce it. What are its options? First, the company could give back the contract, admitting that it made a mistake. This would be a rarity indeed; not many companies are willing to admit they are incompetent, and very few customers would be willing to give that company additional contracts, even if they were for different products. Second, the company could buy the technology or subcontract to another firm that has the necessary know-how. This might solve the problem with the customer but it would be frightfully expensive, especially considering that the companies having the correct technology were probably competitors for the contract in the first place. These are two very pragmatic reasons for accurately knowing your company's technical capacity.

When speaking of technical capacity, I am referring primarily to manufacturing capability, comprising the proper facilities and the technical people to oversee and maintain them. The secondary meaning of technical capacity concerns the design engineering function—whether it has the proper resources to design the product, so that the product can be produced by the company's own factory. This part of capacity analysis is the domain of the producibility engineer and will be explored in Chapter 6.

## PHYSICAL CAPACITY

Besides technical capacity we must consider physical capacity: how much product can be produced in a given period of time. From an economic viewpoint, it is vital to be able to accurately determine this facet of capacity. Firms that habitually produce and deliver products later than promised do not accurately know the capacity of their factories. This costs the producing company money, because the company is usually not paid until the product is delivered, or if progress payments are allowed, not until certain benchmark production phases have been completed. In addition, if making the product is falling behind schedule, more of the company's funds arc tied up in inventory and not earning interest, which is like putting money in non-interest-bearing bank accounts. This also means that the company may have to borrow funds to keep operating until it can collect for sale of its products.

The ability to accurately determine capacity is important for other reasons as well. One is to avoid penalty clauses in contracts for nondeliverance by a specified date. This type of contract is fairly common for large capital goods, and if the capacity is only vaguely known, the company is at considerable risk of falling into an adverse penalty situation. Finally, accurately knowing the physical capacity of the factory is a factor in preserving the company's good reputation and avoiding the situation where other companies are awarded contracts because the customer cannot chance late delivery from a firm that has a reputation for erratic deliveries.

## TECHNIQUES FOR DEVELOPING
## WORKSTATION CAPACITIES

Unless a factory is made up of all skilled craftsmen using hand tools, it must be considered an amalgamation of many human-machine units. Therefore when we talk about determining workstation capacities, we must take into account the designed capability of the machine tool as well as the capability of the operator to run the workstation.

Because humans are not machines or robots, and their performance varies from day to day, we cannot expect exact repetition of effort every time a workpiece goes across the workstation. Hence, determining workstation capacity depends on finding the lower bound of the effort range to be expected of the average person. This is a complex task. It may be done by complex work measurements for the most analytically correct answer, by estimation by knowledgeable people for a much quicker answer. The result is an efficiency factor by which the theoretical design capability of the workstation is modified downward to obtain a capacity factor for the workstation.

**Example:** Let us use a semiautomatic welding station utilizing a submerged arc continuous-feed welding process and a positioner to rotate the parts to be welded. In this example, we are producing circumferential welds with each weld being applied in the vertical downward position. The basic outline of the workstation is shown in Fig. 3.1.

The known parameters of the workstation are (1) the speed and direction of rotation of the positioner, and (2) the deposition rate of the weld process. Therefore, it is possible to calculate the deposition rate on the workpiece. Also knowing the volumetric cross section of the weld, it is possible to determine how long it should take to make the completed circumferential weld. This is the theoretical capacity of the weld machine workstation. Unfortunately, it is useless in developing workstation capacity.

The machine can work forever, but it needs human intelligence to control and guide it. For this example we will state that the machine is not computer controlled in whole or part. Hence, the problem of workstation capacity ultimately reduces to how much time out of the 24 hours in a day an operator will guide the machine. Let us assume that this particular submerged arc process can deposit 30 pounds of weld per hour, or 720 pounds per day. This

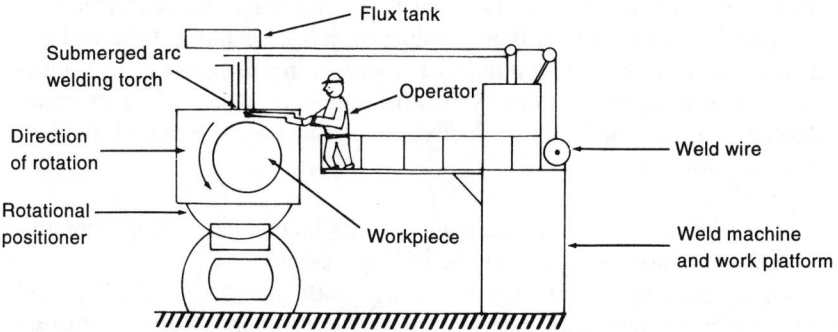

**Figure 3.1** Semiautomatic welding workstation.

is the ultimate theoretical capacity of this workstation. To determine a practical capacity, the following factors must be taken into account:

1. Setup time. The workpiece must be set up on the positioner. Experience dictates that roughly 10% of the time to do any job is the setup time. This is equivalent to a reduction of 3 pounds per hour in the deposition rate.
2. Personal needs. An operator cannot work continuously for an entire 8-hour shift because of fatigue, the need to consume food, and the need to take care of personal hygiene. As a rule, people need 30 minutes for a lunch break and another 30 minutes per 8-hour shift for rest and personal hygiene. Therefore, out of 8 hours we must subtract 1 hour for personal time, or 12.5% of the work time available per shift. This corresponds to a reduction of 12.5% of the theoretical capacity, or 3.75 pounds per hour.
3. Absenteeism. The usual absenteeism rate in a job shop industry is about 2%. This means that the typical worker will not show up for work approximately 5 days per year because of illness or personal needs. The absenteeism factor of 2% equates to a 0.6 pound per hour reduction in deposition rate.
4. Interference time. This is nonproductive time when the operator is available to work but cannot. For a positioner weld workstation interferences might be due to refilling the flux tank, replacing an expired weld wire reel, performing machine maintenance, performing in-process inspection checks, reading process instructions, and fixing the machine when it breaks down. The normal practice is to assign 10% of available time for the interference factor. In terms of deposition rates, this amounts to 3 pounds per hour.
5. Mistakes. Operator mistakes happen and must be accounted for in any realistic capacity evaluation. Typically, the reduction in capacity allowed for operator mistakes is 3%, which in this example is equivalent to a 0.90 pound per hour reduction in deposition rate.
6. Efficiency. This is a measure of how the operator performs the task compared to how the task was designed to be done. Continuous performance at 100% efficiency would require all the operator's moves and the placement of all tools to be continuous exactly as originally planned by methods engineering. This is virtually impossible for job shop applications and very difficult to achieve for continuous manufacturing operations. In semiautomated welding, an efficiency of 80% could be expected. This means that 20% of the time is nonproductive, or a reduction of 6 pounds per hour in deposition rate.

Now let us see what a practical capacity would be compared to a theoretical capacity. Table 3.1 shows the computation for the example.

In rating the capacity of this workstation we would have to use 12.75 pounds per hour for the weld deposition rate instead of the 30 pounds per hour theoretically possible. This is the figure that will be used whenever we are calculating

**Table 3.1 Computation of practical capacity for a submerged arc positioner circumferential welding process**

| | |
|---|---|
| Theoretical capacity | 30 lb/hr |
| Deductions | |
|   1. Setup time | −3.00 lb/hr |
|   2. Personal time | −3.75 lb/hr |
|   3. Absenteeism | −0.60 lb/hr |
|   4. Interference time | −3.00 lb/hr |
|   5. Manufacturing losses | −0.90 lb/hr |
|   6. Efficiency | −6.00 lb/hr |
| Total deductions | −17.25 lb/hr |
| Practical capacity | 12.75 lb/hr |

the amount of work going across this particular workstation and determining how long it will take to perform it.

Efficiency is the largest of the deduction factors. For the purpose of capacity analysis, the advanced manufacturing engineering unit must know the level of efficiency the factory segments are working at. This information is obtained from the normal measurements of efficiency, which is continuously being evaluated. It does not matter what the basis of the measurement is, as long as it is consistent. If it is consistent, the efficiency ratings will be usable as deduction factors to obtain practical capacity values for each workstation. As the productivity performance of the factory improves, its ultimate capacity improves. Thus, the exercise of determining capacity is dynamic, and factory capacity evaluations must be an ongoing activity.

Once we have a file of workstation practical capacities, we can begin to construct a matrix of factory capacities that can be used to load the shop. For a single product line this is relatively simple.

Figure 3.2 illustrates the sequence of operations used to produce a clutch plate. There are three basic operations and one inspection that must be completed. If we know the cycle time for each operation, we can determine the capacity to make clutch plates in the factory. These operations are listed in Table 3.2, where the factored times shown are practical times; that is, all adjustments have been made to increase from ideal time to practical time.

Table 3.2 gives a great deal of useful information. It shows that the complete cycle time for making a clutch plate is $1/2$ hour (time factor), while the longest cycle time is for the Numerically Controlled (CNC) lathe and the shortest for the forge operation. Thus, in order to have no queuing of parts, we need the equivalent of 1.8 lathes, 1.2 Horizontal Boring Mills (HBMs), and 1.0 inspectors for each forge. These are the ratios of workstations required if a continuous-flow process is desired. Otherwise, a batch processing mode would be used. The decision would depend entirely on how many clutch plates are to be made. If the

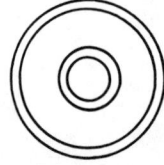 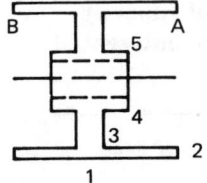

Make from ASTM C2204 grade steel

Step 1: forge basic shape

Step 2: use CNC chucking lathe to machine surfaces
1, 2, 3, 4, 5 on side A, turn around, and
machine side B

Step 3: use horizontal boring mill to mill bore

Step 4: inspect for dimensional accuracy

**Figure 3.2** Operations used to produce a clutch plate.

number is only a few thousand and another product will then be made with the same equipment, the batch mode would be chosen. If the number is in the hundreds of thousands or more, the continuous mode would be necessary.

Other useful information is obtained from this example. For instance, we can determine how many people are required (people factor) and the labor cost to produce the product (cost factor). The labor cost is determined by simply multiplying the labor rate in dollars per hour by the cost factor to obtain labor cost per part.

Now what is the capacity to make clutch plates? It takes $1/2$ hour to make one if we have only one set of equipment. Since we already have factored times, then we know in an 8-hour shift 16 clutch plates will be produced, or a maximum of 48 per day. Deducting vacation and holiday times, the average factory operates for 46 weeks per year. Therefore we have 230 productive days per year at 5 days per week, which corresponds to a maximum capacity of 11,040 clutch plates per

### Table 3.2 Clutch plate operations analysis

| Operation | Factored time to perform (hr) | Number of operators | Man-hours |
|---|---|---|---|
| Forge | 0.10 | 2 | 0.20 |
| CNC chucker lathe | 0.18 | 1 | 0.18 |
| Horizontal boring mill | 0.12 | 1 | 0.12 |
| Inspection | 0.10 | 1 | 0.10 |
| Total | 0.50[a] | 5[b] | 0.60[c] |

[a]Time factor.
[b]People required.
[c]Cost factor.

year. With this information, the master scheduler can now load the shop and give promise dates for parts to be delivered.

The example above is for a part that is made with machines dedicated to making that part. When we have many different parts of varying complexity to make, the technique must be extrapolated and expanded. If the parts are many and varied, and especially if they are unfamiliar, the first thing the advanced manufacturing engineering unit must do is determine whether the part can be made in the factory as presently constituted. This leads us to the basic definitions and relations between capability and capacity:

Capability: The physical limitations on what can be produced in the factory including size, weight, processes, and materials

Capacity: The number of items that the factory has the capability to produce that can be produced in a given period of time

Before we can determine factory capacity in a multiproduct or component shop, it is necessary to determine whether the factory has the capability to make the product. Figure 3.2 outlined the steps necessary to make a clutch plate. If the factory does not possess any of this equipment—forge, CNC chucker lathe, horizontal boring mill—then it does not have the capability to produce the product. The capability review then becomes a matter of investigating whether the product can be made.

When the factory knows exactly what it is going to make, the capability review is a review of fitting equipment to the known product. Often a factory is set up to make a range of similar products or products made by similar manufacturing techniques. In this case capability is defined by the ranges the equipment can handle. Figure 3.3 shows a capability matrix for a company that manufactures special-purpose motors. Under each category the largest sizes and tightest dimensions the factory can produce are listed. The processes available and the materials the factory is prepared to use are also shown. The capability matrix would be used by design engineering so they do not design a motor that cannot be made in the factory, and by marketing as a guide as to what they can sell to customers.

## EVALUATING FOR PRACTICAL CAPACITY

Now that we have discussed the techniques of evaluating individual workstation and individual product capacities, it is possible to consider the techniques for evaluating total factory capacity. It is interesting to note here that the only independent variable is the capability of the factory to produce, that is, the physical and process limitations identified in the capability matrix. The capability matrix shows what can be made but does not attempt to quantify how much of each known product can be made. It is very important, therefore, to be certain that a contemplated mix of products is not to be in violation of the bounds prescribed by the capability matrix. If it were, it would necessitate either investments for

## SPECIALTY MOTORS
## MANUFACTURING CAPABILITIES & CAPACITIES

| Item | Physical characteristics ranges | | | | Processes available for mfg. | Equip. capacity shift | Comments |
|---|---|---|---|---|---|---|---|
| | Dimensions | Weight | Tolerances | Materials | | | |
| Rotors | | | | | | | |
| Shafts | 18′ × 36″ dia. | 30.0 tons | ±0.002″ | Alloy steel | Lathe, slotter | 3 | |
| Punchings | 0.025″ thick | 30 lbs | ±0.0001″ | Silicon steel | Punch press | 3 | Min. size = 0.014″ thick |
| Coils | 2″ × 2″ cross section | — | ±0.0625″ | Copper | Form, mold, tape | 2 | max. length = 48″ |
| Bars | 2″ × 2″ cross section | — | ±0.0625″ | Copper | Form, mold, tape | 2 | Max. length = 62″ |
| Rotor bodies | 7′ × 54″ O.D. × 20″ I.D. | 25 tons | ±0.001″ | Alloy steel | Lathe, slotter | 3 | Laminated construction |
| End rings | 4″ × 1″ cross section 60″ O.D. | — | ±0.003″ | Alum., alloy steel | Roll form | 3 | |
| End plates | 8″ × 0.25″, 60″ O.D. | — | ±0.003″ | Alum., alloy steel | Mill | 3 | |
| Amort rings | 2″ × 2″ cross section 60″ wide | — | ±0.003″ | Copper, brass | Roll form | 3 | |
| Poles | 7′ long, 8″ high, 6″ wide | 1.5 tons | ±0.010″ | Steel, copper | P/P, positioner wind | 2 | |
| Exciters | 2′ × 22″ dia. | 0.4 tons | ±0.004″ | Steel, copper | P/P, handried | 2 | |
| Stators | | | | | | | |
| Frames, fabricated | 14′ × 8′ × 6′ × 0.75″ plate | 7.5 tons | ±0.005″ | Steel | Submerged arc, HBM, VBM | 3 | |

| Part | Dimensions | Weight | Tolerance | Material | Process | | Notes |
|---|---|---|---|---|---|---|---|
| Frames, double shell | 12' × 6' × 4' | 7.5 tons | ±0.005" | Steel | Cast, HBM, VBM | 3 | |
| Top hats | 14' × 8' × 4' × 0.25" plate | 3.0 tons | ±0.030" | Steel | Brake, drill | 2 | |
| Bearings, sleeve | 24" × 25" | — | ±0.0005" | Steel, tin, lead | Spin cast, lathe, HBM | 2 | |
| cylindrical | 14" × 15" | — | ±0.0005" | Steel, tin, lead | Spin cast, lathe, HBM | 2 | |
| spherical | 14" × 15" | — | ±0.0005" | Steel, tin, lead | Spin cast, lathe | 2 | |
| ball | 14" × 15" | — | ±0.0005" | Steel, tin, lead | Purchase | 2 | Purchase to spec. |
| roller | 14" × 15" | — | ±0.0005" | Steel, tin, lead | Purchase | 2 | Purchase to spec. |
| Coils VPI | 2" × 2" cross section | — | ±0.0625" | Copper | Form, mold, tape | 2 | |
| | 19' × 160" dia. | 50.0 tons | — | Epoxy resin | Pressure tank, bake over | 3 | |
| Varnish | — | — | — | Epoxy resin | Dip tank, bake oven | 3 | |
| Punchings | 0.019" thick | 0.5 lb | ±0.0001" | Silicon steel | Punch press | 3 | Min. size = 0.014" thick |
| Cores | 8' × 20" I.D. × 70 O.D. | 45.0 tons | ±0.001" | Steel, silicon steel | Stack press | 3 | Strap weld |
| End shields | 6" thick × 72" O.D. | 2.0 tons | ±0.002" | Steel | Cast, fabricate, VBM, HBM | 3 | |

Figure 3.3  A capability matrix.

new equipment and personnel training, or subcontracting of phases of manufacturing to other companies.

Evaluating for practical capacity in a multiproduct company requires predicting how much of the resources will be devoted to each product. This is an all but impossible task for manufacturing engineering. In fact, the marketing function is the only segment of the company that has the proper resources for estimating the market. For this reason manufacturing engineering does not try to determine market characteristics and trends, but develops basic data showing a specific mix of products with percentage quantities that can be made. Often this is not acceptable to all functions of the business because it specifies a fixed percentage for each product. Clearly, when market opportunities change, the set percentage does not remain useful. Therefore, the most recent trend in capacity analyses is to develop a matrix of all capacities and then let the user plot the effect on capacities if the set of percentages are varied. To do this, manufacturing engineering develops capacities per product as if no other product existed, and then drafts capacity charts to enable the marketing or master scheduling personnel to determine how much capacity is still available after a number of specific products is booked into the schedule.

This is a mortgaging of facilities activities. If we have 10 hours available at workstation A, and product X requires 1 hour, product Y requires 2 hours, and product Z requires 5 hours, the number of each product to go through workstation A will depend primarily on the priorities given to the products, with the result that there will be a large number of possible variations. In practice, the number of product variations to go through the workstation will depend on customer preference and orders on the book or potential orders to be received. This information comes from the master planner via the marketing function. Hence, very little depends on mathematical manipulations. An example of a capacity chart prepared by advanced manufacturing engineering is shown in Table 3.3.

The analysis for each product is done independently in the manner described above. Each workstation is evaluated for practical capacity to make each specific product as if no other product existed. Each workstation will consist of itself alone

**Table 3.3 Workstation capacity per year**

| Workstation | Product | | | | | | | |
|---|---|---|---|---|---|---|---|---|
| | 1 | 2 | 3 | 4 | 5 | 6 | 7 | 8 |
| Forge | 19,872 | 628 | 387 | 12,082 | — | 11,683 | — | — |
| CNC lathe | 11,040 | 721 | 172 | 9,212 | 10,563 | 12,692 | 2,163 | 3,812 |
| Horizontal boring mill | 16,560 | 823 | 279 | — | 9,732 | 14,678 | — | 1,962 |
| Vertical boring mill | — | 562 | 202 | 8,961 | 9,978 | — | 1,672 | 3,177 |
| Inspection | 18,065 | 670 | 311 | 9,743 | 9,853 | 12,672 | 1,833 | 2,468 |
| Total | 11,040 | 628 | 172 | 8,961 | 9,732 | 11,683 | 1,672 | 1,962 |

or in total if multiples of the workstation exist. Note that the total yearly capacity is always equal to that of the lowest-capacity workstation. This means that the lowest-capacity workstation is 100% booked for that product while the other workstations, having a higher capacity to make the product, are less than 100% booked.

> **Example:** For product 1, the yearly capacity is 11,040 pieces, while the forge can make 19,872 pieces. Therefore, only 55.6% of the yearly capacity of the forge is required to make product 1, and 44.4% of capacity is available for other products. (Product 1 is the same product used in the example illustrated by Table 3.2.)

A practical way to evaluate capacity for a proposed multiproduct run would be to look at the critical workstations only. If capacity exists at the critical work-stations, the likelihood of making the products at the other workstations is virtually 100%. Therefore, by focusing on the critical workstations, it is possible to reduce capacity evaluation exercises significantly. In this case, the term critical workstation is synonymous with lowest-output workstation, sometimes called the bottleneck workstation.

Some very important but perhaps unquantifiable data must be subjectively evaluated by the area planners in determining capacities. These data are often critical in factoring the theoretical capacity to a pragmatic capacity. Unquantifiable data must be rated in importance and empirical correction factors used to modify the mathematical results the area planner produces as the capacity per product. Among the hosts of unquantifiable items that must be considered in determining capacities, some of the more common ones are:

1. Key operator retirement. Assuming that the key operator is the most important factor in determining workstation efficiency, his or her retirement will have a detrimental effect on capacity. Usually a 5 to 10% drop in efficiency for 3 months occurs when this happens. Therefore, in the weld example (Fig. 3.1 and Table 3.1) an additional 3 pounds per hour or 10% reductions could result for a full operating quarter.
2. Union contract due for renewal. If the labor climate is favorable, this will have no effect. If it is one of continuous adversarial combat, a serious reduction in output will occur in all six areas of reductions of theoretical capacity.
3. New equipment installed. A learning curve must be anticipated going from lower to higher (designed) output. The length of the learning curve must be estimated on the basis of previous experiences and the complexity of the new equipment compared to the old.
4. Tolerance changes. Any tolerance change will affect capacity. The closer the tolerance is to the practical capability of the machine, the less product will be produced in the given time period. My experience shows that a reduction of 3:1 in output occurs whenever the required tolerance in machining goes below ±0.015 inch.

5. Weather factors. The more uncomfortable it is to work in the factory, the higher the manufacturing losses become. I have observed 1.5:1 increases in losses in foundry-type operations from winter to summer in a northeastern US location.
6. Workstation maintenance. Lack of maintenance not only results in more machines being down, it also affects daily operations through failure to attend to minor items. This causes operators to make corrections that may or may not be correct; that is, the "Kentucky windage"* applied to the engage handle on a worn gear train may not have the result the operator expects. This leads to increased manufacturing losses and lower productivity. The area planner should use a little Kentucky windage to determine what this means in terms of capacity.
7. Excessive planned overtime. Rarely does a continuous 6-day schedule equal 6 days output. Usually, through fatigue, we end up with perhaps $5^1/_2$ days output instead. Causes are excess absenteeism and lower efficiency.
8. Design changes in materials specification. This is one of the most far-reaching types of changes that can affect capacity. If the change results from a manufacturing request, it is usually done to improve manufacturability of the product; hence capacity increases. If it results from new design engineering requirements, the result will be materials that are more difficult to machine; hence capacity decreases. The area planner must thoroughly research any pending changes in materials.

There are countless more unquantifiable or intangible factors related to people, equipment, and material that affect practical capacities. The important point here is that the pertinent unquantifiable items should be identified by advanced manufacturing engineering so that empirical factors can be assigned to subjectively adjust practical capacities upward or downward.

## EFFECT OF LEVEL LOADING ON CAPACITY ANALYSIS

Strictly speaking, level loading is a production control function. Manufacturing engineering becomes involved only to ensure that all technical considerations are observed. For example, to level load for a factory to run at 90% of designed theoretical capacity for an extended period, greater than 3 months, is pragmatically impossible. No factory can run continuously at a rate greater than 67% of design theoretical capacity. Hence, it is important for manufacturing engineering to evaluate the level load plans to ensure that no inherent damage to the factory can occur.

Level loading is a technique used to smooth out production, that is, to produce

---

*A sense of the correct adjustment that an individual gains after many repetitions of the same thing. Originally used to refer to a windage correction made by aiming a firearm to the left or right of a target rather than adjusting the sights.

the same amount of product over each measurement period, usually a week or a month. It calls for the plant to work at a certain level of output regardless of the amount of finished product required for any given period of time. If a workstation can make 12 units per month and the requirements for the next 3 months are 9 units, 16 units, and 11 units, respectively, then more than required will be made in the first and third months and less than required will be made in the second month. The total produced will be 36 for the 3 months, equaling the requirement for the period. The problems for the business caused by this level loading are that carrying costs for inventory are increased in months one and three, and there is a late shipment by one unit in month two. These are typical of the considerations the manager of manufacturing must make when deciding whether to use level loading.

In a more complex system, more than one product would go through a work-station or a series of workstations. The loads at the workstations are the sums of the various products. Figure 3.4 illustrates a level loading chart for a 3-month period at a multiproduct workstation or center (common workstation). In this case we can make 9000 units, assuming that products 1, 2, and 3 require the same operation time at the workstation. The problem is that the requirement is 9900 units, or 10% more than can be produced. Level loading would ensure that 3000 units are produced in all 3 months. This means that in May we would be short 800 units; in June short 300 units for the month, 1100 units cumulative; and in July produce an excess of 200 units for the month, resulting in a cumulative short-age of 900 units. Therefore, if the factory could produce 10% more for each of the 3 months, the total of 9900 units could be produced. In May, trying to achieve

| Output required | Product 1 | Product 2 | Product 3 | Total |
|---|---|---|---|---|
| May | 1000 | 1600 | 1200 | 3800 |
| June | 900 | 1200 | 1200 | 3300 |
| July | 500 | 1700 | 600 | 2800 |
| Total | 2400 | 4500 | 3000 | 9900 |

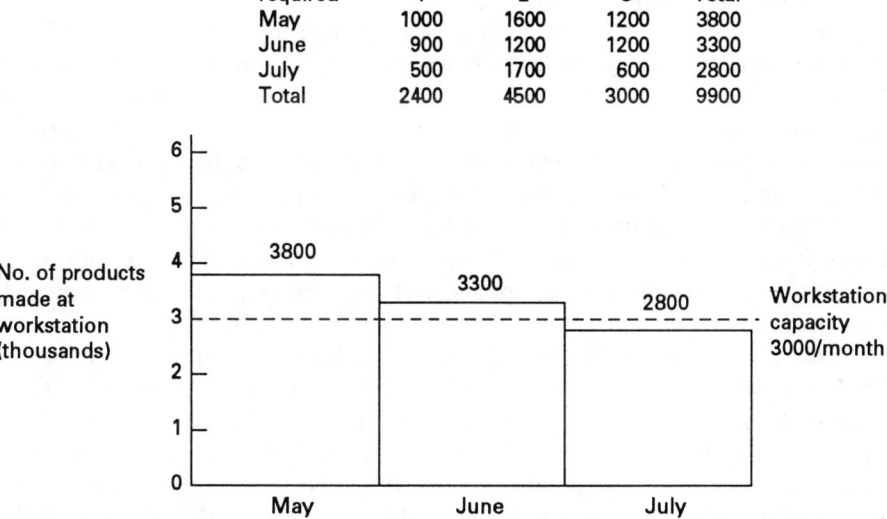

Figure 3.4 Level loading chart for a multiproduct workstation.

3800 units (800 units above capacity) may be impossible, but 3300 units could probably be made. If this can be done in May, it should be possible to do it in June and July. Whether it is possible must be decided by the master scheduler. Manufacturing engineering will have to be able to answer the technical part of this question, but before we deal with the technical part, let us look further into the philosophy of level loading.

Level loading is a way of achieving a steady state. No business wants crisis after crisis, or to go from highs to lows. This fatigues people and wears out machines, which leads to mistakes and a lower life expectancy for equipment. In order to achieve success in an environment of highs and lows, a corporation must have enormous reserves to deploy at the apogee and then place them in storage at the perigee. Industrial concerns do not have such unlimited reserves, and even if they did, employing them in such a manner would be wasteful and nonoptimal. In a company with a profit motive, every dollar spent in a nonoptimal mode increases overhead and thus decreases the profit margin. The simple cost equation is

Total cost = direct labor cost + material cost + overhead cost

Since overhead costs are usually associated with control costs (in this case maintaining an unoptimized reserve to handle highs and lows), the more the variance from a level operation, the higher the overhead costs. Level loading, then, is the most common way to bring order to a business and to lower overhead costs.

Level loading lowers overhead costs by allowing routines to be established. In essence, routines are solutions that have been significantly optimized, and they tend to be the most efficient way of doing things. Routines depend on predictability; hence, random variables such as large swings in production demands must be minimized for routines to exist.

In reference to manufacturing, the basic goal of level loading is to create a smoothly running factory, with machines and processes running at the most efficient load rate, inventories of raw material at as low a level as possible, and the labor force stable. Therefore, true level loading requires a balance between demands for equipment, labor, and material. This affects manufacturing engineering primarily in the equipment areas, and secondarily in the labor and material areas.

Manufacturing engineering, through its objectives and goals statement, must consider the needs for equipment to maintain a level load. Such considerations as cost to buy, floor space for setup, and utilization (or load) must be continuously evaluated.

Considerations made by shop operations management to maintain level-loaded production are of interest to manufacturing engineering. Shop operations will consider such factors as retaining trained people, morale of the work force, effects of cyclic hiring and layoff on productivity, availability of labor to hire, and overtime. Each of these items can affect the work load of manufacturing engineering in providing technical support to the shop floor, particularly the need for training and the detail of instructions necessary in the planning. A changing work force usually has a detrimental effect on efficiency, which in turn affects workstation capacity.

Materials considerations also affect manufacturing engineering operations. Increases in materials require additional handling capability and storage locations. Manufacturing engineering must deal with these factors as it evaluates factory floor layouts. Having the machine capacity for the desired level of production is worthless if the raw materials cannot reach the workstations on time per schedule.

The ways in which level loading is accomplished have already been alluded to in the examples. Specifically, it is done by:

1. Manipulating schedules through
   a. Early starts
   b. Delayed starts
2. Making supply or stock parts during slack periods
3. Having plant shutdowns (vacations) during the slack season
4. Applying more labor during peak periods through
   a. Overtime for existing personnel
   b. Adding personnel to existing shifts and starting up idle equipment
   c. Adding shifts (i.e., if only one shift is employed, starting up a second)

The overall aim is to maintain a steady output per hour per workstation, and to regulate the total output by how long the workstation is run. This is supplemented by adding additional workstations only as a last resort. As a rule, additional work stations are added for a short duration by utilizing idle, less than optimum equipment and for a long duration by purchasing additional efficient equipment.

Now let us look at the technical portion of the answer for the problem posed in Fig. 3.4. The factory wants to run at a 10% overload for 3 months. Manufacturing engineering will have to determine whether the physical plant can handle the overload. They will do that through the following steps:

1. Recognize that no factory can run above 67% of theoretical capacity for more than short periods of time. This usually means up to 3 months.
2. Determine what facilities will be required to run at or above the 67% theoretical capacity.
   a. Review the additional maintenance needs to do so.
   b. Evaluate the spare parts inventory available and reaction times needed to obtain additional spares if necessary.
3. Evaluate the cost of maintaining the factory in an accelerated mode.
4. Evaluate the labor costs, considering
   a. Availability, vacation plans, and so on;
   b. Effect of the fatigue factor on efficiency of maintenance and repairs.
5. Determine what factors in the capacity analysis for each workstation can be optimized for a 3-month period to produce 10% more output.
   a. Reschedule vacations so that key operators are available, ensuring better efficiency, attendance, and less manufacturing losses.

    b. Consider designing new setup fixtures and improving methods to reduce setup time.

    c. Consider assigning manufacturing engineers to reducing interference times by giving specific attention to workstations that are potential bottlenecks.

These are common items that the advanced manufacturing engineering unit would look into to answer the technical question of whether a 10% increase in output can be achieved temporarily for 3 months. The answer, of course, would consist of facts and figures detailing what could be done and how that would minimize the six deduction factors used to derive practical capacity from theoretical capacity. The Advanced Manufacturing Engineering (AME) unit would also include their evaluation of the costs of such an exercise and how long they thought such an accelerated program could continue. It would then be up to the manager of manufacturing to decide whether the risks are worth the potential benefits.

## BOTTLENECK IDENTIFICATION AND ELIMINATION

A discussion of factory capacity and loading would not be complete without devoting a section to the subject of bottlenecks. Bottlenecks, or choke points, in the manufacturing flow are the bedevilment of management. They cause countless frustrations, delays in production, extra costs, and conceivably lost orders. Simply stated, bottlenecks need to be identified and eliminated. If bottlenecks go undiscovered, they can and often do wreak havoc with the capacity analysis.

### Bottleneck Definition

What is a bottleneck? The simplest definition is that a bottleneck is a point in the production flow where the workstation is working at capacity, yet work load volume grows faster than the completed work leaves the workstation. The situation is analogous to a three-lane highway being restricted to two lanes at a time when two lanes are inadequate to handle all the traffic. Hence we have a backup of vehicles before the restriction. Recall that when one encounters this situation vehicular velocity decreases considerably. The same happens in a factory. When a bottleneck occurs, the flow of parts traversing the factory also slows down. As in the highway analogy, bottlenecks are not permanent situations. They come and go depending on the dynamics of the situation. Let's now explore how a bottleneck can be identified and then how it is minimized and often eliminated completely.

### Identifying Bottlenecks

Rates of production reveal bottlenecks. A bottleneck will be uncovered if and only if the overall rate of production is beyond the capacity of the workstation to maintain. This means that if the capacity built into the workstation is never exceeded, that workstation will not become a bottleneck. This in turn means that bottlenecks are not permanent. They exist only at specific workstations at specific rates of production. They arise and disappear as rates of production increase or decrease, if left to their own devices and not interfered with by the factory's controlling function. The manufacturing engineer's role vis-à-vis bottlenecks is predicting where bottlenecks will occur at what levels of production volume. This is done via the capacity and capability determination techniques previously discussed in this chapter.

A bottleneck, when it appears, will have the following characteristics:

1. There will be a large amount of work queued in front of the workstation waiting to be processed.
2. The workstation will be working at its maximum capacity, as presently defined.
3. Succeeding workstations will be underutilized.
4. Factory output will most likely be less than the desired amount per time period.

## Possible Solution Alternatives for Eliminating Bottlenecks

Several methods are available for relieving bottlenecks. Let's examine them. In the section on level loading we explored ways to balance production over longer periods of time. We saw that it was possible to get production through a factory that had bottleneck conditions as described above, although we didn't call them bottlenecks. They were, and it becomes apparent that bottlenecks are not unusual situations but, unfortunately, common. The methods of elimination resolve down to three:

1. Schedule manipulation
2. Methods improvement
3. Facilities addition

**Schedule Manipulation.** The basic techniques for accomplishing balance of production were discussed in the level loading section of this chapter. Essentially, we maneuvered the schedule so that the threshold for creating a bottleneck was raised. This meant there was no bottleneck because we made sure all workstations were working at a level that did not cause a bottleneck. Work flowed freely from one workstation to another at the desired rate. Admittedly, some workstations may have had to work longer hours than others for enough work to flow at the proper rate throughout the factory. But this is the heart of this bottleneck elimination technique. The terminology for creating this situation is "creating a buffer."

A buffer is material waiting to be worked on that is stocked beyond the bottleneck in order to feed the workstations down line that work at higher rates. Faster workstations consume inventory faster than bottleneck workstations—like a Gatling gun consuming ammunition faster than a muzzle-loading rifle. We purposely create a large inventory queue in front of these faster workstations so that they can work at their most efficient rates and still have an adequate production line balance (e.g., the fire bucket brigade synchronous flow). The level loading techniques create buffers as needed. By working overtime, for example, we are creating a buffer for the faster but idle workstations. Thus, when all the workstations are working they can work at their individual optimum speeds (which need not be the same) and the factory is level loaded. It's interesting that this technique of bottleneck elimination revolves about working the bottleneck workstation longer by use of overtime, additional shifts, and perhaps working on days the factory is normally closed. We do this to create the downstream buffer, but we never attempt to relieve the root cause of the bottleneck. The next two bottleneck solution techniques do.

**Methods Improvement.** Bottlenecks are caused by workstations not being able to keep up, flow-wise, with faster workstations ahead of them. The schedule manipulation technique solves the flow problem by making sure there are adequate supplies of inventory at the bottleneck (a prebottleneck buffer, so to speak, but one that forms naturally) so that it can work when the faster stations are tactically idle, thus creating a suitably large buffer for them to work on at superior rates when they do work. If we can get the bottleneck to work at the same rate as the downstream and upstream faster workstations, we do not have to employ the schedule manipulation technique. Instead we have solved the root cause, not maneuvered around it.

One technique for doing this is the methods improvement technique. Here we examine how work is carried out at the workstations, look for ways to eliminate any wasteful activities, and also look for applications of better jigs and fixturing. These are the stuff of the Just In Time (JIT) procedures, which will be covered in a later chapter. Techniques for improving methods will also be presented in the chapter on methods, planning, and work measurements.

**Facilities Addition.** A second way to eliminate bottlenecks, and not just bypass them, is through facilities additions. If methods improvements cannot get the bottleneck workstation up to sufficient flow rates, the company may want to purchase improved or simply duplicate (or other multiple) workstation equipment in order to produce output at a rate that eliminates the bottleneck designation.

> **Example:** We need to deposit 22.5 lb/hr in a submerged arc positioner circumferential welding process. This will allow us to supply valve bodies in sufficient quantity to balance the downstream Computer Numerical Control (CNC) machining centers. With one head we have a practical capacity of only 12.75 lb/hr (see Table 3.1). If another head of equal capacity is added to the weld positioner, the deposition rate capacity will be 25.5 lb/hr. This is a higher capacity than needed, thus the bottleneck is eliminated.

Typically this is the most expensive and longest cycle time technique for eliminating bottlenecks. Consequently, it is the last technique to be considered. Effective approaches for purchasing capital equipment are discussed in the chapters on capital equipment programs and machine tool and equipment selection and implementation.

## USE OF COMPUTERS FOR CAPACITY AND LOADING SIMULATIONS

Experience has taught me the folly of trying to develop and use a computer simulation for capacity and loading independent of total manufacturing needs. A system developed specifically for manufacturing engineering usually lacks items that can affect actual capacity. For one thing, manufacturing engineering systems usually disregard the materials aspect of capacity; they tend to make the erroneous assumption that materials will always be available and in accordance with ordered specifications. They also tend not to factor in lead times for replacement materials to arrive. Finally, independent manufacturing engineering capacity and loading systems rarely include all the workstations. A judgment is often made to cull down the volume of workstations to a number that is easier to work with, which is euphemistically called the set of critical workstations. A manufacturing engineering capacity and loading system lacks expert judgments and inputs from the other functions and subfunctions. Hence, it should be abandoned in favor of a system developed for the whole company, usually called the Manufacturing Production Control System (MPCS). Figure 3.5 shows an MPCS. By studying the various elements of this system we can understand the flow of a useful computer system and, more important, understand how manufacturing engineering's responsibilities for determining factory capacity and loading parameters fit into the overall manufacturing pattern.

The MPCS system shows the flows of information in a typical factory. This flow then becomes the basis of the Manufacturing Resources Planning (MRP II) system. MRP II takes the logic of the MPCS system and explodes it in chronological need sequence for each and every individual part and subassembly. This becomes the chronological factory load that has to be factored for workstation capacity. By running an MRP II scenario, manufacturing engineers can simulate the capacity of the factory to make the intended numbers of products, by type and volume.

By following the central core of Fig. 3.5 from top to bottom, we can see that it is a simple chronological sequence from receiving a customer's order to dispatching the work to the proper workstation. Along the line of the central core, we further refine the data from the customer's broad-based request for a finished product to the necessary discrete parts breakdown and planning required to make the parts at the various workstations.

Let us now look at each block of the flowchart to understand what is accomplished there. The basic flow diagram is presented so that the interrelations between the various manufacturing subfunctions and other functions can be appre-

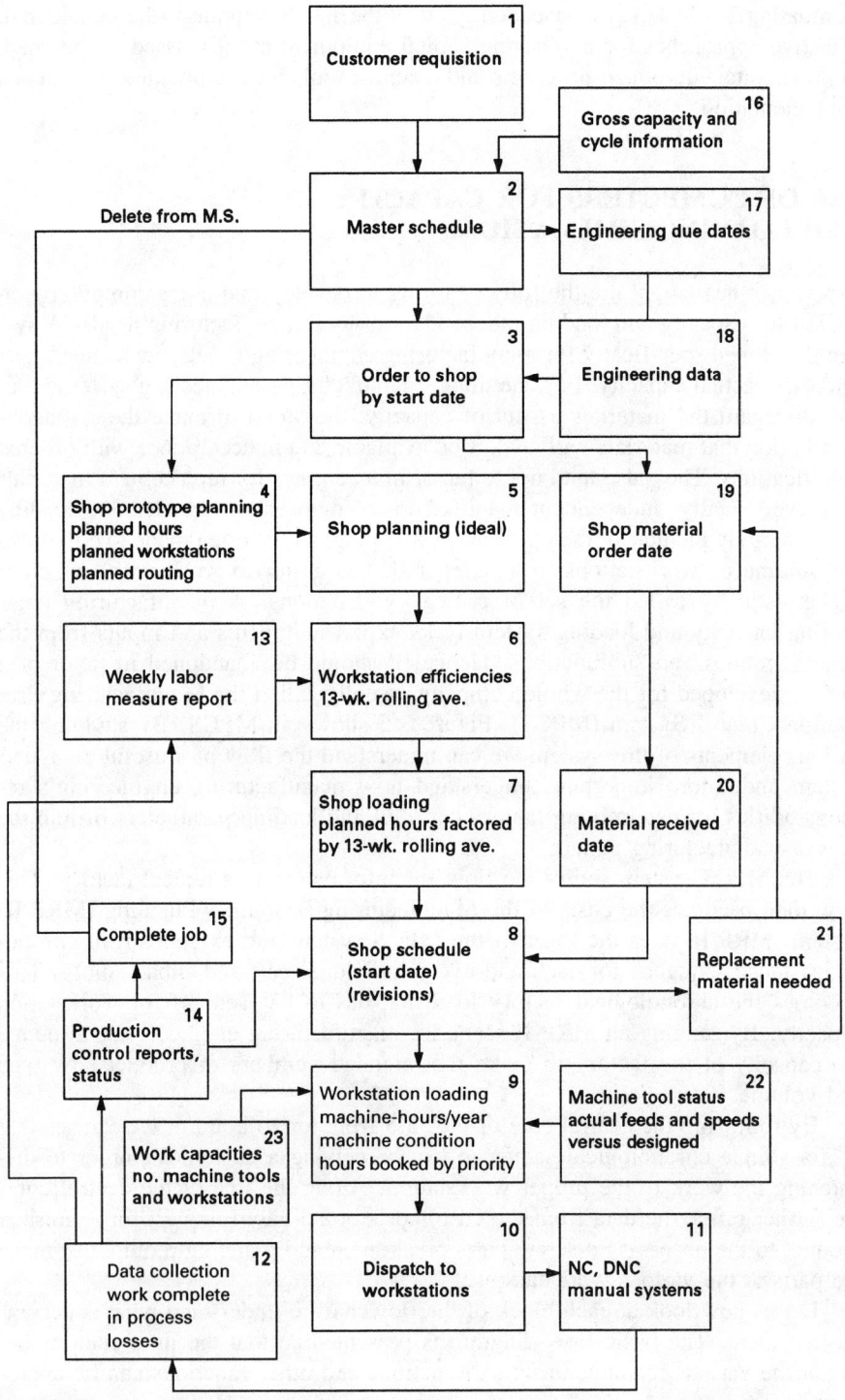

**Figure 3.5** Manufacturing production control system.

ciated. However, by understanding the nature of the work for each block we can visualize what the programmer will have to express in mathematical terms to create a useful MRP II system.

1. Customer requisition: The customer's order defined in terms the factory can understand, that is, by model number, drawing number, rating of output, manufacturing procedure number, and so on. This is a responsibility of the marketing function.

2. Master schedule: The basic sequencing of orders for the shop to produce. Here the customer's order is placed in the queue for sequence of production. Time slots for the various operations that must occur are reserved or mortgaged. This means that the design engineering cycle, the materials procurement cycle, and the major manufacturing cycles are fitted into an overall time schedule. This is a responsibility of the materials subfunction.

3. Order to shop by start date: The basic output to shop operations showing when major manufacturing operations must commence. It eventually triggers the production control schedule once specific planning has been completed. This is a responsibility of the materials subfunction.

4. Shop prototype planning: The preplanning of common models of company product offerings done to minimize the need for detailed planning for each customer requisition. Prototype planning is used to simulate capacities and hold places in the shop schedule when an order is expected but not yet received. It is also used to aid in the mechanization and reduce the time necessary to complete specific order planning. This is accomplished by storing the prototype planning in a data base and making changes only for exceptions related to the specific order. This activity is a responsibility of the AME and methods, planning, and work measurement (MP&WM) units of manufacturing engineering.

5. Shop planning: The breakdown of the order into its component parts and the sequences of manufacture of those parts. Shop planning details the exact workstations to be used and how long it will take to complete each specific part. It is based on prototypes extracted from the data base (step 4). At this stage it is ideal planning, not factored for current efficiencies at the various workstations. However, all the other five deductions from theoretical to practical workstation capacity have been made. This is a responsibility of the MP&WM unit.

6. Workstation efficiencies: The collating of workstation efficiency data in order to factor the shop planning by the efficiencies currently being experienced in the factory. In order to smooth out week-to-week variations in the reporting of efficiencies, a 13-week rolling average is commonly used. This is a responsibility of the MP&WM unit.

7. Shop loading: The planned hours and instructions for each operation that must be performed for all the components, assemblies, and tests that make up the completed product. These are the practical times that shop operations should be able to meet, barring unforeseen circumstances. This is a responsibility of the MP&WM unit.

8. Shop schedule: The detailed production control scheduling of the workstations, showing starting dates and expected completions. Revisions due to less

than expected performance caused by poorer than anticipated efficiencies and variations in the other five practical capacity deductions are also scheduled at this step. This constitutes the actual promise date that shop operations publishes. It is a responsibility of the shop operations subfunction.

9. Workstation loading: The assignment of specific jobs to specific workstations. Here the production control specialists make reservations for workstation time based on machine tool and equipment conditions and the hours during the year when the workstation is available for production. This is a responsibility of the shop operations subfunction.

10. Dispatch to workstations: A time-delayed release to the designated workstations of work to be performed in chronological order. The planning packages consisting of instruction sequences, cycle times, drawings, N/C tapes, quality control forms, and so on are stored in a file based on scheduled release to workstation sequence dates. Dispatches are then made to the workstations based on current promised starting dates. This is a responsibility of the materials subfunction. This commonly is done via the MRP II program.

11. N/C, CNC, and manual systems: The actual workstations where work is performed. The designation N/C and CNC refer to computer-automated subsystems that may be required to initiate and terminate actions at those types of workstations. The design of N/C and CNC subsystems is the responsibility of the AME and MP&WM units of manufacturing engineering. Operations at the N/C, CNC, and manual systems workstations are the responsibility of the shop operations subfunction.

12. Data collection: The manual or computerized collection of data reflecting the current status of work being performed at each workstation. Such things as work completed, work still in progress with percentage complete, and extra work having to be (or having been) performed to correct mistakes are reported on. Since this is a time-dependent step, the information may be readily used to compute efficiency reports. This is a responsibility of the shop operations subfunction.

13. Weekly labor measure report: The compilation of the data from step 12 into management reports related to efficiency of the workstations, and the percentage of the total work done to correct errors. The data on workstation efficiency are fed back into step 6 to correct the shop loading and scheduling. This is a responsibility of the MP&WM unit.

14. Production control reports, status: The compilation of data from step 12 that updates the shop schedule, step 8. In step 8 these data will be used to revise the schedule, if necessary, for early completion or incomplete work. This also forms the basis for production control status reports to management. It is a responsibility of the shop operations subfunction.

15. Complete job: A special input from step 14 when the last operation of a customer requisition subcomponent is finished. Completion of such components as a stator for a large AC motor or the stop valves of a steam turbine would normally require payment by the customer. Hence, a report of their completion is required. Output from this step is fed into step 2 to delete the item from the master schedule. This is a responsibility of the materials subfunction.

16. Gross capacity and cycle information: The workstation capacity data developed by AME as described in previous sections of this chapter. These data are used by the master scheduler to determine manufacturing cycle time for scheduling use of the critical equipment. This is a responsibility of the AME unit.

17. Engineering due dates: The schedule imposed on design engineering by the master scheduler in order to meet customer commitments. Determining the due dates is an iterative process based on marketing function needs, the complexity of the project, customer requirements, and manufacturing cycle time requirements. This is a responsibility of the materials subfunction.

18. Engineering data: The design data used by MP&WM to produce the planning for the specific customer order. This is a responsibility of the design engineering function.

19. Shop material order date: The date by which material must be ordered to support the shop schedule, step 8. Material order due dates are based on design engineering requirements as expressed in step 18. Purchasing develops lead times based on inquiries and current economic conditions. This is a responsibility of the materials subfunction.

20. Material received date: A confirmation that material is available to support production. This is a vital input by purchasing to the preparation of the shop schedule, step 8. Very extensive purchasing, inventory control, vendor performance, and receiving computer systems have been developed to assist in managing this important function and are normally part of the MRP II system. No material means no production, and late material means at best compressed cycles to meet completion dates. Hence, there is a significant risk of increasing costs through mistakes and inefficiency. This is a responsibility of the materials subfunction.

21. Replacement material needed: A result of either defective material being supplied or manufacturing mistakes causing scrapping of otherwise good material. The information is generated at the data collection stage, step 12, and is fed back to the shop schedule, step 8, where an order for replacement material is generated. The replacement material step, in turn, generates a requirement for a shop material order data, step 19, to reorder the material. This is a responsibility of the materials subfunction.

22. Machine tool status: Information on the current capability status vis-à-vis the actual feeds and speeds of the production equipment versus the designed capability. It is a moderating factor in workstation loading, step 9, and would be a legitimate reason for shop operations not to meet the times specified by the shop schedule, step 8. The maintenance unit keeps very close tabs on equipment capability as compared to its original "as new" capability. This information is vital for AME data to justify new equipment. It is a responsibility of the maintenance unit of manufacturing engineering.

23. Workstation capacities: The modifications to the gross capacities as tabulated in the gross capacity and cycle information, step 16. They are the intangibles described in an earlier section of this chapter that temporarily affect the capacity of individual workstations to produce. This information is fed into the

workstation loading, step 9, so that modifications to mortgaging of time on a workstation can take place. This, too, is a legitimate reason for shop operations to miss a scheduled completion date. It is a responsibility of the AME unit.

These are the 23 steps of a typical manufacturing production control system. Such a system is a powerful integration of most functions of a business and certainly all of the direct subfunctions of the manufacturing function. It is also interesting to note that AME can easily perform capacity simulations by inputting its data to this system, and making assumptions based on experience for other function, subfunction, and unit inputs. This is inherently better and more believable by management than a capacity simulation program developed and run independently by AME.

## SUMMARY

In this chapter the concepts of capacity, capability, and level loading have been defined and discussed. Companies must concern themselves with two types of capacities: technical capacity to offer a product line, and physical capacity to determine how much product could be produced in a given period of time. Level loading has been defined as a means of using capacity as a tool for optimum production.

Applying the six deductions from theoretical capacity to compute practical capacity is an important aspect of setting factory output expectations. The physical capacity has been used to explain critical workstations, a monitoring tool useful for level loading. Level loading philosophy and the work of the manufacturing engineer demonstrate the integrated nature of manufacturing. The concept of bottleneck identification and elimination is closely associated with level loading. Here we are analyzing causes that prevent successful level loading. Finally, the concept of using computers as a tool to assist in capacity analysis has been introduced.

Manufacturing engineering only plays a role, albeit a significant one, in factory capacity and loading decisions. This chapter points out the synergism between manufacturing engineering, the rest of the manufacturing function, and the other functions of the industrial organization.

## REVIEW QUESTIONS

1. Discuss the differences between technical capacity and physical capacity with reference to a manufacturing facility.
2. What is meant by the term "lower bound of the typical effort range to be expected of an average worker"? How is it determined?
3. A machine tool has a theoretical capacity to produce 620 units per week. Using the values for deductions employed in Table 3.1, calculate the practical capacity of the machine tool.
4. In question 3, the setup time is improved by 15% and the interference time

by 5%, but the manufacturing losses increase by 8%. What is the practical capacity of the machine tool now?

5. The following is an incomplete analysis of a manufacturing operation. (a) Fill in the missing information.

| Operation | Factored time (hours) to perform per 100 feet of wire | Number of operators | Man-hours |
|---|---|---|---|
| Draw wire | 0.23 | 2 | — |
| Rotational bend wire 122° | — | 4 | 0.32 |
| Anneal wire | 0.12 | — | 0.36 |
| Spool | 0.05 | 2 | — |

(b) What is the complete cycle time for the operation? (c) What is the most likely bottleneck operation that would have to be improved in order to have no queuing of product. (d) What would the direct labor cost be if the hourly rate is $8.50 per hour for the draw wire operation and $8.25 per hour for all other operations?

6. Based on the definitions of capability and capacity, give reasons why the manufacturing operation of question 5 should be considered incapable or capable of performing if the following would have to be assigned to another factory: (a) draw wire, (b) rotational bend wire 122°, (c) anneal wire, and (d) spool.

7. Eight unquantifiable items have been given in this chapter as examples of intangibles that affect times to perform work. Using the four operations of question 5, determine whether the eight items are pluses or minuses for those operations.

8. Give reasons why a factory cannot sustain a 90% capacity operating condition for a long period of time. What is considered a steady-state running condition?

9. Discuss the plus and minus factors from a management viewpoint if level loading is achieved.

10. The following data are for a factory output plan for 3 months: September, 12,830 units; October, 4,620 units; and November, 8,240 units. The standard capacity is 8000 units per month. Determine whether the factory can make the units over the 3-month period and list the specific recommendations manufacturing engineering would make to ensure that the factory has the best chance to produce the desired quantity and yet preserve the factory's equipment.

11. Referring to Table 3.2, suppose the factored time to perform (hr) for all four operations were 0.18 hr. Does a bottleneck exist in this situation? Explain why or why not.

12. Explain why schedule manipulation tends to be the primary method for relieving bottlenecks even though it normally does not eliminate the root cause of the bottleneck.

# FOUR

## CAPITAL EQUIPMENT PROGRAMS

Capital equipment programs are the most visible manifestation of manufacturing engineering activities to senior management. They represent a portion of a corporation's investment funds directed toward the future. Hence they require a precise analysis of needs coupled with the always present constraint of funds available to meet those needs.

In this chapter we will investigate the process of establishing the facilities program, the process of justifying the choices, the development of the budget for purchase of the equipment, and the management of capital equipment projects.

## FACILITIES PROGRAM

A facilities program cannot exist in a vacuum. It must be closely coordinated with the objectives and goals, the business plan, and the product scope of the company. In fact, there can be no facilities program until these items are fully explored and understood. It is an ongoing task of the Advanced Manufacturing Engineering (AME) unit to make sure that the facilities program they develop and pursue is compatible with the desires of management. Of course, the desires of management can and should be influenced by knowledgeable AME engineers.

We will start this discussion of facilities programs at the point of having an approved objectives and goals statement for manufacturing engineering. However, it must be pointed out that objectives and goals, business plans, product scope evaluations, and facilities plans are not strictly linear in their development. For convenience, we define them as linear, but in actuality they exist in a state of flux. For example, the objectives and goals statement may have been prepared on the basis of a previous strategic plan and may now have to be changed. Or the business plan may change due to marketing pressures or perceived new opportunities. A facilities plan developed on the basis of an objectives and goals statement that is now out of date may no longer be valid. To further complicate things, the new business plan may require a quick reaction affecting facilities. This could mean cancellation of previously reserved but not yet purchased equipment from a vendor or placement of a hold order on a vendor's production schedule, all major changes made before thoroughly modifying the objectives and goals statement. Consequently, the system is nonlinear even though it is described as if step A must occur before step B.

Most AME units have someone assigned to the secretary/librarian function of updating and correcting the objectives and goals statement to agree with the current version of the business plan. This keeps the facilities plan consistent with its higher-order management documents. The frequency of such changes can vary from once per quarter to once per year, depending on the volatility of the businesses the company is engaged in.

A facilities plan is a well thought out procedure for purchasing either new equipment or rebuilt equipment in order to meet the capability or capacity requirements of the company. In the objectives and goals system the facilities plan is a specialized action plan for the AME engineers who are assigned the task of finding, selecting, and purchasing the needed equipment. Simple facilities plans are the same as the objectives and goals project statements. For more complex facilities plans the project statements constitute only the current year's portion of the plan. The remainder of the plan would be extracted from the current five-year or long-range forecasts of equipment needs to meet the defined capability/capacity statements.

A facilities plan could also be defined as the extraction of capital equipment purchases from the current objectives and goals statements and the long-range forecasts arranged in chronological order of intended purchase, with action steps outlined to accomplish the purchase on time and to meet the needs requirements. This definition certainly implies a need for purposeful evaluation of equipment. Without such a step we could never be sure that the equipment purchased would perform its intended function. Therefore, selection of the equipment from a long-range forecast is not a first step, but a summation step. The equipment must be on the current projects and long-range forecast document as the result of forecasts of needs made by the area planners for their particular areas of responsibility. These are the capability/capacity needs determined by analysis to meet customer demands and produce desired new products. The compilations of all entries pertaining to capital equipment by all the area planners constitute the list from which the facilities plan is drawn.

The steps to be taken in developing the facilities plan are:

1. Consolidate facilities needs lists.
2. Evaluate estimated expenditures compared to budget for each budget period.

3. Rank equipment needs.
4. Place equipment in year-of-purchase groups.
5. Establish due dates for major action items through implementation of the equipment.
6. Establish a second-tier list of equipment projects to be pursued if additional funds become available.

Let us now explore these six activities in greater detail.

## Consolidate Facilities Needs Lists

In this step the AME engineer collects data on needs from all the area planners and other cognizant personnel and drafts a chronological list of equipment needs by purchase date. This gives advanced manufacturing engineering a total listing of all equipment necessary to support the company's business plan justified by the objectives and goals statement.

## Evaluate Estimated Expenditures
## Compared to Budget for Each Budget Period

Normally, a budget for capital expenditures is established after all operating costs have been firmed up and profitably analyzed. What is left over becomes the capital available for investment in new equipment. This is true even if the firm expects to borrow funds to purchase equipment; the money is then allocated for loan repayment. Unfortunately, money allocated for capital equipment may not be sufficient to purchase the equipment because an investment budget is rarely made up after the equipment needs are identified. This means that the worth of specific equipment purchase projects must be evaluated against that of other projects. But the severity of the evaluation cannot be set until the total funds shortfall is known. Therefore, during this step, equipment is listed by year of intended purchase and expected funds available for that year, the results being shortfall figures that will require judgments of how severe the comparison between projects must be. The result of this step is a breakdown of the list of needed facilities into years of purchase matched against expected costs of each project versus expected funds available.

One thing to be aware of during this step is the level of accuracy of the expected cost of each project and the expected funds available. Since engineers want their projects to succeed and therefore want them to appear as attractive as possible to management, they invariably underestimate the cost of a project and overestimate the funds to be available. It is incumbent upon the AME manager to ensure that costs are assessed correctly. This can be done by requiring engineers to obtain cost estimates from potential vendors and by comparing costs with previous, similar programs. The AME manager must also assess whether the engineers have thoroughly researched the project and uncovered all possible cost areas. On top of this, the prudent AME manager will also add a contingency of 10% to

## Table 4.1 1995 facilities list

| Equipment | Cost estimate ($000) | Budget estimate 1995 |
|---|---|---|
| 1. Vertical machining center, CNC<br>    24 × 60 in. table<br>    3 in. spindle<br>    x travel = 60 in.; y travel = 24 in.;<br>        z travel = 24 in.<br>    18-place tool changer | 750.0 | |
| 2. Phoenix-type horizontal turret lathe<br>    40 in. swing, automatic control | 300.0 | |
| 3. Notching press, 10 ton, manual control | 85.0 | |
| 4. Notching press, 20 ton, automatic control | 110.0 | |
| 5. Blanking press, 90 ton, manual control | 225.0 | |
| 6. Electric lift truck | 10.0 | |
| 7. Pick and place electronic programmable robot | 85.0 | |
| 8. 10-ton panel truck, automatic transmission | 14.0 | |
| Total | 1579.0 | 1000.0 |
| Projected shortfall | 579.0 | |

20% to the project cost for unknowns. The amount of the contingency is a judgment factor depending on the type of project. For projects similar in kind to those recently accomplished, it will be closer to 10%. Finally, the AME manager will never utilize 100% of the budget available for facilities list projects. Usually only 90% will be used, leaving 10% for minor projects that are necessary for the operation of the company but not necessarily documented in the objectives and goals statement.

The problem of not having firm budgets, especially for the long-range forecast period, is one of ongoing concern for manufacturing engineering. The farther the future year is from the present, the less precise the budget figure for capital investment will be. Nevertheless, so that planning will be possible, manufacturing engineering must operate as if the budget projections are real and will be maintained. To aid senior management it must make sure the facilities plan is tied precisely to a point in time, that is, date the plan and identify it to the current version of the current year budget and five-year long-range forecast.

## Rank Equipment Needs

Ranking is a subjective activity that must be done as objectively as possible. Before it is attempted, advanced manufacturing engineering must know how severe the constraints will have to be. Can the company do projects with priorities down to 8 out of 10, or must it restrict itself to projects ranked 1 and 2 out of 10? This funding level constraint comes from step 2. Table 4.1 is a typical listing by year of equipment purchases, cost estimates, and the budget estimate prepared by AME during step 2. Clearly, the budget estimate for capital equipment investment during the budget year will not support all eight facilities projects. We must assume that the eight projects have survived the objectives and goals selection process and support the current business plan. Therefore, all the projects are desirable and would benefit the company. A choice must be made as to what will and will not be allowed to proceed.

**Example:** Using data from Table 4.1, let us determine the severity level that the ranking must satisfy. The budget estimate is $1000K but, using the 90% rule, the AME manager would use $900K for the calculations, preserving the remaining $100K for unforeseen and miscellaneous investment needs. This means that the severity level will be 900/1579 for 0.569. Therefore 56.9% of the total number of projects can be funded, or 4.5 projects. Another way of looking at this would be: any number of projects totaling $900K can be funded. The number 4.5 may sound illogical, but it means that a project may be partially funded in two years, usually successive.

Knowing the severity level is necessary in order to select projects, let us look at a useful scheme for ranking them. Like so many other activities within the overall manufacturing discipline, there is no best way of accomplishing this task. The method proposed here is one that I know is effective.

All the projects listed supposedly do benefit the company. First, make sure this is true; that is, make sure that the benefit does not turn out to be detrimental to another segment of the company.

**Example:** A new drawing coding system may save design engineering 10 to 15% filing time, but requires additions to the manufacturing engineering planning staff to continue to find drawings.

Second, put every proposed facility project to the following test. Can the company survive—that is, literally still be in existence—if the project is deferred? If the answer is no, then the project must be done.

**Example:** The requirement for a bearing finish tolerance is beyond the capability of the equipment presently owned by the company. If the equipment is not purchased, the company cannot compete in the present market and must leave the market, possibly forfeiting current contracts.

Third, for every facilities project, ask whether its completion is highly desirable, moderately desirable, or of low desirability vis-à-vis improving the profitability potential of the company. This is a subjective evaluation, but it can be made fairly objective by comparing the engineering factors of the facilities project with those of its predecessor.

**Example:** If a new lathe is four times faster than the previous way of making a product and 4000 parts are to be made, it is more desirable than a vertical boring mill that is six times faster than an older VBM used in making 600 parts of equivalent value per part. The lathe is worth more than the VBM to the company because more parts go through that particular process.

Logic similar to this should be used to make this subjective assessment as objective as possible.

Knowing the rankings of the facilities lists projects by year group, it is now possible to assign funds to these projects until the funds are exhausted. Obviously, the essential projects will be funded first. The number of projects to be funded depends on the severity level calculated. In Table 4.1 we can surmise that none of the projects are essential for company survival. Not knowing the particulars of the company involved, it would be difficult to determine the levels of desirability, except to say that projects exhibiting trends toward newer technologies would probably be higher in desirability. Projects 1 and 7 would be in this category; they cost $835K, which leaves only $65K. The notching press, project 3, might be selected for a total investment of $920K; this is only $20K higher than the goal of $900K, which is a reasonably good match.

To summarize, the steps in ranking are:

1. Calculate the severity factor.
2. Ensure that the project listings are universally beneficial.
3. Determine what projects, if any, are essential.
4. Rate the desirability of the remaining projects.
5. Match projects with available funds according to the severity factor.
6. Determine which projects will not be funded.

## Place Equipment in Year-of-Purchase Groups

Once the ranking is complete we have made the basic decision about which projects from the facilities lists will proceed and which will be deferred. Commonly, good projects may have to be deferred or canceled outright because of lack of funds. Sometimes excellent projects are simply moved one or two years into the future, when additional funds may be available. Therefore, not only items that survive the ranking process described above but also projects that did not meet the previous years's cutoff criteria are considered. Often the AME manager will add projects that were not funded to the next year's total list, and will consider the deferred project against projects initially considered for that budget year. This is not an automatic listing of all projects that were initially not funded; AME will first consider whether the project is still valid in a later year group.

Once the manipulations of nonfunded projects from one year to later years are complete, we end up with a listing of funded projects by year-of-purchase group. This becomes the official project listing, hence the action plan to be accomplished by the AME unit. At this stage unfunded projects are removed from the objectives and goals statement and placed in a reserve category.

## Establish Due Dates for Major Action
## Items through Implementation of the Equipment

In this step the due dates for all the important steps of a project are assigned in order to ensure successful completion of the project. The AME manager usually establishes these dates in consultation with the engineers and area planners, as described in Chapter 2. This becomes the action plan to be monitored during work planning sessions and other reviews.

Figure 4.1 is a typical weekly report showing the steps that are monitored during a facilities project. The steps must be taken in the order shown as the project progresses from a generalized concept to a specifically tailored application of equipment. Let us take one line of Fig. 4.1 and go through each item reported on. Figure 4.2 is an extraction of the heading from the total project list depicted by Figure 4.1. We will use the lowercase letters under each category in Fig. 4.2 to explain the purpose of the specific columns.

a. Status: This indicates whether a project has obtained final approval for purchase of the equipment. "AP" would indicate approved, while a blank or "pending" would indicate that the purchase of the equipment has not yet been approved.

b. PAR No.: The PAR (plant appropriation request) number is an index by which all functions of the company can identify the particular facilities project. A PAR number is usually assigned when the first draft of the PAR is written.

c. Description: This is the English word description or title of the project.

d. Plan By: The initials of the AME engineer or area planner responsible for the project are filled in here.

e. Estimated Costs and Savings: These three columns show the financial summary of the project. The investment dollars are the funds that will be spent for capital equipment—the depreciable funds. The expense dollars are those which are associated with the investment but cannot be classified as capital investment; for example, they would include costs for moving existing equipment out of the way to make room for installation of the new equipment. These are nondepreciable expenses related to depreciable funds. Savings are the calculated project savings on an annualized basis. The savings area will come under the closest scrutiny at the various levels of approval.

f. Area Specifications: These are the scope specifications usually prepared by the area planner, outlining the type of equipment that should be acquired to meet a specific business need. The area planner does not specify detailed equipment, but rather what the equipment should be able to do.

**Example:** The area planner specifies the type of equipment, such as a CNC chucking-type lathe capable of producing 1000 items per day.

The heading T/A indicates the target date for completion of the step versus the actual date when the step was completed. Dates are shown in a code where the first digit indicates year and the second and third digits indicate fiscal week.

PAR APPROVAL STATUS

| STATUS | PAR NO | DESCRIPTION | PLAN BY | BUDGET INVEST. | DOLLAR IN EXPENSE | 1000 SAVING | AREA -SP T/A | SPE CS. T/A | QUO TES T/A | NOI T/A | WRI PAR T/A | ME MGR T/A | FIN ANC T/A | MFG MGR T/A | GEN MGR T/A | DIV MGR T/A | ORD EQU T/A |
|---|---|---|---|---|---|---|---|---|---|---|---|---|---|---|---|---|---|
| AP | 228-84 | SERVICE CONTRACT DE CONTROLLE | RR | .0 | 5.8 | 8.1 | O O | O O | O O | O O | O O | O O | O O | O O | O O | O O | O O |
| AP | 228-80 | FLOOR PLATES | FJ | .0 | 7.0 | .0 | O O | O O | O O | O O | O O | O O | O O | O O | O O | O O | O O |
| AP | 228-94 | PENDANT CABLES | JM | .0 | 18.0 | .0 | O O | O O | O O | O O | O O | O O | O O | O O | O O | O O | O O |
| AP | 228-99 | PARTS FOR TOSHIBA | RR | 23.0 | .0 | 13.4 | O O | O O | O O | O O | O O | O O | O O | O O | O O | O O | O O |
| AP | 252-81 | BACK FACE CUTTERS | FJ | 14.0 | .0 | 62.7 | O O | O O | O O | O O | O O | O O | O O | O O | O O | O O | O O |
| AP | 251-80 | REPAIR ROOF | JM | .0 | 30.0 | .0 | O O | O O | O O | O O | O O | O O | O O | O O | O O | O O | O O |
| AP | 251-84 | REPAIR TO SHOT BLAST | JM | .0 | 4.4 | .0 | O O | O O | O O | O O | O O | O O | O O | O O | O O | O O | O O |
| AP | 229-91 | INSTALL BXLX | DS | 4.9 | .0 | .0 | O O | O O | O O | O O | O O | O O | O O | O O | O O | O O | O O |
| AP | 228-81 | SERVICE CONTRACT—BENDIX CONT | RR | .0 | 5.0 | 2.3 | O O | O O | O O | O O | O O | O O | O O | O O | O O | O O | O O |
| AP | 228-86 | FIRE HYDRANTS | JG | .0 | 9.5 | .0 | O O | O O | O O | O O | O O | O O | O O | O O | O O | O O | O O |
| AP | 229-83 | LM-2500 FACILITIES | FJ | 22.4 | .0 | .0 | O O | O O | O O | O O | O O | O O | O O | O O | O O | O O | O O |
| AP | 229-88 | TOSHIBA MS MILL | FJ | 162.0 | 30.6 | .0 | O O | O O | O O | O O | O O | O O | O O | O O | O O | O O | O O |
| AP | 229-86 | STEEL RACKS | RC | 8.6 | 14.4 | .0 | O O | O O | O O | O O | O O | O O | O O | O O | O O | O O | O O |

Figure 4.1 Steps monitored during a facilities project.

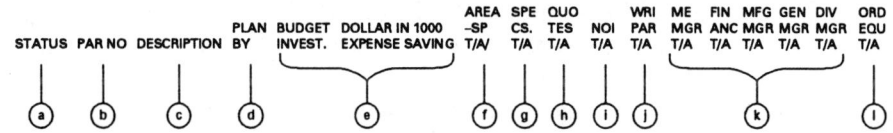

Figure 4.2 Headings from the project list in Fig. 4.1.

Therefore 208 on the target line means year 1992, fiscal week eight. If the step was completed in the tenth week of 1992, then 210 would appear under A.

g. Specifications: In a large AME organization the area planner now turns the project over to an AME engineer. In a smaller organization the area planner and the AME engineer may be the same person. During this step detailed specifications are prepared so that quotes can be solicited from equipment vendors.

**Example:** For the CNC chucking lathe the AME engineer would specify such things as bed length, area of swing, type of controls deemed acceptable, physical size limitations, power requirements, types of tooling to be supplied with the machine, protective screens required, safety requirements, required ranges for feeds and speeds, required tolerance capabilities, required surface finish capabilities, and so on.

All requirements necessary to ensure that the equipment will perform in accordance with the requirements set forth by the area planner during step f are included.

h. Quotes: At this point vendor quotes are solicited and received on the basis of step g. The date shown is the date when the quotes are to be received; this is always specified in the invitation to quote.

i. NOI: A Notification of Intent (NOI) is sometimes required by companies to let everyone in the approval process know that a plant appropriation request is coming. Whether an NOI is required usually depends on the dollar level and/or complexity of the PAR. The purpose of the NOI is to secure speedy and thorough review. If a double zero, 0/0, appears in the schedule, as it does frequently in Fig. 4.1, no NOI is required for the particular project. Notifications of intent are usually short summarized versions of the appropriation requests stating in one or two paragraphs what the project is, why it is being implemented, how much it will cost, and what the savings will be.

j. Write the PAR: The drafting of a PAR consists of thoroughly describing the project so that the reviewing agencies thoroughly understand its purpose and can judge whether to approve or disapprove it based strictly on the merits of the project. PARs can range in length from a single page to a large volume, depending on the complexity of the project. Regardless of length, they all contain the following information in distinct sections: (1) overall description and justification of the project, (2) savings to be achieved, (3) costs to implement, and (4) implementation schedule.

Figures 4.3 and 4.4 are illustrations of typical PARs, one very simple and the other more complex. However, PARs can take many forms other than these. PARs are selling documents; they are used by various levels of management to convince other levels of management that a particular project has significant merit and should be approved. Hence they emphasize whatever the authors believe to be the pertinent or strong points, or the points that the senior reviewer likes to see on requests for capital investment funds. Nevertheless, the four points mentioned above must be covered at the level of detail outlined below:

1. Overall description and justification of the project. This spells out the equipment selected, the alternatives not selected and the reasons for not selecting them, and the reason for going ahead with the project.
2. Savings to be achieved. This section contains the calculations that lead up to and support the savings claimed. Here a very clear and logical buildup of data to calculate the claimed savings is required. This section also states when savings will commence and reach maturity. The return on investment, payback period, and present worth and future worth of calculated funds are also presented in this section.
3. Costs to implement. This is the schedule of expenditures by investment and expense categories, usually showing dates of payments.
4. Implementation schedule. This detailed schedule of purchase through turnover to operations is an important section. If the implementation schedule is considered reasonable, then confidence is gained that the savings are achievable in the years claimed.

In addition to these four generic parts, more complex PARs include an executive summary. This is normally an NOI appended to the front of the appropriations request stating what the request is about. It enables senior managers to decide whether or not it is necessary to review the PAR in detail in order to approve or disapprove it.

k. Approvals: Usually the minimum level for approval of a PAR is at the general manager level. All companies set approval level policies based on dollar expenditures involved. The AME manager must know who will be responsible for the ultimate approval and have the PAR written with that level in mind. In general, the greater the managerial spread between the AME manager and the ultimate approver, the more financial detail and justification will be required. The AME manager must keep in mind that when very high levels of approval are required, managers will not be familiar with the working details of the operation, nor do they need to know them. At the higher levels, the most important thing is to choose investments wisely to optimize the company's profitability. Therefore, good financial justification becomes more critical as the level of the investment increases.

l. Order Equipment: At this point, the AME manager has shepherded the

PLANT APPROPRIATION REQUEST          No. 228-89

---

1.                                          Location _____

   Power Generation _____    Department    Product Line    Turbine & Generators

2. SUMMARY DESCRIPTION OF PROPOSED PROJECT

Relocation of Pandjiris Manipulator for sub-merged arc boom machine and positioner from Bay #2 to Bay #3, Column C31

                                                         Category ____V____

| 3. AMOUNT OF APPROPRIATION | This request | Previously approved | Future requests | Total project |
|---|---|---|---|---|
| Investment | | | | |
| Expense | | | | |
| Total | | | | |

Starting costs (as a memo.)                                          _____

Amounts included in Forecast for total project—Investment _____ Expense _____

6. Date of initial expenditure for project      May 1995

   Date of project completion      May 1995

---

7. STATUS OF PLANNING  80 % complete.
   Remarks (if any):

---

14. PERIOD OF EXPENDITURE

| Month and year | Total | Investment | Expense | Quarter and year | Total | Investment | Expense |
|---|---|---|---|---|---|---|---|
| May 1995 | 3,200 | | 3,200 | 2nd 1995 | 3,200 | | 3,200 |
| | | | | | 3,200 | | 3,200 |
| | | | | Total Contingencies | 350 | | 350 |
| | | | | Grand total | 3,550 | | 3,550 |

**Figure 4.3** A simple plant appropriation request.

15. SUMMARY OF ESTIMATED CHANGES
    IN ANNUAL COSTS

| | Proposed facility | Present facility | Reduction (increase) |
|---|---|---|---|
| Variable costs | | | |
| Direct labor (including employee benefits) _____ | | | 6,650 |
| Direct material _____ | | | |
| Indirect labor (including employee benefits) _____ | | | |
| Maintenance of equipment _____ | | | |
| Expense tools and supplies _____ | | | |
| Special tools (one year's amortization) _____ | | | |
| Transportation (incoming and outgoing) _____ | | | |
|     Relocation | | | (3,550) |
| _____ | | | |
| Fixed costs: | | | |
| Taxes and insurance _____ | | | |
| Normal depreciation _____ | | | |
| Product engineering costs and expenses _____ | | | |
| | | | |
|     Total | | | 3,100 |

PROJECT DESCRIPTION. REASONS FOR EXPENDITURES AND RESULTS TO BE
ACCOMPLISHED.
*Expenditure* of $3,550 is requested to relocate the Pandjiris Manipulator for sub-merge boom
machine and a positioner from Bay #2 to Bay #3, Column C31.

The Pandjiris manipulator was dismantled 2 years ago because of Plant layout rearrangement
and was never used. This proposal is to reinstall it together with a positioner at the west end
of Bay #3 to do the welding of Air Lock shells and L.M.2500 short beams.

*Savings* are calculated as follows:

Based on 8 shells of Air Lock produced per year, total length of weld seams = 505 ft.
Use sub-merged welding process with the Boom Machine it will take 100 hours.
With other methods it will take 400 hours including crane service and idle time waiting for
crane lift.
Defect rate will be 85 hours more.
Savings of labour = (400 − 100) × $10.788 = $3,236.
Savings in rework = 85 × $10.788 = $917.

Total labour saving = $3,236 + $917. = $4,153
Add 60% perquisites = $2,492.
Yearly Saving     = $6,650.

---

| | |
|---|---|
| Estimates and data prepared by _____ | Appropriation approved by _____ |
| Appropriation recommended by _____ | Manager _____ Section |
| _____ | _____ |
| _____ | Manager-Finance |
| Appropriation approved and supporting data certified by_____ | Department General Manager    Date |
| | President-CGE Company Limited    Date |
|     Manager _____ Accounting | |

**Figure 4.3** (*Continued*) A simple plant appropriation request.

PLANT APPROPRIATION REQUEST       No.___229-80___

---

1.                                              Location _____

___Power Generation___       Department    Product Line ___Turbines & Hydro Generators___

---

2. SUMMARY DESCRIPTION OF PROPOSED PROJECT

To purchase and install a CNC Vertical machining center (OKK MCV 800) and the necessary tooling.

Category _____V_____

3. AMOUNT OF APPROPRIATION

| | This request | Previously approved | Future requests | Total project |
|---|---|---|---|---|
| Investment | 351,800 | | | 351,800 |
| Expense | 18,200 | | | 18,200 |
| Total | 370,000 | | | 370,000 |

Starting costs (as a memo.)                                            _____

Amounts included in Forecast for total project—Investment $420,000   Expense $10,000

6. Date of initial expenditure for project    ___August 1995 Amount in 1995 budget___

Date of project completion    ___31st December 1995___

7. STATUS OF PLANNING ___75___ % complete.
   Remarks (if any):

---

14. PERIOD OF EXPENDITURE

| Month and year | Total | Investment | Expense | Quarter and year | Total | Investment | Expense |
|---|---|---|---|---|---|---|---|
| Aug 1995 | 33,000 | 33,000 | 0 | 3rd Q | 33,000 | 33,000 | 0 |
| Sept 95 | 206,500 | 190,000 | 16,500 | 4th Q | 206,500 | 190,000 | 16,500 |
| Oct 95 | 110,000 | 110,000 | 0 | 4th Q | 110,000 | 110,000 | 0 |
| | | | | Total | 349,500 | 333,000 | 16,500 |
| (5.5% inves. 10% expense) | | | | Contingencies | 20,500 | 18,800 | 1,700 |
| | | | | Grand total | 370,000 | 351,800 | 18,200 |

---

**Figure 4.4** A more complicated plant appropriation request.

15. SUMMARY OF ESTIMATED CHANGES
    IN ANNUAL COSTS (1996)

| | Proposed facility | Present facility | Reduction (increase) |
|---|---|---|---|
| Variable costs | | | |
| Direct labor (including employee benefits) | $ | | $ |
| Direct material | (95452) | | (95452) |
| Indirect labor (including employee benefits) | | | |
| Maintenance of equipment | | | |
| Expense tools and supplies | | | |
| Special tools (one year's amortization) | | | |
| Transportation (incoming and outgoing) | | | |
| | | | |
| Fixed costs: | | | |
| Taxes and insurance | | | |
| Normal depreciation | | | |
| Product engineering costs and expenses | | | |
| Subcontracts Costs | | 223177 | 223177 |
| Total | | | 127725 |

Estimated change based on annual production of    *Exclusive of depreciation

Budgeted production—current year      Actual production last year

## PROJECT DESCRIPTION. REASONS FOR EXPENDITURES AND RESULTS TO BE ACCOMPLISHED.

Permission is requested to purchase and install a CNC vertical machining center (OKK-MCV-80) and the necessary tooling.

This new machine tool is required to provide continuous support to the operation as defined in the 1993 Long Range Forecast. With the puchase of this machine tool, three old worn-out milling machines will be scrapped. This machine will relieve the overload situation on the existing OKK-MCV-500 N/C machine as well as handle the work now placed on the TOS horizontal miller as vertical capacity is not purchased.

See Appendix A for description of the machine and a summary of the proposed expenditure.

*Project Necessities*

(1) To replace old and worn-out machine tools. Plant at the present time has the following small milling machines.

    (a) #4705 Kearney & Trecker (horiz) – 27 years old, to be scrapped.
    (b) #4801 Cincinnati (vert) – 34 years old, to be scrapped.
    (c) #4503 Milwaukee (horiz) – 28 years old, to be scrapped.
    (d) #4584 TOS (horiz) – 14 years old, operable.
    (e) #3050 OKK (N/C vert) – 1 year old, excellent.

These machine tools, with the exception of #4584 and #3050, are in extremely poor operating condition, to the point that they are considered hazardous.

Based on a recent survey there is approximately 10,000 non-N/C hours of small vertical milling annual load on these machines. This load will have to be transferred to other machines after these three are scrapped. The only machines available are #4584 (TOS) and #3050 (OKK). #3050 is already overloaded heavily and #4584 can take up only a small portion of the 10,000 hours.

It will be necessary to subcontract work to outside vendors if no replacement machine is purchased.

**Figure 4.4** (*Continued*) A more complicated plant appropriation request.

*Alternatives:*
1) Continue as it is. The retirement of #4801, #4705, and #4503 is way overdue. They should be scrapped. Without the machines, and with the projected overload condition of #3050 (OKK) and the vulnerable condition of #4584 (TOS) if it has to absorb all this load, it is necessary to subcontract work to outside vendors if production schedule is to be met. This will result in high subcontract cost plus extra lead time. For this reason, this alternative is rejected.
2) Buy a conventional machine. No productivity gain is achieved by going for a conventional machine.
3) Rebuilding the old machine tools. The machines have an average age of 30 years. It is not worthwhile to commit a major overhaul on these machines, especially almost no productivity improvement and N/C features could be achieved after such overhaul. In fact, the conditions of the machines are so bad that they are limited to a very restricted type of operation or the safety of the operators will be jeopardized. They will be scrapped as soon as the replacement machine for the workload is secured. For this reason, this alternative is rejected.

*Major Results to be Achieved:*

1) Achieve a payback of 2.7 yrs and an annual savings in excess of $10,000.00.
2) Flexibility of production control because of the availability of a machine with capabilities similar to the existing OKK MCV-500. It will also serve as a backup machine in case maintenance service is required by the MCV-500. Quality of product will also improve.
3) Replacing old and worn-out machine tools with new advanced machine tools will enhance productivity and morale of the shop labor force.
4) Achieve part of the CAD/CAM application of manufacturing technology

See attached appendices for reference.

Appendix A — The proposed machine and expenditure.

Appendix B — Loads and Savings Summary

Appendix C — Cash Flow Analysis

| | |
|---|---|
| Estimates and data prepared by _____ | Appropriation approved by _____ |
| Appropriation recommended by _____ | Manager _____ Section |
| _____ | Manager-Finance |
| | |
| Appropriation approved and supporting data certified by _____ | Department General Manager      Date |
| Manager _____ Accounting | President-CGE Company Limited      Date |

*APPENDIX A*

*Proposed Machine*

The proposed machine is OKK-MCV-800 made by Osaka Kiko Co. Ltd. of Japan. The machine fulfills all our requirements.

One advantage of MCV-800 over other candidates is that the experience gained from the MCV-500 could easily be carried over to this machine. This will include experience in N/C programming, operation and maintenance. This will drastically facilitate the installation, start-up, and operators training on the machine. Another advantage is the universality of tooling, operators and spare parts. Past experience on the OKK MCV-800 has been very satisfactory with respect to its features and performance.

**Figure 4.4** *(Continued)* A more complicated plant appropriation request.

*Proposed Expenditure*

|  | Investment | Expense | Total |
|---|---|---|---|
| (1) CNC Machining Center | | | |
| Foundation | | | |
| Transportation & Insurance | $303,000 | $ — | $303,000 |
| (2) Installation | 2,000 | — | 2,000 |
| (3) N/C Post-Processor | — | 5,000 | 5,000 |
| (4) Set-up Equipment | 3,000 | — | 3,000 |
| (5) Tooling: Cabinets | 5,000 | — | 5,000 |
| Durable Tooling | 20,000 | — | 20,000 |
| Non-durable Tooling | — | 5,000 | 5,000 |
| (6) Area Preparation | — | 2,000 | 2,000 |
| (7) Training: Operators | — | 2,000 | 2,000 |
| Maintenance | — | 2,000 | 2,000 |
| Programmer | — | 500 | 500 |
| Sub Total— | $333,000 | $16,500 | $349,500 |
| Contingency | (5.5%) 18,800 | (10%) 1,700 | 20,500 |
| TOTAL— | $351,800 | $18,200 | $370,000 |

## APPENDIX B

*Savings Calculations:*

Savings is based on comparing the costs of subcontracting the load to outside suppliers vs costs of manufacturing using the proposed machine.

| Costs of Subcontracting | 96 | 97 | 98 | 99 | 00 | 01 |
|---|---|---|---|---|---|---|
| Load (Conv. hr) | 1363* | 4095 | 4095 | 4095 | 4095 | 4095 |
| Load (N/C Hr) | 546 | 1638 | 1638 | 1638 | 1638 | 1638 |
| N/C rate ($/hr) | 49.5 | 54.5 | 59.9 | 65.8 | 72.3 | 79.7 |
| Costs of Subcontracting to N/C Vendors | $27027 | $89271 | $98116 | $107780 | $118427 | $130549 |
| Load (Conv. hr) | 1363 | 4095 | 4095 | 4095 | 4095 | 4095 |
| Conv. Rate ($/hr) | 29.7 | 32.7 | 35.9 | 39.5 | 43.4 | 47.8 |
| Costs of Subcontracting to Non-N/C Vendors | $40481 | $133906 | $147010 | $161753 | $177723 | $195741 |
| Total Subcontract Costs | $67508 | $223177 | $245126 | $269533 | $296150 | $326290 |
| Costs of Mfg. | | | | | | |
| Programming Costs | $5000 | $16500 | $18150 | $19950 | $21900 | $24150 |
| Operating hours | 1090* | 3276 | 3276 | 3276 | 3276 | 3276 |
| Operating Rate ($/hr) | 21.9 | 24.1 | 26.5 | 29.2 | 32.1 | 35.2 |
| Operating Costs | $23871 | $78952 | $86814 | $95659 | $105160 | $115315 |
| Total Mfg. Costs | $24961 | $95452 | $104964 | $115609 | $127060 | $139465 |
| Net Savings | $42547 | $127725 | $140162 | $153924 | $169090 | $186825 |

**Figure 4.4** (*Continued*) A more complicated plant appropriation request.

The work load is derived from a survey of work-station load samples over a period of 12 to 20 weeks and extrapolate over a year (47 weeks) to obtain the annual load.

| Machine | Status | Period Surveyed | Load in Conv. Hrs. | Annual Load in Conv. Hrs. |
|---------|--------|-----------------|--------------------|---------------------------|
| #4705 | Scrap in 1996 | 12 weeks | 114 | 445 |
| #4801 | Scrap in 1996 | 12 " | 470 | 1840 |
| #4503 | Scrap in 1996 | 12 " | 1169 | 4580 |
| #4584 (TOS) | Operable | 12 " | 274 | 1073 |
| #3050 (OKK) | Excellent | 20 " | 900 | 2115 |
| | Projected Load for 1995 – | | | 10053 Hours |

Note: (1) Load from #3050 (OKK) is the overflow load of 360 CNC hours, converted to conventional hours by applying an improvement factor 2.5:1.

(2) The load from #4584 (TOS) is the load that should be processed on the proposed machine to achieve a cost improvement. It does not represent the complete load on #4584 over the period surveyed.

The machine is presently operating in two shifts.

(3) For #4503, #4705, #4801, the load represents the complete load on the machine surveyed in 12 weeks period.

Anticipated load beyond 1995 will be adjusted by a projection factor based on the Long Range Forecast as follows:

| LRF | 95 | 96 | 97 | 98 | 99 | 00 | |
|-----|-----|-----|-----|-----|-----|-----|-----|
| Total D.L. Hours | 375 | 390 | 317 | 327 | 396 | 450 | (Hours '000) |
| Load Projection Factor Using 1995 as Base Year | 1.00 | 1.04 | .85 | .88 | 1.06 | 1.20 | |
| Projected Load On Proposed Machine in Conv. Hours | 3351* | 10455 | 8545 | 8847 | 10656 | 12064 | |

*Adjusted for production to start in September 1995. (10053 x 4/12 = 3351).

Machine capacity based on 2 shifts operation

46.8 wks/yr x 35 hrwk x 2 shifts = 3276 CNC hours
= 8190 Non-CNC hours

Since the available load is greater than machine capacity, 8190 non-CNC hours will be used in savings calculations. (8190 non-CNC hr. is equivalent to 3276 CNC hr. when a 2.5:1 improvement ratio is applied to it)

*Cost of Subcontracting*

When subcontracting, the initial attempt is to find vendors with CNC capability to obtain the lowest possible costs. This is not possible for the practical reason of availability. It is estimated that the load will be subcontacted to CNC and non-CNC vendors on a 50/50 basis. An improvement factor of 2.5:1 is used to convert non-CNC hours to CNC hours. Also, a 10% per year inflation rate is applied to the rates. Subcontract rate is based on quotes from vendors around the Toronto area (see Appendix D) for 1994, $45./hour is used for the CNC rate and $27./hour for non-CNC rate. A 10% handling cost is added to the rates.

*Cost of Manufacturing*

Programming costs based on past experience is estimated to be half a CNC Programmer at $30,000/yr/Programmer with 10% inflation per year.

**Figure 4.4** (*Continued*) A more complicated plant appropriation request.

Operating costs are based on 1995 rate of $9.738/hr plus 125% burden rate (half full rate) with 10% inflation per year. Operating hours are based on 2 shifts operation.

## APPENDIX C

### SHORT FORM CASH FLOW

| MONTH | YEAR | INVESTMENT | EXPENSE | DEPRECIATION |
|---|---|---|---|---|
| Aug. | 1995 | −33000. | 0. | 20.% of Book Value |
| Oct. | 1995 | −318800. | −18200. | 20.% of Book Value |
| Dec. | 1995 | 92222. | 0. | Residual Investment |

| Tax Rate | | 44.0% | 1995–2000 | |
|---|---|---|---|---|

| Year 19-- | Book Value | Inventory Reduction | Savings | Net Income | Deferred Tax | −Cash Flow− Net | Cumulative |
|---|---|---|---|---|---|---|---|
| 95 | 281440. | 0. | 42547. | −25767. | 0. | −307207. | −307207. |
| 96 | 225152. | 0. | 127725. | 40005. | 0. | 96293. | −210915. |
| 97 | 180122. | 0. | 140162. | 53274. | 0. | 98304. | −112610. |
| 98 | 144097. | 0. | 153924. | 66024. | 0. | 102048. | −10562. |
| 99 | 115278. | 0. | 169090. | 78552. | 0. | 107371. | 96809. |
| 00 | 92222. | 0. | 186825. | 91711. | 0. | 114766. | 211575. |

Zero Interest Breakeven Point 3.43 Elapsed Years

Discounted Rate of Return based on 6 year life
Per B.A.B. No. 414 (Yield method) 28.80%

Average Increm. Return on Increm. Investment per B.A.B. No. 413
22.42% based on 5 year life

**Figure 4.4** (*Continued*) A more complicated plant appropriation request.

project from generalized concept through the PAR documentation and through the approval of various diverse management authorities. Now the funds are approved and the equipment can be ordered. This is the beginning of the intensive job of ensuring that the equipment is delivered on time and to specification. Essentially, it is the end of the planning stage and the beginning of the implementation stage. (The implementation activity will be discussed in Chapter 5.) Since obtaining equipment is time-dependent, the prudent AME manager will place the order with the successful bidder as soon as approval is received to do so; hence the short time cycle shown between approval and placement order. This is possible because it is common to work out the contractual details with the vendor while the project is going through the approval stages.

## Establish a Second-Tier List of Equipment Projects to be Pursued if Funds Become Available

Since AME will almost always have more worthwhile projects than funds available to implement them, it is prudent to have second-tier projects ready to go through the project planning phases described in the preceding section at very short notice. Budgets are dynamic entities, and it is not uncommon to have additional funds released part of the way through a budget year. Hence many AME managers will take several of these second-tier projects through the PAR writing stage and hold them there waiting for additional funds. This is one aspect of the manager's responsibility to equip the shop optimally to produce its products.

## FINANCIAL JUSTIFICATION TECHNIQUES

Capital equipment purchases usually involve large sums of money; therefore, management must quantify the decision process to justify its affirmative or negative actions. In a business environment all actions affect the company's ability to make a profit, and all management decisions are intended to result in increased profit and not the reverse. Most decisions are based on common sense and experience, and this is usually sufficient. However, capital equipment decisions are complex and very costly, and a wrong decision can cost the company a great deal for many years to come. For this reason, quantitative decision-making (or justification) techniques have evolved for capital equipment projects.

Before we discuss these techniques, a word of caution is due. Like all other mathematical analysis techniques, financial justification techniques are just tools. They use as inputs the best possible data that the manufacturing engineer can gather. But they are only as good as the data used. The results of a financial justification analysis must be evaluated for their quality, that is, how well they fit the real-world situation. If the data clearly define all the variables, then the solution will have merit and the analysis will be believable. If compiling the data required considerable assumptions, the analysis will be only as good as those assumptions and may not be worth much. Engineers should not use financial justification techniques as a substitute for exercising professional judgment.

Three methods commonly used for evaluating the financial worth of a capital equipment project are:

1. Payback period method
2. Return on investment method
3. Present worth method

They all have in common that they attempt to quantify capital equipment project decisions. The theory is that if we can objectively place all projects on a selected rating scale, we will pick the best ones. The drawback is that it is usually not possible to quantify all aspects of the decision.

**Example:** A decision to purchase a CNC chucking lathe depends on being able to make the most parts in a given period of time. The production rate of the machine can be quantified. However, the cost to maintain it versus that of a manual lathe cannot be quantified. Estimates of maintenance costs can be made but hard data are not available. Therefore, to quantify maintenance costs we are forced to make an assumption, which may or may not be accurate.

Unfortunately, all financial justification techniques treat assumptions as if they were facts. As can be seen, the possibility of an erroneous analysis is not trivial. Therefore, it must be reiterated that financial justification techniques are only tools. In the hands of an experienced, competent engineer, they can be very useful, but when used by someone unfamiliar with the company's operations, they can easily be misused. Let us now consider these three techniques.

## Payback Period Method

This is the simplest method, hence the most frequently used. However, since the advent of stronger financial controls, it is most often used in conjunction with the other methods. Referring to Fig. 4.4, under the heading "Major Results to Be Achieved," it is stated that a payback of 2.7 years will be achieved. This payback is determined by dividing the cost of the project by the expected annual savings, which gives the number of years it will take to reach the break-even point. The formula for determining payback period is:

$$\text{Payback period (years)} = \frac{\text{cost of project}}{\text{incremental project savings/year}}$$

**Example:** Figure 4.4 shows that the cost of the project is $370.0K while the average savings over the life of the equipment is $136.7K. Therefore the payback period is 2.7 years.

This number has no significance unless the company has a specific goal in mind for the time it is willing to allow until a project returns a profit. Usually, during prosperous times a 3.0- to 4.0-year payback period is acceptable, while in recessionary times 2.0 years is often required. This calculation can give an indication of the merit of a project compared to the company's business plan parameters.

## Return on Investment Method

This technique takes the payback period method one step further. In the payback period method the effective life of the equipment is not mentioned; it is assumed that the equipment has eternal life and that there will be a payback break-even period. However, if the equipment lasts only 18 months, or the project is only operational for 18 months, the 2.7-year payback period means that break-even can never be achieved. To take into account the effective useful life of capital equipment, the Return On Investment (ROI) calculation was developed. The formula for calculating the ROI is:

$$\text{ROI} = \frac{\text{average savings/year} - \text{cost of project/years of life}}{\text{cost}}$$

**Example:** In Fig. 4.4 the average savings per year is $155.5K for a full 5 years. The fractional year savings of $42.5K was not used because we are using a 5-year life rather than $5^{1}/_{2}$ years. The cost of the project remains $370.0K. Using the ROI formula, we calculate a value of 22.0%. Note that in Appendix C of Fig. 4.4, the average incremental return on incremental investment is 22.42%. This is the same calculation modified for tax write-offs.

The ROI claims to measure investments. An ROI of 22.0% can be compared to any other type of investment that earns interest. In this case, to find an equivalent investment the company would have to find one that would gain interest of 22.0% per year.

## Present Worth Method

This is the time value of money method. It is also sometimes called the Discounted Cash Flow (DCF), Internal Rate of Return (IRR), or Net Present Value (NPV) method. The concept of the method is that all projects should be compared with an alternative financial investment yield at anticipated interest rates. That is, if we invested a sum of money equal to the cost of the project, would it yield less than, as much as, or more than the incremental savings generated by the project? If an investment at currently available interest rates yields a greater return than the proposed capital equipment project, then the project is unacceptable and should not be pursued. That is the theory of the method. In practice, we are comparing dissimilarities and are not taking into account intangible benefits that may accrue to the company.

If we have a project that yields $200.0K and the comparison investment yields $250.0K, both over the same time period, the present worth method concludes that the project should not be approved. But what if the project yielding $200.0K is an entry into a new market or expands opportunities for present markets? Then if the company settled for an additional $50.0K profit, it might miss a significantly larger opportunity. Therefore, depending on the present worth method to reach a decision can be very shortsighted, and such shortsightedness has resulted in the demise of many companies. What they did not use was good manufacturing engineering and business judgment.

This method uses the classic time value of money formula,

$$P = \frac{F}{(1 + i)^{n}}$$

where $P$ is the present value of money, $F$ the future value of money after $n$ interest periods, $n$ the number of interest periods, usually years, and $i$ the interest rate. We can now use this equation to start with the expected savings over the life of the project and calculate its equivalent present worth today. This tells us how much money we would have to invest today to earn as much money as we would if the capital equipment project was completed. We will use the example of the project shown in Fig. 4.4.

> **Example:** The projected 5-year savings for the capital investment project is $155.5K average per year or $777.5K. What would the present worth of an investment have to be to earn that much money if the interest rate available for an investment is 10%?

$$P = \frac{F}{(1 + i)^n} = \frac{\$777.5K}{(1 + 0.10)^5} = \$482.8K$$

Once we have an alternative investment value—that is, how much it would cost in the financial markets to earn as much money as the capital equipment investment—we can use the present worth method to see whether the project is acceptable.

> **Example:** Data from Fig. 4.4 show a cost of $370.0K to earn $777.5K over 5 years. The alternative is to invest $482.8K at 10% interest to earn $777.5K. Since only $370.0K must be invested through the capital equipment project to earn the same amount of money, the present worth method shows that it is less expensive to invest in the equipment than the alternative. Thus the capital equipment project should be approved.

## MANUFACTURING ENGINEERING BUDGETS

The facilities program description touches on many aspects classified as budget matters and project management matters. Now let us look more closely at the budget aspect of capital equipment programs; in the next section we will consider project management.

The budget represents the funds allocated for a specific purpose, in our case for the purchase of capital equipment. It also shows by its size and adequacy the priority level given by senior management to this aspect of the business. A budget that is set low by comparison to previous budgets indicates that management is not interested in business expansion, but will be satisfied with maintaining the company's current position vis-à-vis market penetration and profits. A budget that is set high indicates that management wants to take market share away from competitors and place the company in position to become more profitable. Therefore, by evaluating the size of the budget as compared to previous years, manufacturing engineering can easily gauge the intent of senior management and act accordingly. This is always true and provides guidelines for how manufacturing engineering should act once the budget is presented.

However, this is only half the story. Manufacturing engineering has considerable opportunity to influence management in deciding whether to hold the line or to set an aggressive budget. As the technical arm of manufacturing, it must make sure that senior management understands what funds are required for the company to maintain its present capability or capacity, and what would be required in addition to improve capability or capacity. Therefore, it must translate matters from a technical sense to a financial sense, that is, the cost versus opportunity relationship, keeping in mind that technical improvements must result in improved profit-making opportunities to be of use to an industrial enterprise. The algorithm the manager of manufacturing engineering must keep in mind is that technical improvement equals opportunity for profit minus costs:

$$T = OP - C$$

and that opportunity for profit minus costs must be greater than zero:

$$OP - C > 0$$

With these principles in mind, manufacturing engineering can proceed to put together proposed budgets, for both the current period and the forecast period. Even such items as upkeep of plant and facility—for example, painting offices— must satisfy the algorithm; that is, walls should not be painted unless the benefits outweigh the costs. Therefore manufacturing engineering must go beyond the reasoning that "it would be a nice thing to do" to justify a proposed budget item. In the case of painting offices, a justification that can be related to improved opportunity for profit must be shown. A reason to paint a sales office—to influence potential customers favorably—would not be sufficient for a manufacturing floor dispatch office. Perhaps the dispatch office should be painted to alleviate a safety problem that, if not solved, could cost the company funds in lost production, fines, and so forth. These mundane examples point out that manufacturing engineering must be aware that every dollar they propose to be included in the capital equipment budget must relate directly to improved profit opportunity.

When manufacturing engineering goes through the objectives and goals process, it is automatically going through a capital budget proposal process. Each project finally approved by the objectives and goals system satisfies the algorithms presented above. Whether a specific project reaches the approved budget depends largely on the degree of positiveness with which it satisfies the expression $OP - C > 0$. In the budget process manufacturing engineering is competing with all other company functions and subfunctions for limited funds. Therefore, it must strive to place only projects with large positive values of $OP - C$ calculations in its submittal of a proposed budget. Of course, it must not lose sight of the fact that projects must be consistent with overall goals. If manufacturing engineering follows the simple procedure of developing capital expenditure projects as part of the objectives and goals procedure, and then assuring that the profitability algorithms are adhered to, it will have substantial success in achieving funding for its programs.

In making a facilities program budget it is desirable to classify projects in accordance with the major category to be accomplished by implementing the project. The classifications show what the major thrust of the company is for any given budget period, that is, whether the company is attempting to enlarge its market share or trying to maintain the status quo.

Classifications vary among companies, but as a minimum, for an adequate description of all capital equipment projects, they should include the following categories:

1. Rebuild—used for projects concerned with taking existing equipment out of service and rebuilding it as new or with slightly enhanced capabilities.
2. Replace—used for direct replacement projects, replacing one piece of equipment with another, usually of the same type and capability.
3. Cost improvement—addition of equipment that allows the company to produce its products at lower costs, hence improved productivity.
4. Capacity addition—addition of equipment that allows more product to be produced than before or a new type of product to be produced.
5. Safety and environment—equipment purchased primarily to meet company, industry, or government regulations.
6. Miscellaneous—used for capital equipment projects that do not fall into any other categories.

Classification of all projects helps senior management to visualize where funds will be spent. It also makes the intent of the company visible. If the company states in its business plan that it wants to increase its market share, and all of its capital equipment projects are in the replace and rebuild categories, it is clear that the actual plans and the strategy are not in agreement. These types of mismatches occur frequently, and the check afforded by classifying capital equipment projects brings them into the open and may lead to either a change in funding levels for capital equipment projects or an honest assessment that the strategy cannot be accomplished and should be changed. Classification, then, can be thought of as a check on the system. If the majority of the equipment project classifications agree with the stated strategy as reflected in the objectives and goals, then the company's planning systems are working well and one would tend to have confidence in its ability to achieve the stated goals. If the converse is true, there is a management problem that should be addressed.

The final budget matter that concerns manufacturing engineering is the question of how much to propose. A proposal for spending at a level that is impossible to achieve will be totally discounted, and vital projects may inadvertently be discarded. To maintain its credibility, manufacturing engineering must know what level of expenditure is reasonable and what is unreasonable.

What is reasonable, of course, depends on the situation. Fortunately, there is a way to check whether a budget proposal is in the realm of reasonableness. Assuming that a company intends to stay in business and intends at least to maintain its current market level, it is fair to say that the capital equipment investment budget should be equal to or greater than the depreciation amount. This means that to stay healthy over a three- to five-year period, a company should invest as much money in capital equipment as it depreciates for bookkeeping and tax purposes. If it does less, the company is shrinking in size and its net worth is decreasing.

With this fact in mind the manager of manufacturing engineering can set budget targets. If the company intends to attempt to increase its market share for its product line, then the manager would be justified in proposing a budget that is higher than depreciation. I believe that the increment should be a one-for-one increase based on the amount of market penetration increase aimed at. At the other end of the spectrum, the manager of manufacturing engineering should protest if investments are kept below depreciation for the entire long-range forecast period. Such a situation is a de facto slow exiting from the business, and it is the manager's duty to point this out. Therefore the general budget target should be equal to capital equipment depreciation plus or minus amounts corresponding to current year situations, with concern for consecutive negative years.

Depreciation is a very simple yardstick for budget proposals. Depreciation accounts are calculable well in advance, since they are based entirely on existing capital equipment and firm depreciation rules. This makes it easy for manufacturing engineering to obtain the figures and set visible targets for budget preparation.

## CAPITAL EQUIPMENT PROJECT MANAGEMENT

The key factor here is due date control. Figures 4.1 and 4.2 show the elements needed to ensure that capital equipment projects reach fruition by establishing dates for all activities. The schedule shows when each activity must be completed in order to meet the end date, in this case the placement of a purchase order with the selected vendor.

The purpose of capital equipment project management is to make sure that the schedule is adhered to. It involves good management techniques as previously described, including work planning, action plans, and so forth. We will describe here the aspects of project management that apply specifically to capital equipment programs.

The key management tool for capital investment projects is the PAR status meeting. This is a specialized production control meeting where the product is the plant appropriation request, its writing and its approval through the authorization to purchase stage. The PAR status meeting is a weekly meeting held by the manager of manufacturing engineering but presided over by the AME manager and attended by all the area planners, AME engineers, and other manufacturing engineering unit managers. The purpose of the meeting is twofold: (1) to review the status of all capital equipment projects, and (2) to inform the manager of manufacturing engineering of the status and solicit the manager's support and help in approval routines. The latter purpose is important because the manufacturing engineering manager is the first managerial approver and will be instrumental in having complex projects approved at higher levels of management.

The meeting is conducted by the AME manager, who reviews the status of all projects. For each project the subordinates must state whether or not a target date is met, explain why if the date is not met, and state what is being done to achieve completion of the steps. If the target date is not being met, the effect on the overall project and what can be done to get it back on schedule are also discussed. One of the main functions of the meeting is to impart a sense of urgency to the organization. Since virtually all the target dates are the result of inputs from the area planners and the AME engineers, date compliance is not an unreasonable burden.

The AME manager will also preview PARs that will be submitted to the manager of manufacturing engineering for approval in the near future, pointing out any unusual or interesting facts that the manager of manufacturing engineering should be aware of. This, of course, is the primary reason for the manager of manufacturing engineering to attend the meeting. In virtually all industrial organizations it is expected that the manager of manufacturing engineering will personally shepherd PARs of significance through the approval procedure.

At the PAR status meeting the manager of manufacturing engineering will want to be assured that the financial justifications for PARs are being adequately attended to, since this will most likely be the point of contention in any approval problems. Most approval agencies are willing to abide by the manufacturing engineering decisions pertaining to technical content but consider it within their prerogative to question the financial justification. Therefore, for capital equipment projects the manager of manufacturing engineering must make sure the financial justification is as sound as the technical justification.

One way to ensure sound financial justification is to invite early participation by the finance function. Most companies recognizing this need will assign a financial analyst to the manufacturing function to assist in this. The prudent manufacturing engineering manager will make use of this resource.

The final outcome of the PAR status meeting is the updated PAR approval status report (Fig. 4.1). This report should be distributed to all who have a need to know—that is, all approval agencies. Those who are behind schedule are officially advised of their lack of performance, and those who are doing well are officially recognized. In addition, the report tells the approval agencies when they can expect to receive PARs for their disposition. This document then becomes a control document as well as an information document. It serves to tie the entire capital equipment program together and give it substance.

## SUMMARY

We have examined the mechanics and philosophies of capital equipment programs from inception to purchase of equipment. This activity is in the forefront of senior management activity and is vital to the success of the company. The range of activities involved in capital equipment programs has been covered from the development of the facilities plan and writing PARs, through financial analysis methods and project management. Of overriding importance, however, is the need for excellence in management, which makes the difference between success and failure.

## REVIEW QUESTIONS

1. A facilities plan is a specialized action plan for purchasing capital equipment. Explain why a five-year plan is desirable for this activity.

2. What is the difference between the facilities list and the facilities plan?

3. Discuss the advantages of preparing an investment budget after the equipment purchase needs are identified. Compare this method with that of allocating funds after all operating costs and profit levels have been determined.

4. Referring to Table 4.1, calculate the severity level if the budget estimate is $1050K and the majority of the projects listed are considered to represent new technologies for the company.

5. Explain the benefits to management of requiring establishment of due dates for action items through the implementation of the equipment in the factory.

6. For the vertical machining center, CNC, shown in Table 4.1, develop a set of specifications, as the area planner would for this capital equipment project. Briefly justify each major entry of the specification.

7. A PAR is to be written for the equipment in question 6. (a) Prepare an overall description and justification of the project. (b) Demonstrate a suitable method for calculating project savings. (c) Prepare a cost to implement schedule. Use 80% of the funds for purchase of equipment and 20% for installation. A three-phase payment to the vendor may be used. Recommend progress accomplishments required before each payment is made. (d) Develop a suitable implementation schedule.

8. Write an executive summary for the PAR in question 7. Include the pertinent facts that the senior operations manager would need to know to approve or disapprove the PAR. Discuss the different information that the chief financial officer would require.

9. A company wishes to increase its market share from 10 to 20% of the total for its basic product within three years. Explain the kind of capital equipment budget strategy it would pursue.

10. Using the algorithms $T = OP - C$ and $OP - C > 0$, justify the following projects: (a) Purchase a fleet of cars for sales personnel rather than rent or lease cars. (b) Purchase a document copier machine rather than rent or lease one. (c) Paint the plant cafeteria. (d) Install a central air-conditioning system for the perishables warehouse.

11. Explain why the classification by categories system for projects is a useful management tool.

12. A project to purchase a new CNC horizontal boring mill has been proposed. It will cost $475,000.00 and save $820,000.00 over the next six years of

operation. The company is operating in a down-turning economy and competition is intense. One of their concerns is that if they cannot improve this particular plant's productivity, they may be forced to close it down. There is also a possibility of entering a new product line in addition to their current offerings. This new line will contribute an additional $100,000.00 per year to the company's profit. Calculate: (a) The payback period. (b) Return on investment. (c) Present worth financial justifications.

Based on the data given: (d) Determine whether this project should be approved. Assume all figures are after-tax figures. The company can invest at 10.5% interest.

# MACHINE TOOL AND EQUIPMENT SELECTION AND IMPLEMENTATION

Once the plant appropriation request has been approved, the facilities program enters into an implementation phase. We leave the domain of strategic planning and enter that of equipment purchase, including negotiation, procurement, and installation of the equipment to perform the intended task. Also, once the PAR has been approved the facilities project ceases to involve all aspects of the company. It continues to have high priority only within manufacturing engineering and, secondarily, within shop operations. To the other functions of the company the project is history; to manufacturing engineering, it is just beginning.

In this chapter we will explore the many facets of a capital equipment project that lead up to successful implementation. Some of the items occur before PAR approval, some after. The sequence described is the optimum way of approaching the accomplishment of capital equipment projects.

## MAKING THE PRELIMINARY LAYOUT

As soon as a facilities project is initiated and a new piece of equipment is agreed upon, an equipment floor plan known as a preliminary layout is prepared. The purpose of this layout is to determine whether the contemplated piece of equipment will fit into the current factory situation and, if it will not, to obtain enough information to estimate the renovations that will be required to make it fit. A preliminary layout is a major factor in determining whether a proposed project is feasible or whether it is too ambitious—that is, whether the changes in the factory floor plan would be too expensive for the project to be accomplished.

Let us now look at a typical preliminary layout and discuss the major aspects. Figure 5.1 represents the block layout of a machining bay for relatively heavy equipment as it exists before the initiation of a facilities project to add a Computer Numerical Control (CNC) horizontal machining center with a multitool turret. The project is designed to increase the capacity for making large steam valve bodies from four per week to eight per week. The existing manufacturing procedure is to receive a forged body, make a machine layout for it, machine the basic interior and exterior shapes on the Vertical Boring Mill (VBM), do the intricate interior work on either the Horizontal Boring Mills (HBMs) or the lathe, depending on the inside diameter of the particular valve body, drill the various holes on the drill presses, and then inspect and ship the product. The project is to add a CNC horizontal machining center capable of doing the HBM or lathe work and the drill work at one workstation, thus saving queue time, setup time, and machining time. The goal for the new piece of equipment to be able to do eight units per week is set because it would match up well with the VBM capacity of ten units per week.

The problem has been that each HBM and each drill press has a capacity of two units per week. Therefore, with the proposed new machining center, an additional capacity of up to ten units per week would be available, if the older HBMs and drill presses were kept in service. The limitation in output now becomes the VBM. It was determined by the area planner, after consultation with the marketing function, that eight units per week would be adequate for the company's needs with an occasional peak load of up to ten units per week. The purpose of the preliminary layout is to see how the CNC horizontal machining center will fit into the production flow and to determine whether it will work in this factory.

Figure 5.2 shows the layout flow with the existing method of manufacture. The flow is essentially U-shaped with one double back to the inspection station. The new layout should at least not make the flow worse.

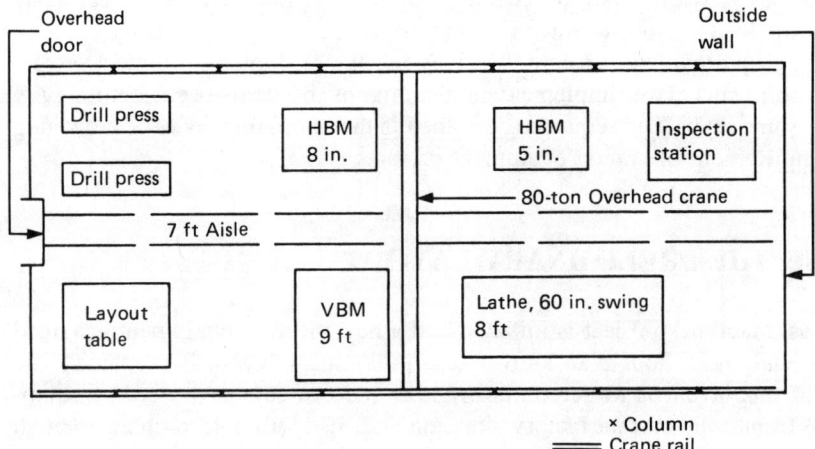

**Figure 5.1** Layout of a machining bay for relatively heavy equipment.

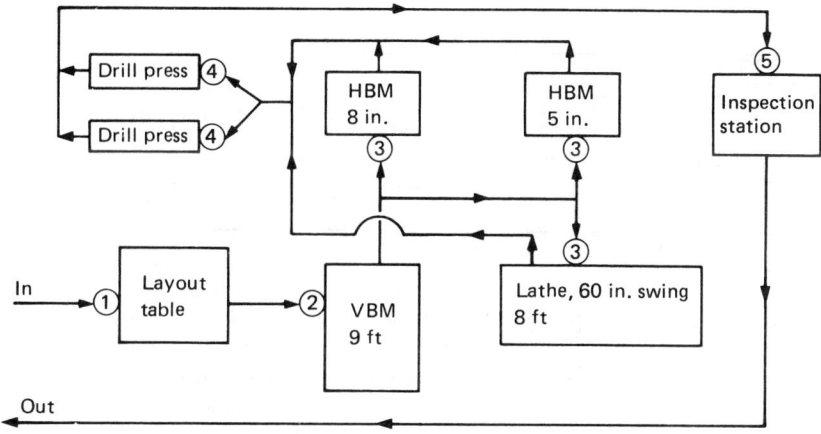

**Figure 5.2** Original flow diagram for the machining bay in Fig. 5.1.

An evaluation of available CNC horizontal machining centers shows that the geometry of the machine will allow placement in the factory to the right of the lathe in Fig. 5.1. However, the area planner must also be concerned with supplying electric power and compressed air to the proposed facility and making sure that it can be serviced by the overhead crane, the only method of transporting the workpiece in this factory. To do this the planner reviews the architect's drawings of the building to ensure that the crane rails extend far enough to accommodate the proposed facility. This type of machine tool will require a below-grade reinforced concrete foundation, which leads to another factor to be evaluated. The area planner will be looking for any underground impediments or other constraints to putting in the required foundation. These and other questions of significance are reviewed and answered only after the preliminary layout is made; an effort to answer them beforehand is not warranted because the answer would not be accurate enough. Figure 5.3 shows the present layout with the CNC horizontal machining center added.

It appears that the proposed addition fits the space. Let us also assume that there are no underground impediments to installing the foundation. Note that the new machine is next to the lathe; therefore the area planner could expect to extend electric power and compressed air from that facility to the proposed facility. However, the preliminary layout does not indicate whether sufficient power or air exists to service an additional machine tool. This will have to be evaluated by investigating the electrical network drawings and the compressed air capacity. The layout does, however, confirm that crane coverage is adequate. If the crane rails did not extend far enough, it might be proposed to extend the rails or to install a jib crane. Fortunately, these options are not necessary.

Knowing that the new machine tool can be physically accommodated, we should now evaluate the product flow again to ensure that it is at least as satisfactory as the original flow. If the flow becomes worse, it will be necessary to

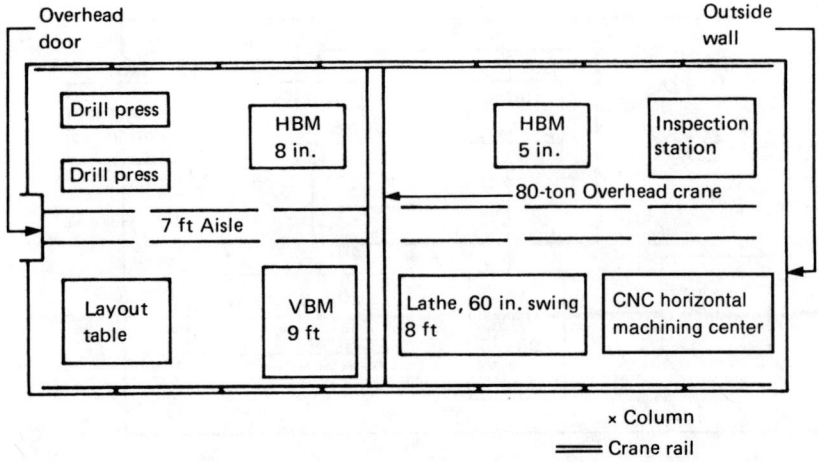

**Figure 5.3** Layout of the machining bay in Fig. 5.1 with a CNC horizontal machining center added.

make another preliminary layout. Figure 5.4 is the flow path resulting from the proposed preliminary layout.

The flow portrayed by the solid line in Fig. 5.4 is essentially a straight-line flow with finished product doubling back out the door. However, this is a flow for producing eight units per week. What will happen when ten units per week are required? In that case we can utilize the older units, as shown by the dashed-line flow in Fig. 5.4, which is the same as the original flow shown in Fig. 5.2. In this case we only need one HBM and one drill press to produce two units per week. This gives us an opportunity to improve the layout further. Note that the additional units require the U-shaped flow and a double back to the inspection station. A review of the layout shows that if the unused drill press were removed, we could relocate the inspection station to its location and end up with a U-shaped flow for both the base load of eight units per week and the peak load of ten units

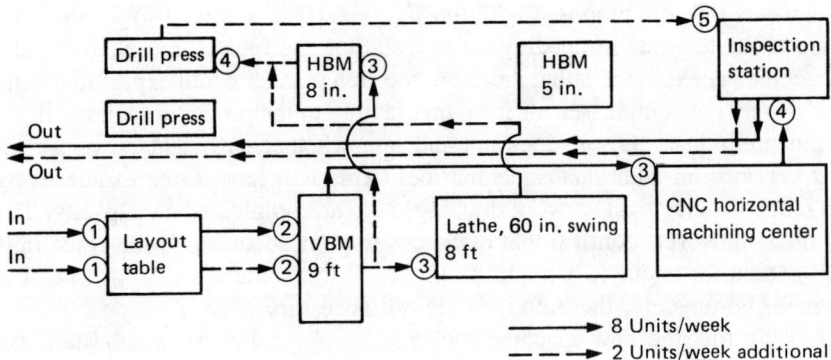

**Figure 5.4** Flow path for the machining bay in Fig. 5.3.

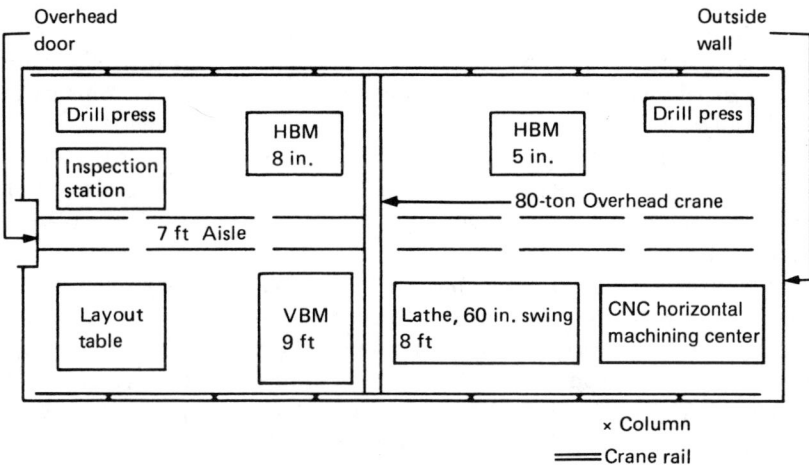

**Figure 5.5** Final preliminary layout for the CNC horizontal machining center project.

per week. Exchanging the unused drill press and the inspection station has another advantage; it places the two unused pieces of equipment adjacent to each other. They can then be used for other product projects or disposed of to make the area available for other uses. The preliminary layout opens up additional opportunities that would probably not have been uncovered if decisions had been made without a preliminary layout. Figure 5.5 shows the final preliminary layout of the CNC horizontal machining center project. This now becomes the working layout that will be used to develop the detailed plans for implementation on the factory floor.

## WORKING WITH VENDORS

Let us now diverge from the chronological path of equipment implementation projects and discuss relations with vendors—a very important subject for the successful practice of manufacturing engineering. Improper relations invariably lead to problems of extra cost, late delivery, and poor equipment performance. There is no need for such problems to develop if the manufacturing engineer understands proper relations with vendors.

First, we must understand that the vendor is the supplier of equipment that allows manufacturing engineering to achieve its stated goal. The manufacturing engineer's project is not to purchase equipment, but to accomplish the stated goal. In the CNC HBM example, the stated goal is to increase weekly output from four to eight units per week with an additional capacity of two units per week for peak periods. Notice that the goal is not to purchase a CNC horizontal machining center. It is important that the manufacturing engineer purchase only what is absolutely necessary and continually keep in mind that the vendor's goal and his goal are not the same. The vendor wants to sell as much as possible. It is not necessarily in the vendor's interest to satisfy the manufacturing engineer's goal.

With these basic facts in mind, the manufacturing engineer can devise a proper relationship with vendors. The engineer must be scrupulously fair with all vendors who have the capability to satisfy the equipment needs of the project. This means giving basic information about requirements equally to all who wish to supply equipment and are deemed capable of supplying it. It is also correct to listen to vendors extol the merits of their equipment and how they would use it to satisfy the company's goal and to incorporate good ideas from vendors in the plans to achieve the goal.

Failure to keep in mind manufacturing engineering's true goal can lead to problems. The manufacturing engineer who becomes enamored with the nuances and elegance of a design may end up specifying and purchasing equipment that does considerably more than is actually required. In this case the manufacturing engineer has fallen into the trap of being swayed through apparent friendship, small favors, technical wizardry, and so forth to spend more of the company's money than required.

It is simple not to fall into this trap. If manufacturing engineers follow a few rules, they can be assured that they will always achieve their company's goal and not the vendor's goal. The rules are as follows:

1. Always keep in mind the goal of the project. For each product presented by vendors, ask whether it is absolutely necessary for the optimum success of the project. If not, reject it.
2. Never accept a favor from a vendor such that the vendor will expect a favor in return. Remember that the only favor the manufacturing engineer can give a vendor that is meaningful is to purchase the vendor's equipment.
3. Never accept gifts of value from a vendor. This puts the manufacturing engineer in a compromising situation when evaluations must be made between competing products. It is legitimate to accept social invitations as long as the value is within common bounds of doing business, and not of such a value that the manufacturing engineer could not conceivably pay for it.
4. Never accept engineering services from vendors other than detailed information necessary to evaluate whether the product offering would fit into achieving the goal. Vendor engineering services could result in a system that would not accept other vendors' products, which is usually detrimental.
5. Always strive to obtain at least two competitive quotes for all equipment purchases. This will ensure that the company is purchasing the equipment at the lowest possible cost.

These rules amount to the exercise of common sense in dealing with persons determined to sell something. By following these rules, manufacturing engineers can pursue their assigned goal in an optimal manner. Whether the sale of equipment is optimal to the vendor is of no concern. Manufacturing engineers must be concerned only with the success of their own company and can do that only if they concentrate on optimizing achievement of their assigned goals.

## SELECTING THE EQUIPMENT

Concurrent with the preparation of the preliminary layout, work starts to specify the equipment to be purchased. In specifying equipment, the more specific or detailed the manufacturing engineer can be, the better the ultimate results will be. This means that the manufacturing engineer must thoroughly understand the purpose of the project, including the technical levels to be achieved as well as the budget limitations. Once a project is close to the PAR writing stage, the area planner will hold a technical briefing session with the assigned manufacturing engineer, representatives from shop operation, the methods engineer, and the process control engineer. Specifically, the overall goal to be achieved and the options identified to achieve that goal will be discussed. During this meeting the area planner will solicit comments from those representing the other areas to ensure that the best possible solution is identified. The manufacturing engineer, on the basis of this information, will begin to determine the specifications for the equipment.

Specifications are prepared in two phases: a preliminary estimate and a detailed purchase phase. The preliminary estimate phase has the main purpose of determining the cost levels involved in the project. Usually this is done by researching current equipment catalogs from various vendors, comparing this project with previous projects, and perhaps contacting vendors to clear up technical points. This information, once compiled, gives the area planner an estimate of what the project would cost if pursued. The estimate is compared with the original budget estimate, and a decision must be made as to whether the project is still attractive and whether the company can afford to go ahead with it. This decision is usually made by the manager of manufacturing engineering, based on recommendations from the AME manager. If the decision is to continue the project, we progress to the next phase.

The detailed specification phase entails sufficient documentation to solicit firm quotes from the interested vendors. It also includes factors to ensure that the company will receive the proper value for the money spent. Therefore, detailed specifications contain requirements for performance of the equipment and requirements to make sure the equipment is manufactured to what are considered satisfactory quality standards. The performance requirements are tied directly to the goals of the project and can be determined by evaluations of engineering needs. The quality requirements are usually designed to ensure that the piece of equipment purchased will not fail in service.

The performance requirement specifications should be straightforward and set the range of what should be paid for the piece of equipment. If a machine tool is to perform to very tight tolerances—for instance, to be accurate within $\pm0.0001$ in.—it will cost more for the vendor to produce than a machine capable of maintaining $\pm0.0005$ in. In these specifications it is important not to overspecify. The manufacturing engineer must be certain to require only what is necessary to perform the task. Another mistake the engineer must be careful to avoid at this point is that of specifying so that a particular vendor's offering will receive favorable consideration. This is called rigging the specification, and it frequently means that the company buys something it does not need. The only certain winner here is the vendor. Performance specifications must be a true reflection of the company's need to achieve its goal and nothing more.

The quality requirements are insurance policies to make sure that the equipment works correctly and does not take extended time to debug and put into useful production. They take the form of compliances that the vendor must meet in manufacturing and testing the equipment. Usually they are tests and inspections that must be satisfactorily completed before the customer accepts the product and pays for it. Typical requirements would be that the materials used in manufacture are of the proper grade, that welding, machining, and assembly are in accordance with the engineering drawings, and that the machine performs in accordance with the specifications. How specific and tight these requirements are will depend on the criticality of the equipment in its intended use. For example, if the equipment is a simple transfer device, the quality requirements would not be as strict as if it were a safety device for a furnace fuel system.

There is a good reason for the manufacturing engineer not to insist on overly stringent quality requirements, and that is cost. The more complex and tight a quality specification is, the more the vendor will charge, so that it is possible for the insurance policy to cost more than a conceivable failure of the equipment. It is important to weigh the cost of tight quality requirements against the cost of failure that the requirements protect against. If failure can destroy the entire investment or significantly hinder production, then high quality requirements costs are justified.

Another consideration in determining whether to place stringent quality requirements in the specification, is whether it is traditional in the industry to do so. If such requirements are entirely alien, it will be difficult to get vendors to quote, and if they do, the costs will be very high.

To summarize, the rules that must be followed in developing detailed specifications are:

## A. Performance Requirements

1. State explicitly what the piece of equipment must do. In our example of the CNC horizontal machining center, the manufacturing engineer would have translated the eight valve bodies per week into machining metal removal rates, that is, feeds and speeds based on acceptable depth of cuts and selection of cutting tools. In addition, the engineer would specify the types of operation the machine must be capable of performing, such as face milling, boring, and drilling, and the dimensional tolerances required for each operation. If the queue times and transfer of material from workstation to workstation are of concern, the engineer would specify limitations in times to perform these functions.
2. State the physical attributes of the machine, that is, geometric size and weight. It is important that the piece of equipment purchased fits in the space available

for it. If the machine is too heavy, a more costly foundation may be required; if it is too big, it will cost considerably more to make space available for it, or it could end up in a location that is insufficient for economic production of the company's products.

3. State the required power sources. The machine must be able to work on the electrical system the company employs. For example, equipment manufactured in Europe is usually designed to run on electricity at 50 hertz, while the North American standard is 60 hertz. It would be embarrassing if a piece of equipment manufactured in Europe could not run in the company's factory until the extra cost of a power conversion was approved. Therefore, the manufacturing engineer must carefully research the company's power situation and write specifications in accordance with it.

4. State other electrical requirements, for instance, for electrical fixtures, controls, switches, and motors. These are usually based on a desire to standardize in order to minimize maintenance costs.

5. State environmental and safety requirements. The purchaser, not the vendor, is responsible for compliance with the various government and insurance requirements. Therefore, it is prudent to state these requirements to be met in the specifications. An example would be hand guards for punch presses or dead-man switches for rotating equipment.

## B. Quality Requirements

1. State performance acceptance testing requirements. The manufacturing engineer must make sure that the piece of equipment will meet all performance requirements before accepting the equipment from the vendor. Even though the vendor is obligated to meet the specifications put forth in the performance requirements, the engineer is in a much stronger position if the machine does not perform as required by the quality requirements during the trials. If the equipment is delivered and paid for prior to actual testing in the company's factory, the vendor can claim that instructions for operation are not being complied with. If the vendor must demonstrate conformance before being paid, the customer is better able to obtain what he contracted for. In addition, the debugging and minor corrections always needed by complex machinery will be on the vendor's account. It is important that the specification include a clause stating exactly what will constitute acceptance of the equipment for delivery and payment. This is usually a test run in the vendor's factory in which the vendor must demonstrate compliance with the equipment specifications, for example, by machining a certain number of parts within a specified period of time under the observation of the customer.

2. State quality requirements during manufacture of the equipment. This usually consists of a set of certifications by the vendor's quality assurance personnel and perhaps key events witnessed by the customer. The exact quality requirements are placed in the specification and become part of the contract between the vendor and the customer.

3. State compliance with code requirements that the vendor must meet. In some cases the vendor must be able to assure the customer that the workers engaged in the manufacture of the equipment are qualified under industry standard or government codes.

Once the specification is prepared, the manufacturing engineer turns it over to the purchasing unit of the materials subfunction to obtain quotes from qualified vendors. Purchasing is responsible for qualifying vendors for all commodities and equipment the company purchases. However, it is common for manufacturing engineers to let the purchasing agents know who they think would be qualified to supply the equipment described in their specifications. In deciding who would be qualified vendors, manufacturing engineers usually confine their judgment to technical matters. They are interested in knowing that the vendor has built similar equipment before and that the vendor can demonstrate successful uses of the equipment. They should not consider nontechnical aspects of vendor qualification with the exception of whether the vendor would be capable of delivering the product on time.

## IMPLEMENTATION SCHEDULE DURING MANUFACTURING OF EQUIPMENT

This is a very busy period for manufacturing engineers. After quotes are solicited and received, the engineers will be called on to make the technical evaluations of quotes and recommend a vendor and will then become involved in placing orders with the successful bidders and monitoring progress through delivery of the equipment. Let us look at the various phases that follow the receipt of bids:

1. Technical evaluation of bids and recommendations
2. Establishment of monitoring schedules with the vendor
3. Vendor surveillance during manufacture
4. Vendor runoff and acceptance

### Technical Evaluation of Bids and Recommendations

Usually vendors will reply to a request for a quote with the same level of detail used in the specification they received. Hence the amount of bid evaluation required will depend on the complexity of the specification.

Before receiving bids, the manufacturing engineer will make out a checklist containing "musts" and "wants." The former are those items of the specification that have to be met. For instance, if a tolerance for a bore inside diameter of 0.001 in., with a concentricity of 0.0001 in., is specified to achieve design engineering requirements, that tolerance capability is essential. A prospective vendor who cannot meet that requirement is automatically excluded from further consideration. "Wants," on the other hand, are desirable rather than critical traits, and inability to provide them does not mean automatic rejection. The manufacturing engineer can establish a point scale for all such items and then rate each bid received on this objective scale. It is important that every item in the specification be declared either a "must" or a "want" and a rating factor (point value) assigned to each item in the second category.

Once bids are received, they are evaluated first for compliance with the "musts." Bids that pass this test are then evaluated for the desirable but noncritical traits they can provide. The manufacturing engineer will then assign point values for the "wants" that are satisfied, giving partial values for items that are only partially satisfied. After all specification "wants" are reviewed against the bid, the manufacturing engineer will compile a score for that particular vendor.

Notice that no credit is given to items in the bid that are not asked for in the specification. Such items represent an attempt to sell something more than required at a higher selling price. It must be the practice of the manufacturing engineering organization to ignore these extras and actually let the vendor penalize himself by having a bid offering at a higher cost than his competitors. Occasionally, a vendor who has a superior offering for all the required items in the specification may be allowed to rebid without the unsolicited extras. However, care must be taken not to engage in an unethical practice vis-à-vis other vendors.

During this evaluation phase it is common for manufacturing engineers to be in contact with the various vendors. This is done to have points clarified so that the engineer thoroughly understands the bid offerings. In each case the manufacturing engineer must obtain written versions of the clarifications, which then become part of the vendor's bid, so that there will be no doubt about the vendor's intentions.

After all bids are evaluated, the manufacturing engineer will make a technical recommendation. If the technical recommendation coincides with the low bidder, everything is simple. The low bidder is awarded the contract. If the technical recommendation is given to a vendor who is not low bidder, the manager of manufacturing engineering will be called upon to decide whether the low bidder or other qualified bidders with lower prices than the technical winner should be awarded the contract. Remember that at this stage the only bidders being evaluated are those who satisfy all the "musts."

There are many rationales that can be used to determine whether a technical winner who is not the low bidder should be awarded a contract. Difference in price is a major consideration. Another consideration will be how many of the high-value "wants" are satisfied by the low bidder. The contract award problem must be resolved to the benefit of the company.

Normally in equipment projects conflicts between the low bidder and the technical winner are resolved in favor of the technical winner. This is true because capital equipment projects are indeed technical projects and are not treated as commodity purchases. Treating capital equipment purchases as commodities leads to more frequent equipment failures and lost profits, and is a decision that is usually regretted.

## Establishing Monitoring Schedules with the Vendor

It is important that manufacturing engineers detail their expectations for monitoring requirements to the successful bidder. They should specify the kind of progress charts the vendor will be expected to produce and review with the customer. However, care must be taken not to become too involved in the vendor's internal procedures. The progress charts are designed to ensure that the equipment is produced to meet the contract date and that the vendor complies with the quality and performance requirements of the contract. Any other requests to the vendor are usually unproductive and may be a factor in late delivery to the customer.

An example of an excessive monitoring item would be one involving review of the vendor's manufacturing drawings. Too many manufacturing engineers fall into the trap of overzealousness in this regard. The engineer may wish to review the engineering drawings to make sure that the design is adequate or that codes are being complied with. However, the bid review should have shown whether the vendor is competent to design the equipment, including code conformance, and the manufacturing engineer should not waste time reviewing drawings. In addition, if reviewing the vendor's drawings becomes part of the monitoring schedule, the engineer is accepting responsibility for the design and in effect voiding the vendor's responsibility to produce a competent design. If the design should prove faulty, the vendor can then claim that the customer approved the design and refuse to accept liability for damages. The manufacturing engineer should adhere closely to schedule and contract performance items only. A typical monitoring schedule is shown in Fig. 5.6.

The monitoring schedule is produced by the vendor from points spelled out in the specification, but the witness points are established by the manufacturing engineer with the concurrence of the vendor. As can be seen from Fig. 5.6, the

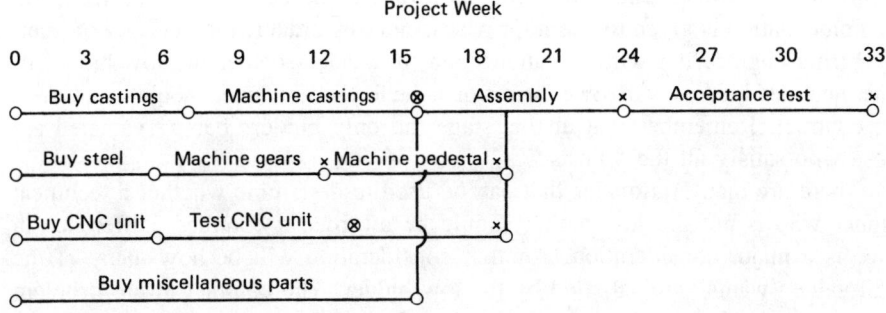

Notes
1. × = customer witness required by contract
2. ⊗ = customer witness allowed by contract
3. Vendor to give customer 72 hours notice prior to each
   contractual witness. Absence of customer witness
   constitutes customer waiver of witness requirement.

**Figure 5.6** Customer monitoring schedule for the CNC horizontal machining center.

manufacturing engineer has placed the contractual witnesses on the schedule; at a minimum this must be done. Optional witness points have also been placed on the schedule so that the engineer can be present at critical manufacturing points. But notice that there are no mandatory holds placed on the vendor that could be used as an excuse for failure to meet the delivery date requirement.

## Vendor Surveillance During Manufacture

The primary objective of vendor surveillance is to ensure that the vendor is doing everything possible to deliver the ordered equipment on time. The manufacturing engineer will be satisfied only if the vendor adheres to the end date of the schedule, that is, delivers the ordered equipment on or before the schedule date. Besides delivery on time, the manufacturing engineer must also be assured that the quality requirements and performance requirements are met. By using the customer monitoring schedule, the engineer can be in a position to do both.

The customer monitoring schedule shows when certain phases of manufacturing, including testing, are to be accomplished. As a minimum, the manufacturing engineer must check with the vendor prior to these scheduled accomplishment dates to ensure that these commitments will be met. If the engineer is confident that the vendor is performing adequately to schedule, nothing else need be done. If it does not appear that the vendor will meet the dates, it becomes necessary to urge the vendor to perform better. Procedures for doing this range from a simple telephone call to express concern, to a high-level management visit with implied threats of no future business if the vendor's performance does not improve. The main point is to communicate displeasure for lack of adequate performance. The vendor will favor customers who are attentive and insistent on timely performance, especially if he is experiencing delivery difficulties. Silence on the part of the customer implies acceptance of the vendor's actions.

Another method of checking a vendor's progress would be through a contract containing provisions for progress payments. On expensive capital investment projects it is common to have provisions in the contract for periodic payments to the vendor. These payments are usually tied to the vendor's achieving agreed on milestones in progressing from the start of the contract through delivery of the equipment—that is, they are tied to the customer monitoring schedule. With this system, a vendor who is late in reaching a milestone is also late in receiving the progress payment. This is a powerful tool because the vendor is probably counting on the progress payment for operating funds, especially in periods of high-interest borrowing costs.

Performance specification and quality specification surveillance are slightly different. Here the manufacturing engineer uses the customer monitoring schedule to determine when to be present to review or observe tests. As in monitoring progress to ensure delivery on time, the customer must know where the equipment stands against the schedule. However, in this case the reason is to ensure that the manufacturing engineer is present when the tests are to be performed, or data are to be approved, before the vendor moves on to the next phase of manufacturing.

It is essential that the manufacturing engineer be available to perform this task, as dictated by the contract, when the vendor requests it. Failure to do so would entitle the vendor to delay delivery of the equipment by at least as much time as the delay caused by the manufacturing engineer. Often this type of delay can be multiplied by factors of two, three, or even five times if it leads to loss of priority in the vendor's sequence of manufacturing flow, so that the required facilities are no longer available because they are now being used on work for other contracts. In addition, if the engineer is not available when required, the vendor can legitimately claim the progress payment due at the milestone, which was not achieved because of a customer-caused delay.

At the opposite end of the spectrum is the case of the vendor trying to eliminate or modify tests in order to make delivery on time. This allows the vendor (1) to reduce manufacturing costs by not having to do the test, and (2) to get back on schedule and transfer the testing to the customer's factory. By delaying the test to the final installation at the customer's site, the vendor has a greater opportunity to blame any lack of performance on improper installation and operator error.

For these reasons, it is never in the manufacturing engineer's interest to waive tests. Remember, the tests were placed in the specification to ensure that the manufacturing engineer's project goal would be successfully achieved. Waiving tests should only be considered in an extreme situation and only after a thorough review of the possible consequences. The only extreme situation that would warrant considering this would be one concerned with severe economic loss to the manufacturing engineer's company if the equipment were not installed on time or within an acceptable late period. Only the manager of manufacturing engineering can adequately judge whether such an extreme situation exists.

## Vendor Runoff and Acceptance

In this phase of the project the equipment will be put through its operations to see whether it does produce as the specifications require. As the company's representative, the manufacturing engineer plays a vital role in the acceptance test. Let us review the sequence of the runoff and discuss the manufacturing engineer's function.

**Assembly and Lineup.** Virtually all equipment used in factories is made up of subassemblies. Putting together the subassemblies is critical for successful operation of the equipment. For bake ovens and furnaces, the walls and doors must fit snugly to avoid unacceptable heat loss. For machine tools, the lineup of shafts and bearings must be precise to prevent rapid bearing wear and ensure accuracy of machined parts, and the lead screws and parts that constitute the transfer roadway of the cutting tool must be plumb and square with each other to very close tolerances, again to ensure accuracy of machined parts. Each type of equipment has specific critical parts that require precise assembly, and for a runoff or acceptance test it is important that the piece of equipment be assembled correctly.

The manufacturing engineer should be present during critical phases of assembly for two reasons. First, the engineer must make sure that the parts constituting the equipment have been made correctly, that is, that they assemble properly. This simple observation indicates that the results of the runoff will be valid, that the machine has been assembled in accordance with design, and that any lack of performance is probably inherent in the design. The second reason for being present during the assembly stage is to learn how the machine goes together and what the difficult phases are. Since the vendor's factory personnel are not likely to be available to assemble the equipment at the customer's factory, it is vital that a customer representative be familiar with the installation procedure for adequate future operation and maintenance of the equipment.

**Test.**   This is the actual operation phase. In most runoffs the vendor runs the equipment in as close a simulation as possible of the customer's situation. During this phase the manufacturing engineer can observe the characteristics of the machine and determine whether it meets the contract specifications.

The test to which the piece of equipment is subjected is specified in the purchase contract. It is the manufacturing engineer's responsibility to see that the vendor conducts the test as required. The engineer must also keep in mind that the vendor cannot be expected to do more than is required by contract. Nuances of interpretation are important here, and the engineer must be certain that the vendor understands the requirements in the same way that he does.

Whenever controversy exists concerning whether equipment meets acceptable test criteria, there are no winners. The vendor's payment may be delayed and his reputation tarnished. The customer is in danger of receiving a piece of equipment that does not perform in accordance with specification, which can result in nonachievement of a goal that depends on the purchased equipment. Such controversy can be avoided by a prerunoff meeting to review the specification for the acceptance test and ensure that there is complete agreement on its meaning.

The manufacturing engineer must explicitly tell the vendor what will constitute passing the test. This is done simply by reviewing the specification and determining the numbers involved in meeting it. In the case of the CNC horizontal machining center of our example, it might mean producing a fully machined valve in the specified time, or preferably two valve bodies in twice that time, since a runoff should demonstrate repeatability. In addition, the specification may require a demonstration of range of capability. In reviewing the details of the test with the vendor to make sure there is mutual understanding, there is always a phase of verbal communication. However, to prevent any misunderstanding it is recommended that this be followed by a written phase, where the manufacturing engineer produces minutes of the meeting and delivers a copy to the vendor. By receiving a copy of the minutes and not disputing any points, the vendor signifies concurrence with the interpretation of what constitutes a successful test.

During the actual running of the tests it is important that the manufacturing engineer observe all key points. If the equipment does not meet specification, the engineer will have to determine with the vendor what can be done to correct the problem. The engineer should review the corrections proposed by the vendor and determine whether they will permanently correct the deficiency and whether the time frame to do so is acceptable. In any event, the manufacturing engineer should not accept equipment that does not meet the specifications.

Occasionally, no matter what a vendor does, a piece of equipment does not meet specification. In this case the manager of manufacturing engineering must decide what should be done regarding acceptance. Several options may be considered:

1. Accept the equipment as is.
2. Negotiate paybacks by the vendor, and then accept it.
3. Reject the equipment and require the vendor to build a product that meets the specification.
4. Reject the equipment, recover progress payments if any, and place an order with another vendor.
5. Reject the equipment, recover progress payments if any, and exercise another option for achieving the goal related to the equipment purchase project.

The option selected depends on the company's situation and the degree to which the vendor misses the specification. If the mismatch is slight or not of primary significance, then option 1 or possibly option 2 is selected. If the mismatch is significant and the company can afford the time delay, option 3 or 4 may be chosen. Finally, if the company's goal must be achieved by a specified time and this time precludes another try by the same vendor, option 5 may have to be selected. In practice, option 5 is rarely used because the purchase of the equipment is typically the only way to achieve the stated goal. Therefore if a piece of equipment fails runoff, options 1 through 4 constitute the practical choices. Most equipment eventually passes the runoff test. Then the data collected by the manufacturing engineer establishes the benchmark for successful operation of the equipment and is used to develop methods and time standards. In the very rare case where equipment fails repeated runoff tests, the data will probably be used as evidence against the vendor for a law suit or settlement.

**Disassemble and Pack for Shipment.** The primary concern of the manufacturing engineer at this point is that the equipment is not damaged during disassembly and is adequately protected for shipment to the company's factory. After successfully completing a runoff, the vendor will naturally wish to deliver the equipment and get on with other business. One way of assuring that the vendor pays proper attention to the disassembly, packing, and shipping is to have this work documented in the specification. This means that manufacturing engineering must be knowledgeable about the construction and assembly procedures for the product prior to purchase, which is usually not the case. Therefore, another method must be found.

The simplest way to ensure that the vendor is not careless during disassembly, packing, and shipping is to insert a clause in the specification stating that the equipment must arrive defect free at the customer's plant. This means the vendor is responsible for the equipment until a receipt is received for it at the customer's factory. In essence, the transfer of title does not occur until the equipment arrives at the customer's site. By not accepting title at the vendor's factory, the manufacturing engineer has only to exercise concerned interest during the teardown procedure and does not have to manage the operation. Often, if the title is transferred immediately after successful completion of the test, the manufacturing engineer is put in the position of being the foreman, planner, and chief expediter over the vendor's hourly work force in order to get the purchase on its way to the company's factory. This potential frustration and confusion is eliminated if the original bid specification states F.O.B. customer's factory for the title transfer and final payment.

## IMPLEMENTING THE EQUIPMENT ON THE FACTORY FLOOR

After this activity is completed, the equipment is installed and ready for operation. The company then can proceed toward achievement of the stated goal. Let us follow our example goal of producing the eight valve bodies per week and examine the steps necessary in implementing the newly arrived CNC horizontal machining center.

Once the contract for the CNC horizontal machining center is let to the successful bidder, work commences to ready the area for acceptance of the equipment. The final preliminary layout (Fig. 5.5) is used as the basis for producing architectural drawings and detailed construction drawings. At this stage of the project we know only the space required for the new machine tool. We do not know its actual shape or where the electrics, hydraulics, air, and so forth will be connected to the equipment. However, we do know what their values will be because they are part of the specification for the machine tool bid. Therefore, work on the architectural drawings can commence; and in fact, unless we are building a new factory, we are only making additions to existing drawings. These additions will constitute bringing the services to the area designated for placement of the new machine tool. After the work on the drawings begins, the locations of the services will be transmitted to the vendor, who needs to know where to place the hookups for the machine. Usually this is an iterative process in which the customer and vendor reach a mutually satisfactory conclusion based on the service location choices dictated by the machine's design and the site situation.

Near the end of the design phase by the vendor, the specifics of the foundation for the equipment usually become known. The vendor supplies this information to the customer, who then completes the architectural drawing by adding in the foundation outline. This architectural drawing then leads to the development of the construction drawings for emplacement of the foundation, completing the design phase of the project. The vendor has designed the equipment and the manufacturing engineering organization has designed the factory to accept it.

The construction phase begins next. The same manufacturing engineer who is responsible for vendor surveillance will usually also be responsible for the installation phase. Here the maximum use is made of project control charts. For very complex projects networking or Critical Path Method (CPM) charts may be employed. For less complex projects gantt-type charts are used. Whatever technique is employed, it is important that it be detailed enough so that all activities are scheduled and can be monitored effectively.

Figure 5.7 is an example of a networking project control chart that is compatible with a project of the magnitude of installing a CNC horizontal machining center. Note that all the basic phases are included: work that must be done by the purchaser, the equipment vendor, and the construction vendors. This chart can be used to track all activities of all organizations involved. It may be as simple as the version shown in Fig. 5.7 or significantly more detailed, as long as it is sufficient for adequate control. Note also that the line corresponding to manufacture of the CNC horizontal machine is void of detail. For our example, this was done for clarity. In actuality that line would be replaced with the detail shown in Fig. 5.6. Hence we would have all the information required to carefully monitor all phases of the project on one chart.

Let us look at the construction phase in chronological order. During the drawing preparation, the manufacturing engineer must make sure that enough information is presented for adequate quotes to be prepared by the contractors. The term contractors can also include in-house staff if the company wishes. Lack of adequate information in drawings usually leads to cost overruns and delays. Therefore, it is important to make sure that architectural drawings are correct, particularly with respect to obstructions. Obstructions are any items that for any reason cause contractors to use more materials than anticipated, which typically

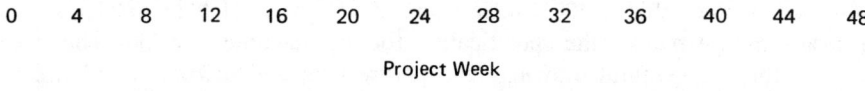

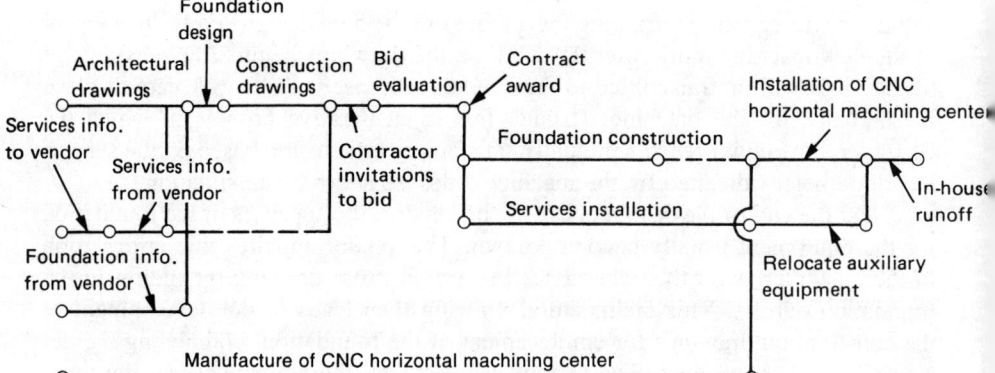

**Figure 5.7** Installation schedule for the CNC horizontal machining center.

leads to extensions to obtain more materials and always adds to the bill. No contractor will allow a client to pass on such hidden costs.

Construction bid activities are very similar to those explained previously in this chapter. They tend to be less complex because they involve a shorter term and a greater direct-work content. The format of construction bids usually follows standard local practices and the legal commercial code will apply. When working with in-house staff there will be no bid, but the manager of that function, the manager of maintenance, will be required to estimate the in-house costs involved, consisting of the time to complete the work and the cost of materials. Based on the time requirement, the manager of manufacturing will determine at an early stage whether to authorize overtime to the contractor, either in-house or external. Once contracts are let, the manufacturing engineer will monitor the work in the factory, paying particular attention to whether performance of the work is at a proper pace and to proper standards.

While the foundation is being built and services installed, the CNC horizontal machining center is going through its acceptance and runoff procedure. This activity will be repeated in the manufacturing engineer's factory after assembly of the machine on its foundation. During assembly the manufacturing engineer will pay close attention to ensure that the machine is being assembled as it was at the vendor's facility. This is necessary to ensure that the equipment operates successfully and that the vendor's warranty is not voided. Usually the vendor's installation service representative will help in this activity. Virtually all large capital equipment contracts include funds for this service, which is very helpful in achieving proper installation. All reputable vendors maintain a staff of experts to assist their customers in installing machines such as a CNC horizontal machining center. Having the machine run correctly is the best form of advertising for the vendor's product.

The culmination of the installation is another runoff. This runoff or acceptance test is quite different from that held at the vendor's location, where the question was whether the machine would work as designed. At the second runoff the only things in question are whether the machine has been assembled correctly and whether it has suffered any damage during shipment. This runoff is also a fine tuning, since there is likely to be some dynamic difference between running on a solid foundation and performing on the vendor's test block foundation. During the in-house runoff, the manufacturing engineer must be cognizant of minor differences that may or may not affect long-term performance. Usually at this runoff things such as shaft balance, hold-down bolt torques, and horizontal and vertical alignment are looked at very critically to ensure that the machine tool is set up in the most stress-free position possible.

The second runoff does not have a limit to the number of pieces to be produced before the equipment is accepted. Manufacturing engineering wants to complete this acceptance test and turn the equipment over to shop operations. Every piece now produced will be used to satisfy the production schedule needs. The only question is how many units per week will be produced and what the rate of increase will be to reach the production level in the stated goal.

Now let us look at the activities involved in transferring the machine from manufacturing engineering control to shop operations control. This is the last phase in the capital equipment purchase project.

The first consideration is training of operators. Operators are assigned to the new machine tool as soon as it arrives in the factory, and it is advisable to have them participate in the assembly of the machine and be able to converse frequently with the vendor's technical support people. It is an old axiom, particularly true in job shops, that good operators know their machine and can work wonders with it. This is a good reason for introducing the operators to the new machine at the earliest possible time.

Training to operate the machine consists of learning how the new piece of equipment will produce the factory's product. It begins with assisting in the assembly and reviewing the operating manuals with the vendor's technical representative and the manufacturing engineer. The training will also cover the method to be followed in operating the equipment, tool holder positions, setup techniques, and how to utilize the CNC controls. Most of this will be accomplished through discussions between the manufacturing engineer and the operators, and some consultation with the vendor's representative. The manufacturing engineer will be keeping a record of the salient points of these sessions to be used later in the methods documentation. The ultimate goal of the training activity is for the operators to be ready to run the machine under guidance of the vendor's representative and the manufacturing engineer during the runoff, and to enable them to progress quickly to the point where they can perform correctly without close supervision.

One of the major tasks of the manufacturing engineer during the installation and runoff phase is to prepare a methods document to be turned over to the methods, planning, and work measurements unit for refinement and implementation within the planning system. This document states the sequence of operations to be followed to machine the valve bodies, such as setup procedure, sequence of tools to be used, feeds and speeds, operator-performed maintenance, and how to operate the machine. Development of this document begins during acceptance testing at the vendor's location, and at least a draft document should exist by the time the second runoff is complete.

During the installation and runoff period, the manufacturing engineer also introduces all key personnel besides the operators to the machine. The engineer familiarizes shop operations management with the machine, paying particular attention to differences from their previous experiences in the factory. Similarly, the engineer briefs process control personnel and maintenance personnel. The purpose of these briefings is to reduce the time required for key personnel to become proficient with this equipment as much as possible.

Finally, we have reached the point of turning the new equipment over to shop operations. Recall that the example of producing eight valve bodies a week plus an overload capacity of ten valve bodies required relocation of the inspection station and a drill press (see Fig. 5.5). In order to turn over the new facilities to shop operations, these other moves will also have to take place. The equipment relocations are relatively simple. It is considerations involving people that must be carefully planned. At worst, people may be relocated and not be capable of performing their jobs. With lack of attention to details, the little things that make factories work smoothly and efficiently will go wrong. Such things as not knowing where the dispatch cage or the tooling lockers are located are reasons for increasing idle time, resulting in lost productivity.

It must be recognized that anything that constitutes change will result in unfamiliarity, hence lost productivity. It is management's responsibility to minimize the level of unfamiliarity. One way to do this involves making a checklist of anticipated problems, such as that shown in Fig. 5.8. A checklist should be made

---

Machine/Process _____

| | | | | Review Complete |
|---|---|---|---|---|

Operators by Shift

First _____ _____ _____ _____
Second _____ _____ _____ _____
Third _____ _____ _____ _____

| Items Discussed | Complete |
|---|---|

- Shift start and finish times _____
- Notification in case of absence _____
  - 1st shift Foreman _____ Phone No. _____ _____
  - 2nd shift Foreman _____ Phone No. _____ _____
  - 3rd shift Foreman _____ Phone No. _____ _____
- Break times _____
- Work rules (general) _____
- Locker assigned and moved _____
- Workstation instructions _____
- Workstation tooling _____
- Bldg. and facility orientation
  - Toolcrib _____
  - Material storage _____
  - Material handling service _____
  - Lubrication storage _____
  - Dispatch cage, paperwork routines _____
  - Inspectors _____
  - Foremans desk _____
  - Rest rooms _____
  - Canteen facilities _____
  - Emergency telephones _____
  - Time clocks _____
  - Parking facilities _____

Foreman Signature _____
Date Complete _____

**Figure 5.8** Checklist for relocated equipment.

up for each workstation that is being relocated. It identifies the operators and requires considerable logistics information to be given to the operators by the foremen. There are places for recording that each step has been covered, so there is an attempt to require management to communicate with the people involved prior to any move. When checklists such as that illustrated in Fig. 5.8 are used, moves usually involve a minimum of disruption.

## CAPITAL EQUIPMENT PROJECT CONTROL CHARTS

Most manufacturing engineers use charts of one type or another to visualize activities to be accomplished in order to achieve their stated goals. Figures 5.6 and 5.7 are examples of networking charts. Charts that take the form of bar graphs are commonly called gantt charts, after the pioneer industrial engineer Henry L. Gantt, who introduced such charts for production control and project control. Regardless of the type of chart used, their purpose is to show the activities necessary to complete a project in a manner that portrays the relationships of the steps to each other. Control charts make it significantly easier to understand the interactions between the various steps. Let us now explain the theory behind these two techniques and demonstrate how they are used.

### Bar Graphs (Gantt Charts)

Bar graphs are a combination of verbal and visual information charts. The technique is to list operations to be performed in sequential order, then use a bar or line next to the items to indicate commencement and completion dates. The length of the bar indicates the time in relation to other steps. We can convert Fig. 5.6, the customer monitoring schedule for the CNC horizontal machining center, into a gantt chart to illustrate this technique. To make this conversion, we (1) list all activities by their starting dates along the left-hand margin, with the first item to be completed at the top and the last at the bottom; (2) construct a time scale along the top of the page; (3) draw a horizontal line to scale, beside each specific activity listed for starting date and completion date; and (4) use a simple key to indicate the actual starting and completion date. Figure 5.9 is the result of this task. In the figure it is assumed that the project is already under way. The filled triangles and circles indicate that the activity has been accomplished. Note that in items 1 and 6 the filled symbols do not coincide with the open ones. This indicates that the activity did not start or end on the scheduled dates.

Now compare Fig. 5.9 with Fig. 5.6. They contain the same information, and both of them show the steps in sequential order and show the relationships of the steps to each other. Which of them better visualizes the project is a matter of opinion. Perhaps the networking chart shows the dependence of later steps on earlier steps more effectively, but again, that is a subjective opinion. Both formats work; the choice of which to use is based on personal preference.

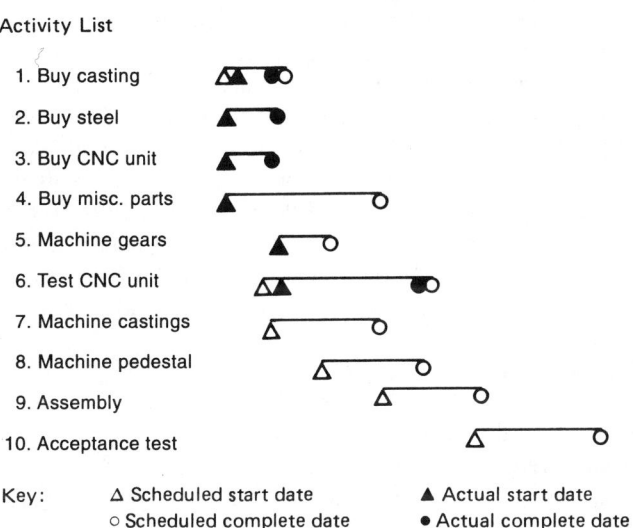

Figure 5.9 Gantt chart: Customer monitoring schedule for the CNC horizontal machining center.

## Networking Charts (Critical Path Method)

For most projects encountered by manufacturing engineers the complexity is such that there is no preferred choice between gantt charts and networking charts. However, occasionally a project is commissioned that is very complex and has a large number of incremental, interrelated steps. If this is the case, a gantt chart may become very long, so that it is difficult to visualize the thrust of the project. For simpler projects, the engineer can visualize the abstract relationships between the steps in a gantt chart and instinctively spot the critical path, or main road, from start to finish. When we have many activities taking place at the same time, as in a very complex project, we must resort to a networking chart.

A networking chart allows us to visually describe the critical path of the project, so that the manufacturing engineer can concentrate on achieving the elements that make up the critical path. Successful conclusion of the critical path elements, which are the items with the longest lead times, means that all other elements are likely to be completed on time.

The most common form of the networking chart is the CPM chart. To illustrate the networking technique, we will develop a CPM chart for the customer monitoring schedule for the CNC horizontal machining center and end up with Fig. 5.6.

First, we may define a critical path as a sequence of chronological events selected out of a universe of events such that that sequence takes the longest time to accomplish. All other sequences will be complete before the critical path sequence.

The logic of the CPM is simple. If we can find the main stream of events and make sure we accomplish them, all the tributary events will be finished on time because they have shorter cycle times. This means that we can tolerate much greater delays for the tributary items than for the main stream or CPM items. The CPM forces manufacturing engineers to focus their primary efforts on the main stream events and not diffuse their energy on comparatively minor items.

The CPM networking technique works, but like every other tool available to manufacturing engineers it must be used with caution. Many projects will not exhibit a clear-cut critical path. One path may be a little longer than all the others and may be designated as the main stream. However, if the tributaries are almost as long, focusing on the critical path may lead to unacceptable delays of the tributaries. When this happens the late tributary becomes the new critical path and a shift in emphasis is required.

The steps for developing a critical path are as follows:

1. List and number the events of the project in chronological order.
2. Determine the immediate event precursor for each event.
3. Determine the time required to accomplish each event, using the same time scale (e.g., weeks and weeks, not weeks and days).
4. Link all the precursor and successor events with lines.
5. Add the times of all events in each of the lines to determine the longest chronological time to complete them. This is the critical path; all other paths require less time.
6. Put the network on a time scale chart similar to a gantt chart. This makes it easier to read and shows the dates when events must occur to maintain the schedule.

Now let us go through the process of constructing a CPM networking chart. Table 5.1 is the CPM work chart with data filled in representing the items pertaining to the CNC horizontal machining center example.

1. List and number the events of the project in chronological order. This list, except for the arbitrary start and finish items, is identical to the left-hand column of the gantt chart (Fig. 5.9).

2. List the immediate precursor for each event. If more than one event must occur before an item is started, then the event item will have more than one precursor. Event 10 in Table 5.1 is an example of this.

3. Determine the time required to accomplish each event. In the CNC horizontal machining center example the time scale will be weeks. This information can be found on the gantt chart.

4. Link all the precursor and successor events together with lines. For clarity, we will use the node convention, where the node is an ellipse containing a shorthand code identifying the event and showing the time it will take to accomplish it. For example, event 4 ("buy CNC unit") is estimated to take 5 weeks to complete; therefore the node would be $\overline{(4.,5)}$ . Nodes for all the events are shown in Table 5.1. The linking of all the nodes is shown in Fig. 5.10.

**Table 5.1 Work sheet for CPM network: customer monitoring schedule CNC horizontal machining center**

| Node | Event no. | Description | Immediate event precursor | Required time (wks.) to accomplish |
|------|-----------|-------------|---------------------------|-----------------------------------|
| 1.,0 | 1. | start | | 0 |
| 2.,7 | 2. | buy castings | 1. | 7 |
| 3.,5 | 3. | buy steel | 1. | 5 |
| 4.,5 | 4. | buy CNC unit | 1. | 5 |
| 5.,16 | 5. | buy misc. parts | 1. | 16 |
| 6.,7 | 6. | machine gears | 3. | 7 |
| 7.,15 | 7. | test CNC unit | 4. | 15 |
| 8.,9 | 8. | machine casting | 2. | 9 |
| 9.,8 | 9. | machine pedestal | 6. | 8 |
| 10.,8 | 10. | assembly | 5,7,8,9 | 8 |
| 11.,9 | 11. | acceptance test | 10. | 9 |
| 12.,0 | 12. | finish | 11. | 0 |
| Item 4 | Item 1 | | Item 2 | Item 3 |

5. Add all event time in each of the lines to determine the longest completion time. This is the critical path. We now add the times of the four paths to find the critical paths:

Path a: 1.,0 + 2.,7 + 8.,9 + 10.,8 + 11.,9 + 12.,0 = 33 weeks
Path b: 1.,0 + 3.,5 + 6.,7 + 9.,8 + 10.,8 + 11.,9 + 12.,0 = 37 weeks
Path c: 1.,0 + 4.,5 + 7.,15 + 10.,8 + 11.,9 + 12.,0 = 37 weeks
Path d: 1.,0 + 5.,16 + 10.,8 + 11.,9 + 12.,0 = 33 weeks

From these additions we find two critical paths and two tributary paths with time values very close to the critical paths. Unfortunately, it is typical of manufacturing engineering projects that the critical path is not very different from the tributary paths.

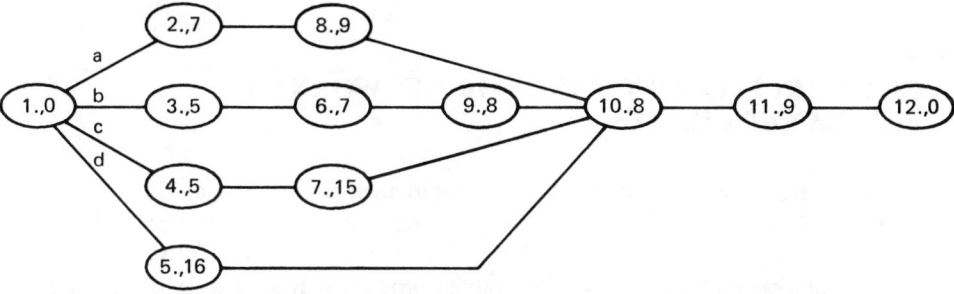

**Figure 5.10** CPM diagram: Customer monitoring schedule for the CNC horizontal machining center.

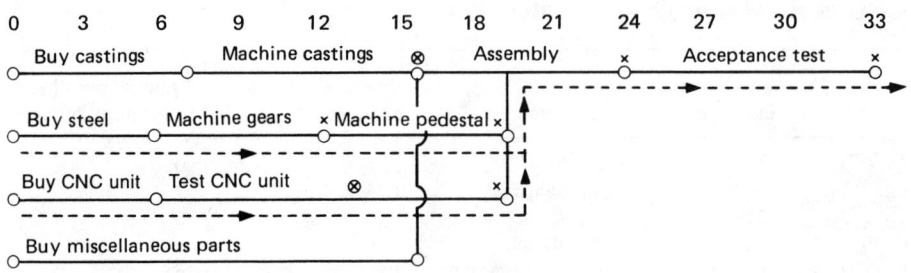

Notes
1. O = customer witness required by contract
2. ⊗ = customer witness allowed by contract
3. Vendor to give customer 72 hours notice prior to each
   contractual witness. Absence of customer witness
   constitutes customer waiver of witness requirement.
4. ----➤ = Critical path.

**Figure 5.11** Customer monitoring schedule for the CNC horizontal machining center: Critical path superimposed on network chart (Fig. 5.6).

6. Put the network on a time scale chart similar to a gantt chart. The CPM diagram is not very revealing; only the engineer who constructed it knows what the node numbers refer to. Therefore, prudent manufacturing engineers will take the next step of constructing a time scale networking chart, such as the one shown in Fig. 5.6, so that all who are interested can follow the progress of the project. Sometimes the critical path will be superimposed on this chart, but this usually occurs only for complex projects. Figure 5.11 is Fig. 5.6 with the two critical paths shown (most often projects of this complexity would not require this further refinement).

Note that the critical path line intersects the assembly line beyond the assembly start point. This is allowable if it is understood that the assembly operation is all-encompassing. In this case it is understood that the assembly of the gears and pedestals cannot begin until the main structure, the casting, has been assembled. The same is true of the CNC unit.

## COMMERCIAL SOFTWARE FOR NETWORKING AND GANTT CHARTS

There are many software offerings that manipulate project data to assist in creating gantt and CPM charts. Most of these software packages run on personal computers, so their costs are minimal. Being low in cost, they definitely are worthy of investigation for use in capital equipment programs. But be wary of their abilities to do the total project management task.

Software programs reputed to be project management programs can be very useful in curtailing the clerical effort needed to create and use the networking and gantt chart techniques—if they work. This means it is essential that the manufacturing engineer carefully investigate the capabilities of the proposed program versus the needs to control the project before attempting to use the program. Software programs that are of insufficient capacity to handle the total data can cause considerable havoc. At worst they can truncate data and give false results, such as erroneously defining the true critical path. It is therefore prudent to take the necessary time to investigate software thoroughly before employing it in a manufacturing engineering project.

Selecting the proper software is nothing more than a "musts" and "wants" exercise, similar to that discussed in the Technical Evaluation of Bids and Recommendations section of this chapter. The engineer has to determine what the "musts" and "wants" are for the intended application. At a minimum the software must be able to construct bar charts. It should also be capable of doing the steps necessary to define the critical path with minimal manipulation by the user. For example, as a minimum, in the steps for developing a critical path, as shown in this chapter, data inputs should be required only for the first three steps. If manual or off-line data manipulation is required, then the software program is probably not sufficient for a manufacturing engineering type of project.

Another point to consider is the number of individual data points the proposed software program can accept. Keep in mind that a networking activity for a large project can have hundreds or even thousands of individual nodes. These data have to be manipulated and gantt charts produced, which could require a significant amount of computer capacity. Hence, the software has to be able to handle the required amount of transactions in a reasonable time frame. This capacity will vary depending on the types of projects the manufacturing engineer will be running. Obviously, constructing a factory to produce valve bodies will require software with a larger capability than that used to design and purchase jigs and fixtures for this chapter's example CNC horizontal boring mill. In general, then, selecting software for program management requires the "musts" to do critical path planning as stated above plus a set of "wants" rated for degrees of desirability.

Commercially available software does make the job of managing large projects easier. But like all other tools available to the manufacturing engineer, it does not supplant the judgment and experience of the engineer. Computer software primarily does the clerical function. The manufacturing engineer still has to supply the creative process.

## SUMMARY

We have discussed the major activities for achieving goals through the purchase of capital equipment. Capital equipment projects are complex ones, with many activities going on in parallel, and require adequate management coverage. Coordination is necessary to maintain adequate coverage and can be achieved only through proper communications techniques.

Manufacturing engineering is a pragmatic science. This chapter describes approaches to working with vendors so that the company's interests are protected while ethical relations are maintained. Another important part of manufacturing engineering is applying good business practice based on sound engineering principles. This blend of engineering principles and business sense has been further explained through the use of project control charts. These are powerful tools for the manufacturing engineer, but professional judgment is necessary for their successful use. The pragmatic approach to business problems has also been demonstrated through the lengthy plant layout description, which shows that successful implementation of capital equipment projects requires attention to practical details as well as theory.

## REVIEW QUESTIONS

1. A preliminary layout is often called a tollgate layout. With reference to the CNC horizontal machining center, discuss what the "go" or "no go" aspects might be.

2. With reference to Fig. 5.2, note the material flow. Determine where the congestion points are that would be areas of opportunity for corrective action for the new layout required by the addition of the CNC horizontal machining center. Explain why these would be areas of opportunity.

3. List in chronological order the major steps required to construct and verify the adequacy of a factory floor layout.

4. Discuss with examples how a vendor's goal and a manufacturing engineer's goal can be complementary with respect to capital equipment purchase.

5. For the CNC horizontal machining center of Fig. 5.5, prepare a list of items that a manufacturing engineer is likely to include in the detailed specifications, and explain how these are related to the CNC horizontal machining center project.

6. For the following items, determine whether the quality requirements would be specific or general and state reasons for your answers. (a) Machine tool electric drive system; (b) layout bench; (c) operator vertical lift (allows the operator to be at same level as the rotating cutting tool of an HBM); (d) machine column horizontal translating rails.

7. Make a list of "musts" and "wants" to evaluate bids for the CNC horizontal machining center. Include rating factors for the "wants" and reasons for the priority selection sequences.

8. The following vendors meet all the "must" requirements for a capital equipment project. The following data represent a summarization of the bid evaluations. The budget allocated for this project is $512K. Delivery is desired by April 15 with an absolute date of June 1. Make a recommendation with reasons for vendor selection.

|                          | Vendor A | Vendor B | Vendor C | Vendor D |
|--------------------------|----------|----------|----------|----------|
| Cost                     | $490K    | $565K    | $505K    | $510K    |
| Delivery date            | May 20   | April 10 | May 1    | May 15   |
| "Wants" score            | 58       | 68       | 56       | 66       |
| Compatible with existing equipment | No | Yes | Yes | Yes |

9. Referring to the example of the CNC horizontal machining center, the following is a list of monitoring schedule items. Explain why each item is proper or excessive. (a) Review material certifications to specified code; (b) determine progress to manufacturing schedule; (c) review design stress calculations for rotating cutting tool shaft; (d) review tolerance selections for tool holder; (e) review performance of machine tool at runoff.

10. Discuss the pros and cons of entering into a progress payment contract for capital equipment.

11. A manufacturing engineer receives a request from the vendor to waive a test pertaining to adequate performance of bearing parts. This is the fourth of five identical machines the vendor is to supply. In each of the previous three, the test was 100% satisfactory. In addition, any defect could be detected during runoff. What should the manufacturing engineer's response to the request be? Give reasons for your decision.

12. A manufacturing engineer is purchasing a CNC lathe to make studs for valve bodies. The specification requires a diametric tolerance of 0.0005 in. and a concentricity of 0.0002 in. TIR (total indicator readout). Develop a runoff acceptance test plan to meet these criteria.

13. Discuss the differences and similarities between invitations to bid for construction services and equipment purchases for projects such as the CNC horizontal machining center.

14. With reference to question 12, what differences, if any, would there be between the runoff acceptance test to be performed after installation and the runoff test after initial assembly at the vendor's plant?

15. Develop a strategy for training operators and familiarizing other personnel with the CNC horizontal machining center so that time is optimized from receipt of the machine in the factory to turning it over to shop operations.

16. Construct a gantt chart, using the data contained in Fig. 5.7.

17. Based on the data of Fig. 5.7, (a) develop the work sheet for constructing a CPM network, (b) construct the CPM network, and (c) determine the critical path.

# PRODUCIBILITY ENGINEERING

Producibility engineering is a coordinative discipline with the design function of the industrial organization. Through this coordinative approach we strive to optimize the process of producing the company's products. We cannot, however, simply isolate the manufacturing engineering subset called producibility engineering and discuss it and its techniques without introducing the concept of concurrent engineering (sometimes referred to as design for manufacturability). The work of a producibility engineer is in many aspects an interfunctional one. It deals with finding the pragmatic limits of feasibility for manufacturing and still meeting the intent of the design. In this case, the relationship is between manufacturing engineering and design engineering. When we expand beyond the manufacturing and design engineering relationship to include the entire industrial organization, we enter the domain of concurrent engineering.

The concurrent engineering concept unites all functions of a company into a team for conceiving, designing, manufacturing, marketing, and distributing a product in an integrated and optimal approach. Just as the name suggests, all this work is done in a concurrent approach. Producibility engineering is at the same time a root of the concurrent engineering philosophy and a subset of it, as a member of the concurrent engineering team. Therefore, in this chapter we explore producibility engineering as a prelude to and within the concurrent engineering concept. We will see how producibility engineering techniques are used in this unifying concept.

In this chapter we define the concurrent engineering role in modern manufacturing and then focus on the producibility engineering aspects of that role. We look at the techniques in which producibility engineering leads the concurrent engineering team, and also identify areas where producibility engineering assists other functions.

## CONCURRENT ENGINEERING CONCEPTS

The definition of concurrent engineering is straightforward. It is the synergistic process of doing all the preproduction, production, and postproduction work in a manner such that efforts are scheduled and done in a parallel interactive fashion, rather than in a series singular manner.

We all know the adage that teamwork is more important than individual achievement in reaching an organization's goal. Traditionally, we applied teamwork concepts to functions of a company and rarely to the company as a whole. The conventional wisdom was that there is no commonality between marketing, engineering, and manufacturing. They all do their thing independently, and there is certainly no similarity in management techniques. No commonality exists between managing Computer Numerical Control (CNC) machines and sales campaigns—or so it was thought 15 to 20 years ago. Hence teamwork existed, it was presumed, only within functions where all the activities were definitely adjacent in the spectrum of work. This meant the walls were up and work was done within a narrow universe. Marketing defined a need, engineering created a concept of a product to meet that need, and manufacturing took the concept and made a real thing out of it—and they all did it mostly independently of each other.

By allowing this thought process to dominate our organizations, we incurred significant additional costs and time in the process of creating and delivering products to customers. Costs were not optimized. Why? Simply because we were creating waste. The left hand did not know what the right hand was doing. In many cases the instructions from marketing to engineering or from engineering to manufacturing were incomplete or just plain wrong. Why was this so? Mainly because of unfamiliarity with what happens when the baton is passed and the next step of the product creation process takes place. When people simply do not know, strange things happen, and the laws of probability state that the vast majority of outcomes will be considerably less than optimum.

Concurrent engineering unites an organization's functions by creating synergy between those functions. The synergism is established by setting up parallel efforts between the functions through cross-functional teams, and the output is definitely greater than the sum of the parts. Via concurrent engineering we actually establish an atmosphere of trust and cooperation that allows all functions to do their jobs with the active and real-time advice of the other functions. Thus, the concurrent engineering process allows companies to serve their customers better by shortening the product cycle time. The parallel effort greatly reduces the cycle, from conceptualization of the product through delivery to the customer.

Concurrent engineering is not a precise and bounded body of knowledge such as thermodynamics or strength of materials. It is a concept, or perhaps some would call it a theory, of how a manufacturing company ought to structure its process of delivering its product to the customer. Along with this structure are techniques that are applied in carrying out the process, which are compatible with the parallel effort. Some are internally directed techniques and some are external. The internal category comprises mainly design engineering– and manufacturing engineering–based techniques. External techniques concern primarily marketing principles.

We explore primarily the internal techniques, first with aspects of manufacturing engineering, then with those of design engineering. These two disciplines of engineering are quite broad, so the focus is only on the content that is applicable to concurrent engineering, that which makes the parallel effort work. In a supportive role, we explore the external techniques, for example, marketing, from the viewpoint of how its principles are applied in the parallel team effort and where the ties to producibility engineering are.

## Techniques of Producibility Engineering Used in the Concurrent Engineering Process

Figure 6.1 illustrates the power of concurrent engineering. But what exactly is the technology involved? What is the interface between the functions when they are together as a team? What do they really do? What is the expected contribution of each member? What is design for producibility? These are but a few of the questions that need to be answered and understood before concurrent engineering can be applied properly. Let's investigate the answers to these questions by starting at the beginning, which is producibility engineering.

Producibility engineering is the discipline responsible for creating producible designs. Note that there is nothing in the definition about teamwork or parallel efforts. Producibility engineering was created as a checking mechanism. The idea was to make sure that the factory was not being asked to make products beyond its capabilities. The original concept was not concerned with fostering parallel efforts to meet customer needs (although this later became a valuable attribute of the discipline, as it was used as part of the concurrent engineering process). It had a very focused approach to minimizing manufacturing losses by imparting information to design engineering vis-à-vis the specific drawing and whether or not the factory would have difficulty in achieving the desired design goals.

Producibility engineering is the beginning of the understanding that there has to be cooperation between functions in order to optimize a manufacturing company's performance. It is a process by which manufacturing-based engineers are assigned as liaison personnel to design engineering. The purpose was primarily to impart "manufacturing know-how" to the designer as part of the design process, which is a big part of what later became the full concurrent engineering process. It is important that we understand these procedures.

One way to think about producibility engineering's place in the concurrent engineering continuum is to think of producibility engineering as the manufacturing technical contribution to the concurrent engineering team. Likewise, the designer would be the engineering technical contributor to the team. We must

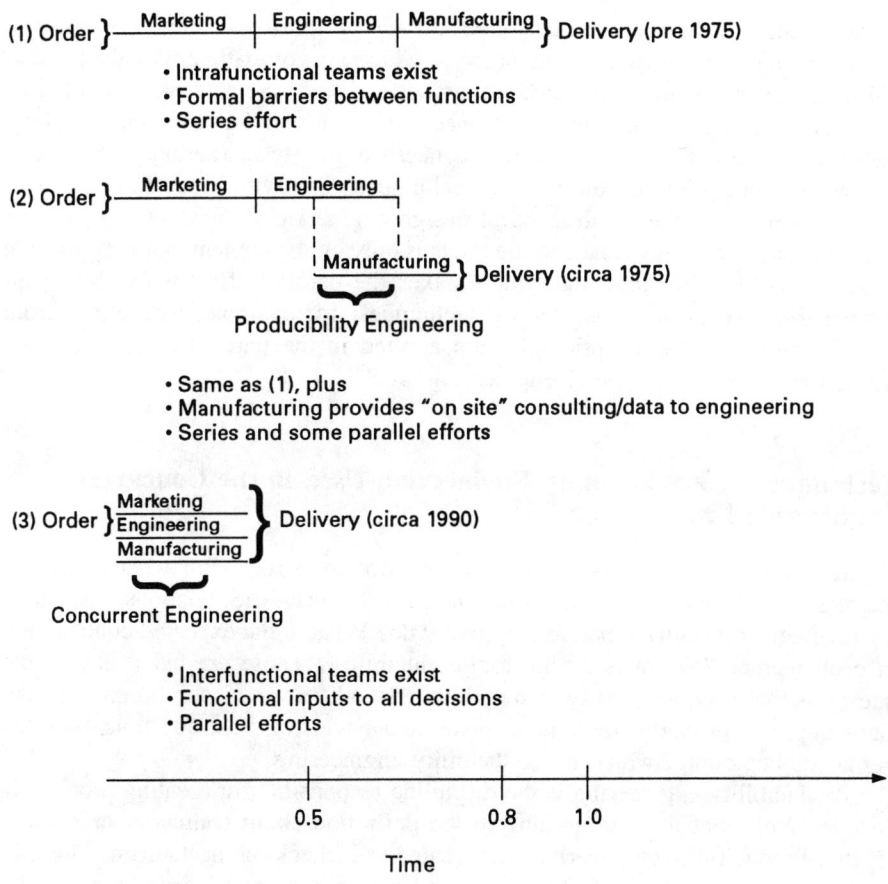

**Figure 6.1** Evolution of concurrent engineering.

understand this manufacturing engineering technical role thoroughly in making concurrent engineering a successful process.

## The Producibility Engineering Process

The producibility engineer is expected to tell the design engineer how to make the design producible, which is commonly known as design for manufacturability. This means that he or she is expected to critique the design from the viewpoint of how to optimize manufacturing process costs. This is a very demanding task. Process costs comprise cycle time and material utilization, and keep in mind that the product still has to perform to the specifications required. Therefore, for the producibility engineer to do the job right, he or she must understand the theory of the design. As the factory's representative, the producibility engineer needs to understand the intent of the design, such as why certain dimensions, tolerances, and materials were selected, before he or she can make credible comments to the designer. Also, as the factory representative, it is absolutely essential that the producibility engineer understands the factory's limitations. So, for the producibility engineer to assure that designs are capable of being produced in the factory, two tasks must be mastered:

1. Understand the intent of the concept design.
2. Understand the factory's capabilities relative to the design.

If this is done successfully, the producibility engineer can offer cogent criticism to the designer.

## When to Accept or Reject a Design for Producibility Criteria

Suppose the producibility engineer found that the intent of the design simply did not match the capability of the factory. Suppose also that the producibility engineer was certain that the intent must be met for the product to be successful. What are the choices? Remember, the producibility engineer's task is to prevent designs from being accepted by manufacturing if they cannot be produced economically. Furthermore, the producibility engineer is morally committed to the success of the cooperative venture between manufacturing and engineering. There are two basic choices:

1. Reject the design for production.
2. Recommend major changes in the factory to accommodate the design.

At first glance, the first choice appears to be a stonewalling strategy. But really it's not. Informing the design team that they must reject what they're proposing because it would be folly for the manufacturing group to attempt is a sufficient answer to give design engineering. This is a hard choice to make but a correct one. The last thing the company could tolerate would be manufacturing trying to do something outside its realm of competence.

## An Example of the Logic of When to Reject a Design for Production

Dilemmas such as this typically occur when companies try to redefine what type of company they are. Let me give an example.

Company A is trying to redefine itself as a musical instrument maker instead of a piano maker. This could lead to serious mismatches. There may be problems in its desire to serve a market versus its capability to do so. When this happens, engineering is apt to respond in a positive manner much more quickly and readily than manufacturing. For example, as a musical instrument maker, instead of only a piano maker, company A might be tempted to offer a line of electronic organs. After all, there appears to be synergy: they are both keyboard instruments and both require fine furniture fabrication. But there the similarity ends, and this might not be a good choice.

Suppose company A were to embark on such a venture. Engineering could acquire the technological expertise by going to the market and hiring bright electronics engineers with organ design backgrounds. They would then produce the design for building organs. But what does manufacturing do? They could contract out the electronic components, to the point of purchasing completed subassemblies. Then they would assemble, test, package, and ship. How confident would manufacturing be in its ability to produce organs at the same quality levels as pianos? Not very. Piano making is 90+% mechanical in design, as the piano is a product dominated by mechanical engineering. Organs, on the other hand, are more than 50% electronic. So there is no synergy. The factory would have to go through a long and arduous learning cycle. It would be difficult to predict realistically a successful business outcome. This being the fact, the producibility engineer would make the proper choice in recommending rejection of the design even though he or she agrees with the intent of the design.

## When It Is Proper to Accept a Design for Manufacturing Even Though Capability to Do So Does Not Exist

The second choice of action puts the onus on manufacturing. In this case the producibility engineer agrees with the intent of the design but does not reject it because manufacturing does not have the capability to produce it. Instead, the producibility engineer says that manufacturing has no choice but to acquire the capability to produce, or the company will suffer the consequence. The last phrase is key. If the producibility engineer is convinced that the design is proper and within the range of products the company must offer, then he or she should recommend that manufacturing take on the challenge of tooling up to make it.

What is the line of demarcation? How does the producibility engineer determine whether the company should be urged to invest capital and human resources to be able to make the product depicted in the design? Sometimes the choice is obvious. For example, in the silicon chip business, the ability to crowd more and more electronic circuits on a silicon wafer is the difference between staying in business or permanent demise. So, if a new chip design requires improvement in photoetching capability, the producibility engineer will probably recommend that his or her manufacturing colleagues develop the necessary equipment. In this case, the need to meet the design requirement is obvious to all.

Sometimes the choice is not so obvious. Remember, we are dealing with cases in which the producibility engineer agrees with the intent of the design but knows the factory cannot make it at satisfactory cost levels. The choice of accepting the design commits the company to spending funds to gain the capability to produce. There is no rule book on how to make the choice. I can only offer some guidelines.

1. If compatibility exists between the proposed design and previous designs, then accept.

2. If tooling requirements to produce are within the realm of state-of-the-art equipment, then accept.
3. If employee experiences are similar to those required for successful manufacture of the proposed design, then accept.
4. If training requirements are extensions of current skills, then accept.

Let me expand a bit on these guidelines.

1. *Compatibility Exists.* Compatibility existing probably means that the new design, while not within the capability of the factory, requires processes similar to existing processes. This is the chip maker example. Getting more transistors on the chip will require new machinery, but in truth it is just more of the same thing. It is somewhat like driving a car and having to go from zero to sixty miles per hour in 7.0 seconds. The process for doing it is the same; however, some automobiles will achieve the goal and others will not. If the goal is critical and you are unfortunate in having a car unable to accelerate fast enough, there is only one choice: trade up to a faster automobile.

2. *Tooling Within the Current State of the Art.* The second condition, tooling within the current state of the art, means that no new technology needs to be developed for the factory to comply. This means that new invention is not required. For example, if the design calls for an electric-powered automobile with a 100-mile radius, this technology exists. The consequence is that the factory can be tooled up to make the product with state-of-the-art technology. On the other hand, if the design requires a range of 1000 miles, the design theory might be satisfactory but the edge of practicality as we presently know it may be breached, and there may be reason to reject.

3 and 4. *Employee and Training Issues.* The third and fourth items are people issues: current capabilities and the ability to learn new technologies. These are the hardest to evaluate. We are dealing with many levels of experience and capability, and the producibility engineer will be hard pressed to come to any conclusions independently. Here the proper course of action is to involve employee relations as well as the various levels of manufacturing management. The goal is to determine the median experience level and compare that with the tasks that will have to be accomplished to produce the intended design. If it is felt that the differences are not too great, the design can be accepted. Likewise, whether training can be successful depends on how close the new skills level is to the current level.

## A Guideline for the Producibility Engineer's Job Performance

Summarizing responsibilities and how the producibility engineer should perform them is a complex matter. I have explained the various factors, but they are not simple. Over the years I have developed what I call the "producibility engineer decision tree," which is shown as Figure 6.2. I trust it will be a useful aid in understanding the role of the producibility engineer. This is a very real and valu-

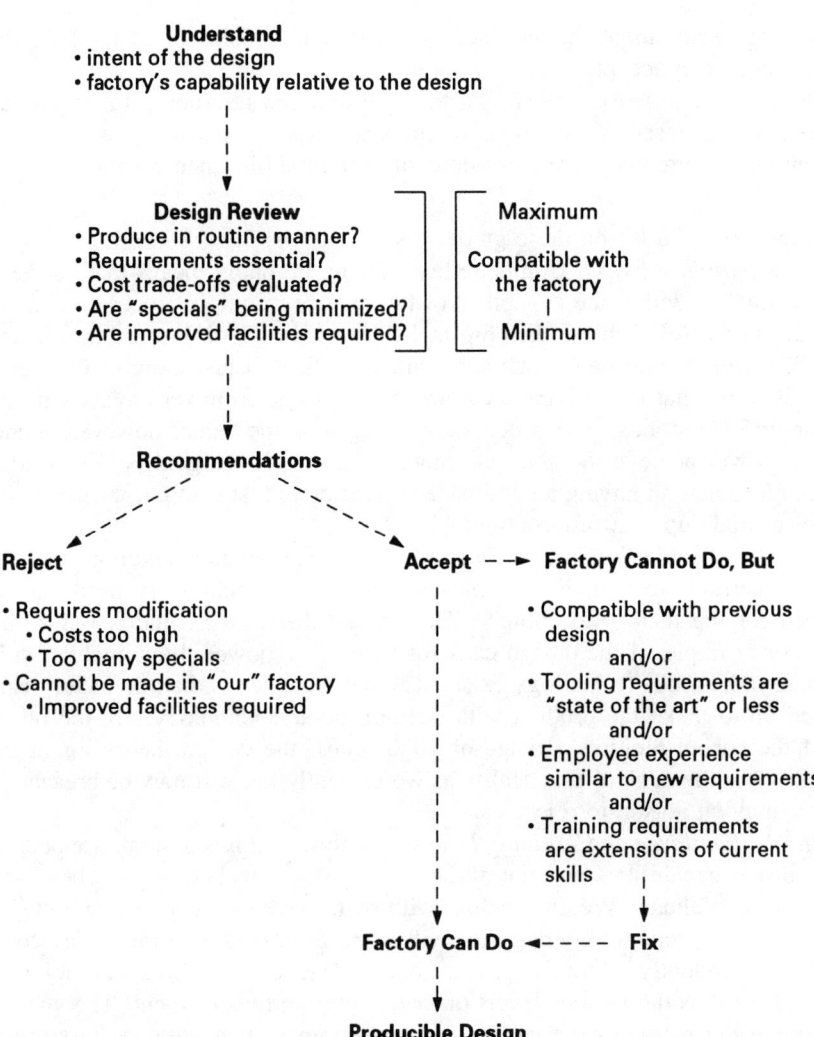

**Figure 6.2** Producibility engineer's decision tree.

able role that needs to be performed within the continuum of concurrent engineering.

So far I have discussed what the producibility engineer's responsibilities are. But how are they accomplished? What methodology is used to do the job? Let's take a look.

## THE PRODUCIBILITY DESIGN REVIEW PROCESS

Producibility engineers perform a liaison role, primarily via the design review. Prior to concurrent engineering this was done at the completion of significant milestones of the design task. With concurrent engineering teams, it is a continuous process. Nevertheless, the process remains the same. The process is not the normal design review for functionality and meeting of the customer specifications. It is a review to discover whether or not a producible design exists.

> A producible design is defined as a design that can be manufactured correctly and economically in the factory for which it was intended.

The producibility design review starts with a look at the basics. Can the design be produced while maintaining the status quo in the factory? If this is so, the design is producible. It gets more difficult from here on. Referring to Figure 6.2, we see that the degree of difficulty increases as the requirements of the design become less compatible with the factory's capability. The trick is to determine how much the factory's ability can be stretched to meet the design requirements. The practitioners of producibility engineering do this by looking at all five items listed under the design review segment of the decision tree. The secret is to list the reasons for incompatibility and then see what can be done to overcome these deficiencies. At the same time as we are looking at the factory side of the equation, the design requirements must be looked at with equal intensity. This ensures that unnecessary costs are not incurred. Creating a producible design is a systems integration activity and one that optimizes the process for the entire company.

To fully understand the producibility design review, let's look at an example. We will use a simple product so that the essence of the technique is not lost in the technology being applied.

## Producibility Design Review Example

The scenario is as follows. We have a typical job shop supplied with normal CNC metalworking and woodworking machines capable of doing sawing, milling drilling, turning, and shaping. The current products are simple hand tools, specifically screwdrivers and chisels. The basic manufacturing process in use is to cut machine tool grade steel to length and then turn it to the desired diameter. Sometimes the process requires drilling holes in the shanks to mount the handles. The final metalworking operation is to mill the flats, either for screwdriver blades or chisel blades. The shop also does some simple wood machining, such as sawing to length, lathe turning, and drilling and shaping to make handles for screwdrivers and chisels. The final operation is an assembly task. Here handles are glued onto shafts. The more deluxe models have a rivetlike pin pressed through the handle and shaft before gluing.

Because of the good quality of work, business has been good and the company is contemplating expanding the hand tool line. In order to do so, a concurrent engineering team has been assembled to bring new products to market. They determined that the market is ripe for a high-quality, simple design needle-nose pliers. The designer has prepared a preliminary product scope to meet the need and is requesting that the producibility engineer review it for practicality of manufacturing in the factory. In addition, the concurrent engineering team is looking at material procurement cost, sales promotions, production run quantities, selling prices, capitalization, and other related and equally important issues for introducing a new product. However, for this example we will make the simplifying assumption that the go/no go decision will be solely a producibility issue. Let's now go through the steps of the producibility design review.

The questions to be answered during the producibility design review are:

1. Produce in a routine manner?
2. Requirements essential?
3. Cost trade-offs evaluated?
4. Are "specials" being minimized?
5. Are improved facilities required?

The needle-nose plier design shown in Figure 6.3 is the item that will be reviewed for producibility.

1. *Produce in a routine manner?* The current production requires straight metal rods turned to a specific diameter. The rod is then milled flat at one end into either a screwdriver or chisel point, and a portion of the production of metal rods is drilled through to accept a pinned handle. The plier will be made of two rods bent approximately one-third of the way along the shaft to a 60° interior angle. This will require milling an undercut near the pivot joint. Also, one side of the short end of the rod beyond the pivot will require tapered milling and knurling to create the needle-nose shape. Finally the two plier halves will have to be drilled at the pivot point to accept the pivot pin.

In the example factory the only thing currently not done is making metal parts with angle shafts. This is different and will be so noted by the producibility engineer.

2. *Requirements essential?* The designer is not requiring anything but bare essentials for the finished product to perform as an adequate needle-nose plier. The producibility engineer would agree that everything specified is required for functionality to be achieved.

3. *Cost trade-offs evaluated?* In this case the answer is yes, they have been. The material to be used is the same tool-grade steel used for the screwdrivers and chisels. Because the company is a quality maker of hand tools, marketing would insist that the same material, which is already perceived as a quality commodity, be continued.

4. *Are specials being minimized?* One special is being considered, that of the bending of the plier half. No bending is currently being done, so this must be considered a special for the factory. Keep in mind that the evaluation is for a producible design, that is, whether the design can be produced in this specific factory. The question is not whether the product can be made in a generic factory. There is no doubt it can. The question to be answered is whether this special will pose an unacceptable burden for the current manufacturing situation.

5. *Are improved facilities required?* The producibility engineer now has to

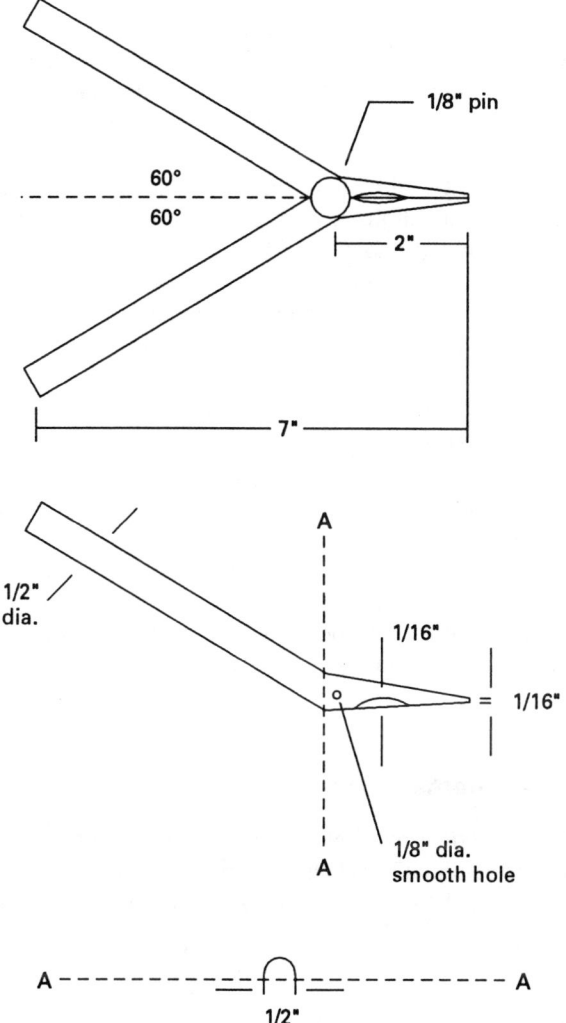

60°
60°
1/8" pin
2"
7"

1/2" dia.
A
A
1/16"
= 1/16"
1/8" dia. smooth hole
o

A --------------- A
1/2"

**Figure 6.3** Sketch: simple needle-nose plier.

decide how the angle bend can be accomplished. There are many ways to do this. It can be cold bent around a mandril. The metal can be heated to reduce the yield strength and then bent much as a blacksmith does. The process of manufacturing can be changed to drop forge the plier halves. The halves can be made out of much larger-diameter bar stock and "hogged out" via lathes and milling machines. There are many possibilities, some with very low initial costs and some with high costs. The answer to whether or not improved facilities are required in this case requires inputs from other members of the concurrent engineering team.

The volume of product being considered and what the target selling price will be must be determined. Once these are known, the team can judge how much capital can be utilized for this product. At this point the producibility engineer, the manufacturing engineer, and the designer need to consult with the marketing and finance team members. If the volume is to be very large, the blacksmith approach will not be viable. If keeping costs to rock-bottom minimum is paramount, the "hog it out" approach is not acceptable—it would have low initial equipment costs but very high unit costs. Similarly, the cold mandril technique can be related to medium- to low-volume choices. The team also has to evaluate the perception of quality. The marketing member must lead this discussion. Certainly, drop forging will be a favorite alternative because market conditioning through advertisements has led the buying public to relate this manufacturing process to high quality. However, it is an acceptable alternative only if the concurrent engineering team is going after a large share of the market, hence can afford the high initial capital costs. Even this solution may have to be tempered with the realities of equipment procurement availability and schedule.

In this example, we can conclude that the producibility engineer will accept the design for manufacturability. It will be adding a burden to the company to do so, but just what that will be requires a decision by the entire concurrent engineering team. The producibility design review is a vital part of the decision process, but it is not done in a vacuum. As shown above, the final decision depends on many issues. However, the strategic issues of competition come to a head faster via the concurrent engineering approach than the obsolete singular stand-alone chain-type method.

## Producibility Design Review, Another Example

Sometimes the producibility design review requires a more extensive review during the second step, "Requirements essential?" When the complexity of the design requires it, the second step is expanded to include more specific questions such as:

1. What is the part for?
2. How does it work?
3. What engineering analysis is important to the design?
4. How can the part be made?
5. What can be done to make it cheaper to produce?

In essence, because the design is more complex, the producibility engineer is looking at the design very closely to understand the factors the design engineer considers important. Then the producibility engineer considers the manufacturing factors to see if normal procedures can be used to make the part or assembly. If not, then recommendations have to be made to the designer to modify the design for producibility concerns. Let's look at an example in which step two has to be expanded to handle a more complex design.

Figure 6.4 is a drawing of a stud to be used by a company with its product,

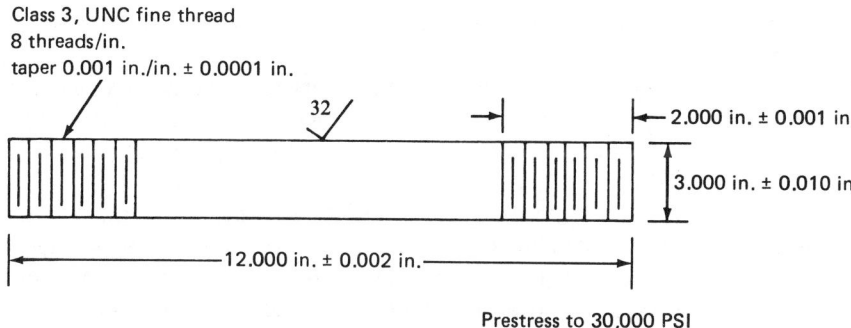

Figure 6.4 Drawing of a stud to be used with a large centrifugal pump.

a large centrifugal pump. The stud is designed to hold the pump base plate to the customer's foundation. The producibility engineer proceeds as follows:

1. What is the part for?
   (a) Holding the base plate to the foundation. It will go through the base plate, then into a structure provided by the customer, secured at both ends by nuts.
2. How does it work?
   (a) Puts plate and foundation in compression; therefore stud goes into tension.
   (b) Taper in thread indicates that the designer intended to get a better distribution of load along the stud and nut threaded length.
   (c) Prestress requirement shows that the designer intended to firmly anchor the pump, probably to guarantee known support conditions for rotor and bearing critical speed and vibration analysis as well as to overcome any possibility of joint loosening.
3. What engineering analysis is important to the design?
   (a) Shear stress calculation for the body of the stud and for the threads.
   (b) Bending stress calculation to ensure workability of the stud if an off-center load is applied due to misalignment during installation or pump axial thrust.
   (c) Fatigue evaluation due to induced vibration when the pump is running in the customer's system.
4. How can the part be made?
   (a) Bar stock cut to length by saw.
   (b) Face mill on Horizontal Boring Mill (HBM) to desired precise length.
   (c) Rough machine in lathe to a finish of 63 RMS.*
   (d) Thread in screw machine; use go/no-go gauge, class 3, for check.
   (e) Grind surface on cylindrical grinder to obtain a 32 RMS finish.
5. What can be done to make it cheaper to produce?

*RMS, root mean square. The value is a measure of surface smoothness.

(a) The part can be made in the usual area or department for such parts; no savings opportunity exists here.

(b) The length is 12.000 ± 0.002 in.; is it necessary to produce such an exact length? *Recommended:* reduce length tolerance to ±0.015 in. ($^1/_{64}$ in.); this is within reason for a metal saw. *Result:* eliminate the face milling step on the HBM.

(c) Is the surface finish requirement of 32 RMS necessary? Obviously, the surface finish was specified to reduce stress risers on a relatively highly stressed part. *Recommended:* reduce finish requirements to 63 RMS; this can be produced easily on a lathe (lathes in good condition consistently produce finishes better than 63, approaching 32). *Result:* eliminate the necessity for surface grinding on a cylindrical grinder.

(d) Is the taper requirement of 0.001 in./in. over a 2-in. length realistic? Probably not for such a short length of thread. *Recommended:* eliminate the taper. Rely on class 3 thread to provide a tight thread mesh. *Result:* eliminate a more complicated setup on the screw machine.

(e) Will the part still do the job if producibility engineer's recommendations are accepted?
  (i) In this example, it obviously will.
  (ii) The design margin is somewhat reduced, but still safely withinany reasonable expectations.

The above example results in less cost to produce the product. The number of steps is reduced from five to three. We have gone from a design to a producible design, from one that could do the job to one that still does the job but at a much lower cost. Notice that the producibility engineer must mentally design the part, then contemplate how it should be made, and then review the method to see whether it can be optimized. Cheaper and quicker ways to do it can be found if the tolerances or processes can be modified. By understanding what the design engineer intended to accomplish, and knowing how the part will be made, including an estimate of the cost, the producibility engineer can discuss his or her findings with the design engineer from a position of knowledge and can beneficially affect overall productivity.

## How to Determine if a Factory Can Meet the Requirements of the Design

How do you evaluate whether or not the factory can meet the requirements of the design? There is no exact answer. The producibility design review example gives some insight into the complexity of determining whether a factory can handle a given design. It is evident that we cannot boil it down to a formula-driven exercise. A lot depends on an understanding of the specific factory. For example, is it capable of following deviations from norm through issuance of revised documentation? Or is it a factory in which change from the routine is abhorred and achievable only through brute-force methods, requiring constant supervision? Where the factory fits within this spectrum will determine the degree of change that can be tolerated. One of the strengths of the concurrent engineering process is that it gets this discussion going up front. If it appears that this will be an issue (more change needed than currently possible), the problem is spotlighted for solution much earlier in the process.

## SETTING DESIGN TOLERANCES

One of the primary functions of producibility engineering is to define to design engineering the factory's ability to hold dimensional accuracy. With this information the concurrent engineering team can set tolerances that meet the design requirements and are compatible with the capabilities of the production facility. How do you determine the proper tolerance to apply? Mistakes in tolerance setting can cost significant amounts of money. The general rule is simple: require what is necessary and nothing more. If design engineering does not know what the tolerance should be, they should work in consort with other members of the company team and learn together what will be required. Here are some basic rules that will help in setting tolerances.

1. *Understand the material you are working with.* To set realistic tolerances, it is absolutely vital to understand the physical and chemical properties of the material we are working with. For example, it makes no sense to require tolerances tighter than $\pm 0.003$ in. for hardwood because of its hygroscopic nature; it absorbs and gives off water vapor depending on the relative humidity of the immediate atmosphere. This means it is constantly undergoing small dimensional changes. So you must understand the limitations of the chosen materials and set tolerances accordingly. Set what is needed to maintain the intent of the design, but be aware of the need to be compatible with the requirements of the materials used.

2. *Use the statistics approach to setting tolerances.* Absolute dimensioning never occurs in the real world. We must understand that we cannot design for either the maximum or minimum side of the dimension. We must design to the given dimension and use statistics and probabilities to set the tolerances. This will give the most realistic values for tolerance ranges. Yes, there will be failures, but the statistics approach will yield fewer failures than any other method. So, when assembly tolerances are critical, use the statistics approach. This technique is demonstrated in the next section of this chapter.

3. *Avoid the need to set tolerances as much as possible by simplifying designs.* Whenever possible, use such techniques as chamfered holes and fastening devices, gaskets, and tapered pins in assemblies to avoid precision assembly requirements. Consciously design for ease of assembly, thereby minimizing the need to set tight tolerances.

These three suggestions are all that is pertinent about tolerances. Set tolerances compatible with the materials being used. Use statistics methods for establishing tolerances instead of arithmetic methods. Finally, avoid the problem as much as possible by designing around the need for tolerances. Other concepts for setting tolerances are often used, such as employing a percentage of the dimension. However, these are not directly related to the primary need, that the part or assembly function as required. They are typically related to administrative procedures; therefore they are not discussed here.

## STATISTICAL APPROACHES TO TOLERANCE SETTING

One of the important functions of producibility engineering is the development of techniques for design engineering that allow product reliability to be maintained while optimizing manufacturability. A significant body of work over the years has been done in the area of statistical tolerance setting. Statistics is used to negate or at least predict the effect of tolerance buildup during assembly operations. The purpose is to allow the widest possible latitude for manufacturing while still maintaining a very high probability that the parts will go together.

Tolerances are set to make sure that the finished part has the attributes required by design. Stress and strain analysis dictate the size and shape of a part. Tolerances are used to ensure that after manufacture, the part retains the proper size and shape. Unfortunately, the most popular method of setting tolerances is that of avoiding interferences. This method dictates that the largest (plus) tolerance must always be smaller than the part it mates to.

**Example:** Noninterference-fit keyway in a shaft; the plus tolerance of the key is always smaller than the minus tolerance of the keyway.

This is fine for a two-part assembly. But for a multipart assembly with many parts that are not directly related, we are concerned with tolerance buildup and we want to know what the probability of interference will be. Therefore, the normal scheme of tolerance setting is too severe, and many parts judged out of tolerance would fit with a high degree of probability. In order to not throw away perfectly usable parts, a statistical approach to tolerance setting is preferred by manufacturing, especially if the probability of failure (using a part that actually does create an interference) is small.

One such way of setting tolerances by statistical approaches is to look at load characteristics versus material strength characteristics. This can be done with a Warner diagram, which is a series of normal distributions on the same set of axes. Let us look at the Warner diagram in Fig. 6.5, using the joint stud of Fig. 6.4 for the necessary data.

The diagram shows an area where the two normal curves overlap. If these curves represent values pertaining to the same physical entity, we can deduce some vital information. From Fig. 6.5 we can see that the maximum load on the part is higher than the minimum strength of the material. That means if we are on the plus tolerance of load and the minus tolerance of strength, there will be failure. In fact, any combination of load and strength that falls into the failure

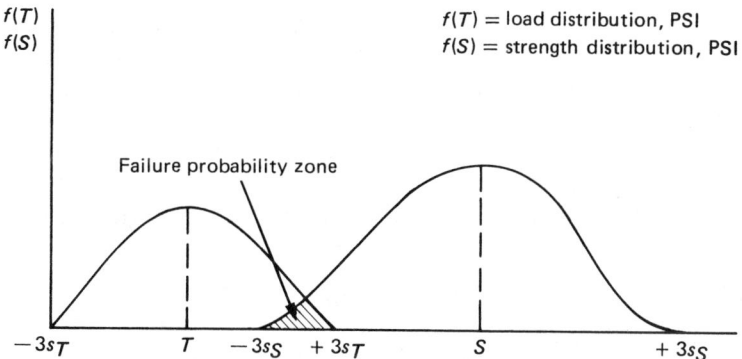

$f(T)$
$f(S)$

$f(T)$ = load distribution, PSI
$f(S)$ = strength distribution, PSI

Failure probability zone

$-3s_T$    $T$    $-3s_S$    $+3s_T$    $S$    $+3s_S$

**Figure 6.5** Warner diagram. Here $T$ is the mean shear stress load on the bolt; $^{-3s}T$ and $^{+3s}T$ the lower and upper 3 sigma spread of the shear stress variation, respectively; $S$ the mean strength of the materials used to make the bolt; and $^{-3s}S$ and $^{+3s}S$ the lower and upper 3 sigma spread of the material strength specification, respectively.

probability zone will result in failure. Normally, when a machining drawing is produced, dimensions rather than stress levels are shown. Therefore, the Warner diagram must be translated into dimensions so that we can set tolerances.

The shear stress, $T$, was calculated by the design engineer to size a stud. In the example given, $T$ is 30,000 pounds per square inch. If no bending occurs, $T = V/A$ or $T = V/\pi r^2$, and there is an axial stress effect only. If bending occurs, $T = 4V/3\pi r^2$, which is the maximum shear stress affecting the part. It is a simple matter to determine the value for $r$ in the equations, hence the diameter. Assuming that bending can occur, referring back to the joint stud example, let $V = 159,000$ lb. Then the diameter is

$$D = 2r = 2\left(\frac{4V}{3\pi T}\right)^{1/2} = 2\left(\frac{4 \times 159,000}{3\pi \times 30,000}\right)^{1/2} = 2.996 \approx 3.000 \text{ in.}$$

Now we have a mean diameter. In order to calculate a reasonable tolerance we must assume that the factor of safety for the material is such that slight variations in $T$ would not ordinarily result in failure. The problem now becomes one of determining the variation in $T$ caused by error in measurement (very likely to occur when prestressing a part). A normal measurement error would be $2\frac{1}{2}\%$. Then $T$ (upper) is 30,750 lb/in.$^2$ and $T$ (lower) is 29,250 lb/in.$^2$ Using the same formula for the diameter that we used above, we now have

$$D \text{ (upper)} = 2\left(\frac{4 \times 159,000}{3\pi \times 29,250}\right)^{1/2} = 3.038 \text{ in.}$$

$$D \text{ (lower)} = 2\left(\frac{4 \times 159,000}{3\pi \times 30,750}\right)^{1/2} = 2.963 \text{ in.}$$

Therefore the mean, upper bound, and lower bound are 3.000, 3.038, and 2.963, respectively, or a $3\sigma$ probabilistic normal curve distribution on the diameter of $3.000 \pm 0.0375$. Now these data are ready to be entered on the Warner diagram.

We now have data for the $f(T)$ distribution but lack data for the $f(S)$ distribution, that is, the material strength distribution. The maximum, minimum, and mean of this distribution will be found in material and test specification data sheets. For our example of the joint stud, let us assume that a steel with a 0.02% yield strength in shear of 34,000 lb/in.$^2$ was selected and that it has a strength range of $\pm 4,000$ lb/in.$^2$ Then

$$S(\text{max}) = 38,000 \text{ lb/in.}^2$$

$$S(\text{mean}) = 34,000 \text{ lb/in.}^2$$

$$S(\text{min}) = 30,000 \text{ lb/in.}^2$$

Now we can enter both load and strength values on the Warner diagram (see Fig. 6.6). Then, using the equation $D = 2(4V/3\pi T)^{1/2}$, where $T = 30.75$ KSI, 29.25 KSI, and 30.00 KSI (KSI $= 1000$ lb/in.$^2$), we have

$$D(\text{min}) = 2.963 \text{ in. } (30.75 \text{ KSI})$$

$$D(\text{max}) = 3.038 \text{ in. } (29.25 \text{ KSI})$$

$$D(\text{mean}) = 3.000 \text{ in. } (30.00 \text{ KSI})$$

We can see there is a definite overlap into the failure zone and the probability of operating in that zone should be calculated. Before that is done, however, let us look at the ranges compared to that shown on the original proposal for the part. The proposal calls for a diameter $D = 3.000 \pm 0.010$ in., or a range of 3.010 to 2.990 in. A standard machine shop tolerance would be $\pm 0.015$ in. ($\frac{1}{64}$ in.). The

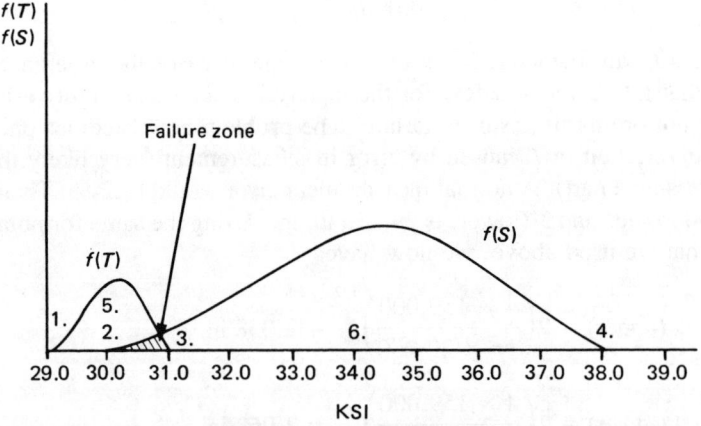

**Figure 6.6** Warner diagram with data for determining diameter tolerances for the studs. 1: $^{-3s}T =$ 29.25 KSI; 2: $^{-3s}S = 30.00$ KSI; 3: $^{+3s}T = 30.75$ KSI; 4: $^{+3s}S - 38.00$ KSI; 5: $T$ mean = 30.00 KSI; 6: $S$ mean = 34.00 KSI (1 KSI = 1000 lb/in.$^2$).

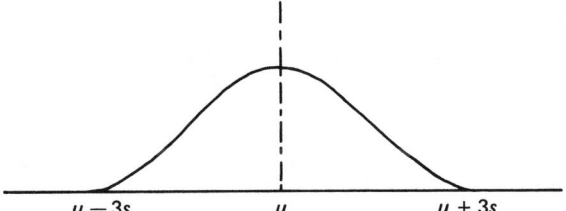

**Figure 6.7** Normal curve.

statistical analysis gives $D = 3.000 \pm 0.0375$ in., or a range of 3.038 to 2.963 in. This is a $\frac{1}{27}$ in. range.

Without any further calculation, it is possible for the producibility engineer to recommend the normal machine shop tolerance of $\frac{1}{64}$ in. for this part. This is correct since the statistically derived tolerance is broader than the normal good machine shop practice. If the outcome had been the reverse, with the original looser than good machine shop practice, the producibility engineer would still have chosen $\frac{1}{64}$ in. If the statistical tolerance had been less than $\frac{1}{64}$ in. but greater than the original selection, the producibility engineer would have selected the statistically derived version in order to prevent needless scrapping of parts with a high probability of being good. In this example, we can see that $\frac{1}{27}$ in. is roughly twice the spread of $\frac{1}{64}$ in.; therefore, noting the very small Warner diagram overlap, it is proper to conclude that the $\frac{1}{64}$ in. tolerance is acceptable and end the problem here. For instructional purposes, however, let us continue and find the probability of failure based on the Warner diagram. The procedure involves finding the percentage of area under the normal probability curves. To do this we use the tabulated areas represented by the number of standard deviations calculated and measured along the abscissa of the normal-curve graph (see Fig. 6.7). Areas under normal curves are given in Table 6.1.

Table 6.1 is used to find the area proportions under the curve by entering with the number of standard deviations, $Z$:

$$Z = \frac{(x - u)}{s}$$

where $x$ is the distance of the variable away from the mean, $u$ is the mean, and $s$ is the standard deviation. This is adequate for problems involving a single normal distribution, but a compatibility relationship must be developed to equate two normal distributions acting together as in the Warner diagram. Edward B. Haugen* has developed such a coupling formula:

$$Z = \frac{u_1 - u_2}{(s_1^2 + s_2^2)^{1/2}}$$

where $u_1$ is the mean of the first distribution, $u_2$ the mean of the second distri-

*E. B. Haugen, *Probabilistic Approaches to Design*, Wiley, New York, 1968.

## Table 6.1 Areas under normal curves

| $s$ | .00 | .01 | .02 | .03 | .04 | .05 | .06 | .07 | .08 | .09 |
|---|---|---|---|---|---|---|---|---|---|---|
| 0.0 | .0000 | .0039 | .0079 | .0110 | .0158 | .0198 | .0237 | .0277 | .0317 | .0356 |
| 0.1 | .0396 | .0436 | .0475 | .0515 | .0555 | .0594 | .0634 | .0673 | .0713 | .0752 |
| 0.2 | .0791 | .0831 | .0870 | .0909 | .0948 | .0987 | .1026 | .1064 | .1103 | .1141 |
| 0.3 | .1180 | .1218 | .1256 | .1294 | .1332 | .1370 | .1407 | .1445 | .1482 | .1519 |
| 0.4 | .1556 | .1593 | .1630 | .1666 | .1703 | .1739 | .1775 | .1811 | .1846 | .1882 |
| 0.5 | .1917 | .1952 | .1987 | .2022 | .2058 | .2091 | .2125 | .2159 | .2192 | .2226 |
| 0.6 | .2259 | .2292 | .2325 | .2358 | .2391 | .2423 | .2455 | .2487 | .2518 | .2550 |
| 0.7 | .2581 | .2612 | .2643 | .2673 | .2704 | .2734 | .2764 | .2793 | .2823 | .2852 |
| 0.8 | .2881 | .2910 | .2938 | .2960 | .2994 | .3022 | .3050 | .3077 | .3104 | .3131 |
| 0.9 | .3158 | .3184 | .3210 | .3236 | .3262 | .3287 | .3313 | .3338 | .3362 | .3387 |
| 1.0 | .3411 | .3435 | .3459 | .3483 | .3506 | .3529 | .3552 | .3575 | .3597 | .3619 |
| 1.1 | .3641 | .3663 | .3684 | .3705 | .3726 | .3747 | .3768 | .3788 | .3808 | .3828 |
| 1.2 | .3847 | .3867 | .3888 | .3905 | .3923 | .3942 | .3960 | .3978 | .3996 | .4014 |
| 1.3 | .4031 | .4048 | .4065 | .4082 | .4098 | .4114 | .4130 | .4146 | .4162 | .4177 |
| 1.4 | .4192 | .4207 | .4222 | .4237 | .4251 | .4265 | .4279 | .4293 | .4306 | .4320 |
| 1.5 | .4333 | .4346 | .4358 | .4371 | .4383 | .4397 | .4407 | .4419 | .4431 | .4442 |
| 1.6 | .4454 | .4465 | .4476 | .4480 | .4497 | .4507 | .4517 | .4527 | .4537 | .4547 |
| 1.7 | .4556 | .4566 | .4575 | .4584 | .4593 | .4602 | .4610 | .4619 | .4627 | .4635 |
| 1.8 | .4643 | .4651 | .4659 | .4606 | .4673 | .4681 | .4688 | .4685 | .4702 | .4708 |
| 1.9 | .4715 | .4722 | .4728 | .4734 | .4740 | .4746 | .4752 | .4755 | .4763 | .4769 |
| 2.0 | .4774 | .4780 | .4785 | .4790 | .4795 | .4800 | .4805 | .4809 | .4814 | .4818 |
| 2.1 | .4823 | .4827 | .4831 | .4835 | .4839 | .4843 | .4847 | .4851 | .4854 | .4858 |
| 2.2 | .4862 | .4865 | .4868 | .4872 | .4875 | .4878 | .4881 | .4884 | .4887 | .4890 |
| 2.3 | .4893 | .4895 | .4898 | .4901 | .4903 | .4906 | .4908 | .4911 | .4913 | .4915 |
| 2.4 | .4917 | .4919 | .4922 | .4924 | .4926 | .4928 | .4929 | .4931 | .4933 | .4935 |
| 2.5 | .4937 | .4938 | .4940 | .4942 | .4943 | .4945 | .4946 | .4948 | .4949 | .4950 |
| 2.6 | .4952 | .4953 | .4954 | .4955 | .4957 | .4958 | .4959 | .4960 | .4961 | .4962 |
| 2.7 | .4963 | .4964 | .4965 | .4966 | .4967 | .4968 | .4969 | .4970 | .4971 | .4971 |
| 2.8 | .4972 | .4973 | .4974 | .4974 | .4975 | .4976 | .4977 | .4977 | .4978 | .4978 |
| 2.9 | .4979 | .4980 | .4980 | .4981 | .4981 | .4982 | .4982 | .4983 | .4983 | .4984 |
| 3.0 | .4984 | .4985 | .4985 | .4986 | .4986 | .4986 | .4987 | .4987 | .4987 | .4988 |
| 3.1 | .4988 | .4988 | .4989 | .4989 | .4989 | .4990 | .4990 | .4990 | .4991 | .4991 |
| 3.2 | .4991 | .4991 | .4992 | .4992 | .4992 | .4992 | .4993 | .4993 | .4993 | .4993 |
| 3.3 | .4993 | .4994 | .4994 | .4994 | .4994 | .4994 | .4994 | .4995 | .4995 | .4995 |
| 3.4 | .4995 | .4995 | .4995 | .4995 | .4996 | .4996 | .4996 | .4996 | .4996 | .4996 |
| 3.5 | .4996 | | | | | | | | | |

bution, $s_1$ the standard deviation of the first distribution, and $s_2$ the standard deviation of the second distribution. One other important fact must be kept in mind in order to use Haugen's equation; the distributions must be of like quantities, for example, both expressed in pounds per square inch. If this is the case, $Z$ is calculated and the tables entered as described above.

We can now use the Haugen equation to find $Z$ for our example.

*Strength curve:*

$$u_1 = 34,000 \text{ lb/in.}^2$$
$$s_1 = (38,000 - 34,000)/3 = 1333.3 \text{ lb/in.}^2*$$

*Load curve:*

$$u_2 = 30,000 \text{ lb/in.}^2$$
$$s_2 = (30,750 - 30,000)/3 = 250 \text{ lb/in.}^2$$

Then:

$$Z = \frac{u_1 - u_2}{(s_1^2 + s_2^2)^{1/2}} = 2.9487$$

and from the normal curve table:

$$\text{Area under curve} = 0.5 + 0.4984 = 0.9984$$

Therefore the probability of failure is

$$1 - 0.9984 = 0.0016 = 0.16\%$$

For this example the probability of failure, even using the entire range determined by the statistical tolerance method, is very small. This means that the tolerance for the diameter can be very liberal.

This example demonstrates the technique of statistical tolerance setting and implies that many normally condemned parts are indeed usable for most situations. The producibility engineer must be aware of this to prevent excessive manufacturing costs.

The method of statistical tolerance setting should be used where the costs of individual parts are high. It is scientifically accurate and will lead to lower scrappage rates. However, to use this technique for all tolerances would be a problem. Not all parts are designed with stress analysis as a part of the procedure. Therefore this technique, like so many others, cannot be applied universally. To try to force-fit the data would detract from the producibility engineer's productivity. The statistical concept has been illustrated here to show that creative technical solutions are possible and can indeed lead to savings if properly applied. As computer numerical techniques become simpler to use, such techniques will probably become more commonplace.

## ATTRIBUTES OF GOOD DESIGN USED IN THE CONCURRENT ENGINEERING PROCESS

In order to perform the producibility function it is necessary to understand design procedure. Concurrent engineering covers many aspects of product development,

---

*The normal curve shows a 3 sigma spread; hence we divide by 3 to obtain the standard deviation.

from conceptualization to delivery of the finished product, but central to this team-work process is the design. Design is absolutely critical to success. It is the plan for producing. If the plan is good, most likely it will succeed and be profitable. If the plan is bad, no amount of heroic effort is sufficient to make it succeed. This is why we need to consider attributes of design as part of any discussion about concurrent engineering and the role of the producibility engineer.

## A Definition of a Design

A design is a combination of many things. First of all, it is a plan for achieving a goal. A goal can be anything. To give an extreme but totally proper example, successfully staging Mozart's opera "The Magic Flute" would be a goal. The producer and director would have to conceive how to do it to achieve the goal. This would be the design—how to meet their interpretation of the artistic merits they wish to portray.

When engineers think of design we narrow the definition. Design to an engineer usually means physical things, such as machines, vehicles, electronic devices, and virtually all the amenities of life. So, in reference to what factories do, a design is a plan that, if followed properly, will result in a physical thing that performs a function the designer intended it to do. In its most common form, the design is recorded in an electronic data base or, more traditionally, in a set of drawings and explanatory notes.

A design is also an expression of creativity. Designers are given a goal to meet and they use their experiences and education, with a good deal of intuitive creativity, to conceive an approach for reaching the goal. If we asked several people to move a 5-pound block from point A to point B, we would probably get as many unique solutions as there are people. This is a manifestation of the uniqueness designs produce. Within the imposed constraints, there will be many "correct" solutions to achieving the desired goal. The designer, using his or her skills and creative abilities, would offer a method for moving the block, and the method would be sufficient. A designer with a mechanical background might offer a solution based on principles of hydraulics. An electrical engineer might opt to move the block using an electric motor geared to a rack-and-pinion screw lift. Both would be appropriate solutions to the block problem. The thing all these solutions would have in common is their uniqueness, and because there are many "right" answers, discussing design is an abstract subject.

When we talk about the constraints, the practicality of the approach must be recognized. Here we begin to think of producibility. There are designs that are extolled for their simplicity and beauty. There are other approaches to design that we berate as "Rube Goldbergs" because they are too complicated and not pleasing to our eye or psyche. We usually praise simplicity because it is makable in the factory; it is producible. We hear terms such as functional, esthetic, and pleasing to the eye. These can be thought of as ways of saying the designer has created a scheme for achieving the goal that can easily be used by those responsible for doing it. In other words, manufacturing can make the product with only reasonable applications of their skills. This we judge to be a good design, a producible one.

## The Set of Design Attributes

A good design has a set of attributes, and all good designs share the same attributes. Let's look at them, and then we'll investigate the various phases of the design process and their compatibility with the attributes. The five recognized attributes of design are:

1. Producibility
2. Simplicity
3. Lowest feasible cost
4. Esthetically pleasing
5. Meet quality requirements

1. *Producibility.* First, we must understand that a design, above all other requirements, has to be producible. Obviously, if the design cannot be produced, all other attributes have no meaning. Attributes 2 through 5 come to naught if the design is insufficient for the factory for which it was intended. The simple fact is that if the design is not producible, we have set up a no-win game.

2. *Simplicity.* Simplicity is the commonsense attribute. The desire to make a design producible virtually dictates that it be as uncomplicated as the physics of the situation allows. In the industrial environment this is affectionately referred to as the KISS principle. KISS means "keep it simple, stupid." This is an admonition to bright people not to demonstrate their technological mastery of their specific segment of engineering theory, but to deliver a design that serves the intended purpose and no more. We do not want to admire the clever nuances the designer employed to solve the design problem, unless they were truly needed to reach a successful conclusion. We do want straightforward understandable instructions that make use of the factory's capabilities. This is simplicity.

3. *Lowest Feasible Cost.* The cost attribute must be met. This means not only the targeted cost to produce but also the cost the company intends to charge the customer for the product. If the design results in a product that is beyond the means of the intended purchaser, it is a failure. Imagine if basic transportation, the "plain" automobile, had a price tag of $100,000.00. If this was the best that could be done, how many of these automobiles would be sold? Precious few indeed! We must understand that taking cost out of a product is not just a nice thing to do, but a necessary thing to do.

4. *Esthetically Pleasing.* Esthetics are also important. We do not live by bread alone. This can be a difficult attribute for engineers to comprehend. For some products we can safely ignore the need to consider esthetics, but it is difficult to say where the dividing line. However, we all know that some classes of products are virtually exempt from this requirement while others are not. For a screwdriver, esthetics is a minor concern. As long as the screwdriver performs its function,

the design is satisfactory. Grand pianos are another matter. Here we require not only supreme functionality but also demand beauty of appearance. Here esthetics are very important. There are no guidelines for determining when esthetics are an important attribute, but failing to consider this can have significant negative consequences.

5. *Meet Quality Requirements.* This attribute sounds deceptively simple, but it is not. That the design must meet the quality requirements is an undisputed fact. However, what is the quality requirement? Is it that manufacturing meets the design specification? Or is it the pleasing appearance of the product? Or is it achieving the goal of the customer? It is all of these requirements—that the product is defect free, works correctly, meets the hopes of the user, and perhaps meets the expectations of admirers. Also, there may be government regulations that have to be complied with, even if the customer really does not care or is oblivious of the need. So we see that quality attributes vary depending on the viewpoints of various constituencies.

## The Process of Creating a Design

Creating a design is very much a multifunctional task. This is something we are just coming to recognize. As recently as fifteen years ago this statement would have been vigorously disputed. Then we would have said that design is simply applying scientific principles to form and function, thus defining what should be produced. This is no longer a valid definition. There are three components of design: predesign, engineering phase, and postdesign. Predesign is the definition of the customer's needs and desires and is coordinated by marketing. The engineering phase is the traditional applying of scientific principles to achieve a workable plan for a product that meets marketing specifications. Postdesign is dominated by the customer service component of marketing and involves helping customers understand and properly use the product they have purchased.

## The Engineering Phase of Creating a Design

The engineering phase consists of three components done in chronological order and concurrently iterating after the preliminary work is done:

1. Concept design
2. Producibility design
3. Manufacturing facilities design

Keep in mind that all phases of design, to be successful, must be compatible with the five attributes we discussed previously. Let's look at the engineering phase in more detail.

1. *Concept Design.* The concept design is the rationalization of an idea in terms of science. Concepts are ideas for product offerings. These ideas can be new, such as when SONY introduced the commercially viable videocassette re-

corder. Or the idea can be an extension of or an improvement on an existing idea. The introduction of the automatic transmission for automobiles by Oldsmobile would be an example of the latter. In the concept design stage, we have to face the reality of scientific facts. This is the defining of the product in accordance with the laws of science. For example, when designing a ship, one must make sure that Archimedes' principle of buoyancy is complied with. We certainly cannot have a hull form that will not float. The task here is to make sure that the concept has been thoroughly vetted. It has to have been well conceived and thoroughly evaluated to make sure the goals of the design are achievable. The ideas employed have to be compatible in all aspects with the relevant science. Once a design is found to be based on science and is well established, we can begin to consider producibility.

2. *Producibility Design.* The primary questions that have to be answered are where will the product be built and what are the capabilities of that source? The designer and the entire concurrent engineering team must make sure that whatever information is transmitted to the source factory is understandable and within the source factory's "normal" capabilities. If not, the probability of the design being executed faithfully is severely reduced. The producibility design phase is the process of "customizing" the design for the production source. Not doing this customization would more than likely lead to substandard production results. The process for achieving a producible design evolves about the design review process as explained previously.

The producibility design is not the end of the engineering phase of the design. Even though we have now tested the design for compatibility with the factory and we know that the requirements meet both the idea of the product and the practicality of building it, we are not done. Now we have to be concerned with fitting the product within the factory. This process is called the manufacturing facilities design.

3. *Manufacturing Facilities Design.* The manufacturing facilities design process is sometimes called methods engineering. You might ask, if the design is producible isn't this a redundant step? Absolutely not. Through the producible design development we have tested the concept design to see if it fits within the envelope of practicality. We have simply evaluated the design to see if it can be done in the factory it was intended for. Now we need to do the nitty-gritty work of tooling up the factory. Because the design is producible, we know we do not have to reinvent the factory. The tooling, although it could be very creative, does not have to be heroic to do the job. The manufacturing facilities design, then, is the specific designs for jigs, fixtures, and sometimes processes that are necessary to implement the producibility design.

## SUMMARY

We have reviewed the role of the producibility engineer in the modern industrial organization. The working relationship between design engineering and manufacturing within the confines of concurrent engineering theory has been demonstrated. The role of the producibility engineer within the overall umbrella of concurrent engineering has been demonstrated to be that of the manufacturing engineering representative in the team process of conceiving and creating a product.

Creating producible designs for the factory is important to a manufacturing company's success. Without this activity, the probability of suboptimal production costs is greatly enhanced. Another important aspect of the producible design is that even though it is approached from the viewpoint of favoring optimum factory processes, it does not sacrifice the designer's basic intentions or increase the quality risk. On the contrary, producible designs tend to improve product quality and, because they require only needed processes, thus a simplifying technique, enhance the elegance of design.

Statistical setting is a method of monitoring and judging usability of assemblies. It evaluates risks objectively so that the engineer can make judgments based on facts. Setting of tolerances cannot be a game played between the designer and the factory to see who gets the greater level of insurance protection against risk of design failure. It has to be a win-win game for all parties; otherwise, the true loser is the entire organization.

The producibility design review is the focus of synergism within the concurrent engineering philosophy. Without it, costs of products could be considerably more than necessary, which would reduce the company's market and lower its profit potential. The producibility engineer is the catalyst for successful design reviews, creation of producible designs through practice of attributes for good designs, and the setting of realistic manufacturing requirements.

## REVIEW QUESTIONS

1. Define concurrent engineering and the role of producibility engineering within the overall concurrent engineering concept.

2. Discuss the concept of producible design. Explain why a design engineer would often require the counsel of the producibility engineer to create a design that can be produced in the factory at reasonable costs.

3. Explain how tight tolerances increase costs. Devise a scenario in which tight tolerances are justified.

4. A design for catalytic converters for automobiles is to be produced in a factory. Make a list of questions that the producibility engineer should ask before reaching any conclusion about whether the design is a workable one.

5. Explain what the statement "minimize the risk, then find a way to do it" means, and how it applies to introducing difficult-to-manufacture projects in the factory.

6. The drawing for a steam valve head joint bolt is shown below. Perform a producibility engineering analysis, answering the following questions:

(a) What is the part for?
(b) How does it work?
(c) What engineering analysis is important to the design?
(d) How can the part be made?
(e) What can be done with the design to make it cheaper to produce?
(f) Will the part still do the job if the producibility engineer's recommendations are accepted?

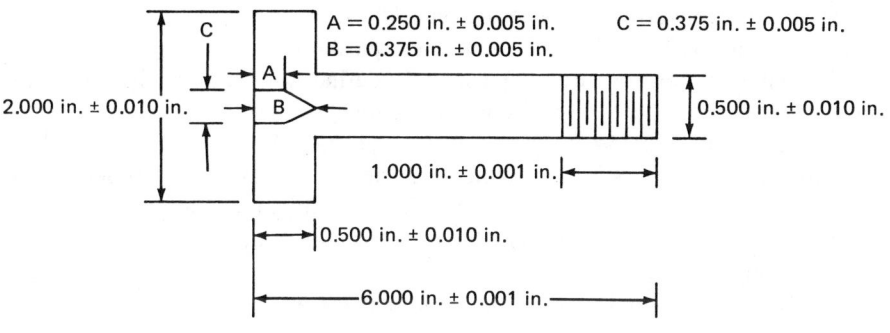

A = 0.250 in. ± 0.005 in.    C = 0.375 in. ± 0.005 in.
B = 0.375 in. ± 0.005 in.

2.000 in. ± 0.010 in.

0.500 in. ± 0.010 in.

1.000 in. ± 0.001 in.

0.500 in. ± 0.010 in.

6.000 in. ± 0.001 in.

Class 3, UNC fine thread, 8 threads/in.
Taper 0.001 in./in. ± 0.00001 in.

7. Perform a producibility engineering analysis of the following part. Use steps (a) through (f) of question 6.

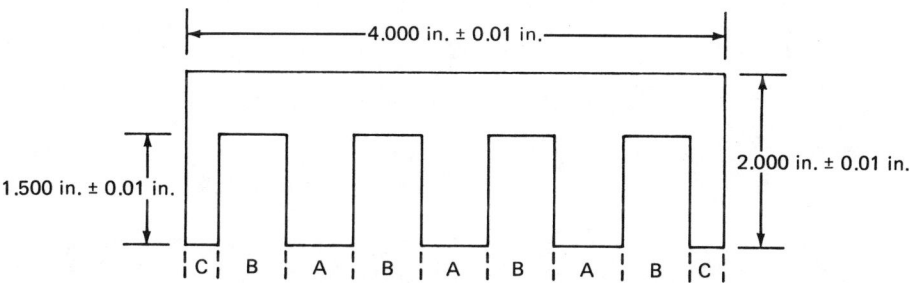

4.000 in. ± 0.01 in.

2.000 in. ± 0.01 in.

1.500 in. ± 0.01 in.

C   B   A   B   A   B   A   B   C

Heating element for toaster oven
Required: 100 watts/in.
Thickness = 0.020 in.
Material: silicone steel
A = B = 0.500 in. ± 0.001 in.; C = 0.250 in. ± 0.001 in.

8. Explain why dimensional tolerances must be converted to strength values in using statistical tolerance techniques.

9. A stud must carry a steady-state load of 25,000 lb/in.$^2$ with a cyclic load of ±1,500 lb/in.$^2$ Assuming that the basic stress mechanism is bending-induced shear, determine the diameter of the stud and the tolerance range. Also determine

the probability of failure if the material used to make the stud has a strength of 30,000 lb/in.$^2$ with a range of 4,000 lb/in.$^2$ and the shear force is 50,000 lb.

10. Repeat question 9 if the steady-state load is 27,500 lb/in.$^2$ and the cyclic load is ±2,500 lb/in.$^2$

11. Discuss limitations to the use of statistical tolerances. State reasons for not using statistical tolerances in all situations.

12. With reference to question 6, outline a procedure via a list of specific items to be covered for a design review to be held between producibility engineering and the concurrent engineering team.

13. Repeat question 12, using data from question 7.

14. Using the design data of question 6, define how each of the five steps of the attributes of good design is complied with.

15. Repeat question 14 using the data of question 7.

16. Referring to question 6, define what elements would be considered concept design, producibility design, and manufacturing facilities design. The design evaluation questions (a) through (f) of question 6 may apply to more than one of the three design subsets.

17. Repeat question 16. using the data of question 7.

## METHODS, PLANNING,
## AND WORK MEASUREMENT

Methods, Planning, and Work Measurement (MP&WM) is the design of work performance on the factory floor to optimize output of the product. The function of methods engineering is to convert broad-based methods or procedures into detailed, easy to follow plans for the workstation operator. These plans must be detailed enough to specify the tools to be used, the materials to be used, the sequence of events, and even the time to complete each event.

The MP&WM unit consists of methods engineers, planning specialists, and time standard analysts. Methods engineers create the broad-based sequence for producing the part. The planning specialists then create the detailed instruction sheet from which the operator will do the work. The time standard analysts work with the method sheets to determine the time it should take to perform each operation. This involves study of the task to be performed at the workstations to determine what the optimum body movements should be to minimize the time and effort required. Finally, the MP&WM unit will have methods engineers who are totally involved in measuring the performance of the factory against objectively set standards. For an MP&WM unit to work effectively, smooth transitions must be made from phase to phase in this sequence from broad-based to particular. How these transitions are made and how the disciplines work with each other to produce manufacturing instructions for shop operations are the subjects of this chapter.

# THE SCIENTIFIC METHOD APPLIED
# TO MANUFACTURING ENGINEERING

The purpose of the MP&WM unit is to produce a manufacturing plan, utilizing the existing facilities, to optimize production of the company's products. This is not a "by guess by golly" procedure. The philosophy of the activity is to use the scientific method:

1. Make observations.
2. Develop a hypothesis.
3. Test the hypothesis.
4. Make revisions to the hypothesis based on the test.
5. Test the revised hypothesis.
6. Reach a workable conclusion.

Let us demonstrate the scientific method in terms of a factory operation problem.

A new producible design is transmitted to manufacturing to produce. The methods engineer devises a sequence of events, called a method, to produce the product. This method is turned over to the planning specialist to create a detailed written plan for the shop to follow. The detailed plan will include times to complete each step, which will be obtained from the time standard equations or computer programs developed for the workstations by the time standard analyst. This series of activities constitutes steps 1 and 2 of the scientific method; observations are akin to studying the design, and developing a hypothesis is the work of creating the method and a written plan.

The written plan is given to shop operations, which will produce the part. Since it is the first time the part is being made in the factory, the methods engineer will be present at the workstation as a combination observer and consultant. At this point the method is being tested to make sure it is the best one possible and optimally produces the part. This corresponds to step 3 of the scientific method.

Having carefully observed the prototype production, the methods engineer now has practical experience to compare to the method and can make necessary adjustments (step 4 of the scientific method). Adjustments are necessary because, with prototypes, it is usually the case that several items are unknown and assumptions must be made.

**Example 1:** The method is derived on the basis of feeds and speeds being delivered precisely by the machine tool. Quite often, machine tool performance varies with the condition of the machine. Therefore modifications could be required in the method.

**Example 2:** The part is produced with a cutting tool from a different vendor, which was thought to be generically identical to a previous supplier's tool, but is not. A change in method is needed to accommodate the different results.

**Example 3:** The cutting tool is not properly sharpened by the operator. This could cause significant variation, especially in stationary or spade drilling activities, where the drill blade is held stationary and the part is rotated. If the methods engineer considered a mean time to perform the operation based on a particular level of drill blade sharpness and the actual mean time was different, a modification could be called for.

Once the revisions to the written planning document are made, the factory is ready to try again on a second workpiece or series of workpieces. Again, the methods engineer will observe the manufacturing activity and offer consultation where required. This is step 5 of the scientific method. The methods engineer is testing the revised hypothesis, that is, the method, and will continue to revise the method until an acceptable manufacturing procedure is developed. This iterative process is continued as long as it is economically justifiable to do so. For instance, adjusting the method to bring it within 99% of optimal may not be economically justifiable if it is already at 95%. The termination of the experiments with the method is similar to step 6 of the scientific method.

This procedure is carried out continuously by the MP&WM staff. It is done so routinely that its practitioners often forget that it is an application of the scientific method. It is the optimizing technique used by the methods engineer and the supporting staff of planning specialists and time standard analysts to find the best possible way to produce the product.

## METHODS ENGINEERING

Methods engineering encompasses all the activities of the MP&WM unit: tool design, fixtures, setup optimization, time standards, feeds and speeds, detailed planning, operator training, and many other related activities. In this section we will discuss only the activities of the methods engineer as defined earlier. The work of the planning specialist and time standard analyst will be covered in later sections of this chapter.

### The Methods Sheet

Methods engineers are experts in what can be accomplished at workstations, including machine tools and processes. They design the manufacturing technique. They do this by carefully reviewing the design information, ferreting out specific details that are critical for the functionality of the completed part, and specifying the individual steps to be taken to complete the manufacture. Figure 7.1 is an example of a broad-based methods sheet. This is considered a broad-based method for two reasons: it assumes that the operator is familiar with the workstation and the typical parts going across the workstation, and no time to perform or allowance is calculated for each step. Nevertheless, it includes such detailed instructions for the operator as what type of tools to use and even where to get the tools. There-

---

Methods Sheet
Sealing Rings Setup and Metal Cutting Data

1. Change spindle if required, use wherever possible Setco no. 3615 spindle for increased rigidity.
2. Clean table of machine and take from cabinet (4) 3-in. square parallels. Wipe clean and position on table at 90°.
3. Put in table T-slot nuts. Assemble studs to nuts and clamps.
4. Pick up sealing ring. Wipe ground surface, clean, and place on parallels.
5. Assemble 3-in. travel indicator to magnetic base. Mount on inside of machine splash pan, positioning indicator to contact sealing ring.
6. Rotate table by hand. Note dial reading on indicator. Tap sealing rings with no. 4 Compethane mallet as indicator reading increases to plus.
7. Repeat above 3–4 times until indicator is zeroed in.
   7A. Should sealing ring be distorted or out of round, center ring until readings at 180° are same.
8. Position clamps on sealing ring as per sketch and secure snug only.
9. Recheck runout. Adjust if required. If part did not move, secure tight.
10. Recheck runout.
11. Mount borozon wheel to spindle if required.
12. Move spindle toward left side of sealing ring.
13. Set stops on vertical slide, top and bottom to clear portion being ground.
14. Set bottom positive stop and check stroke of vertical head.
15. Set rotation of table clockwise.
16. Move spindle toward left side of scaling ring so borozon wheel is within .020 of sealing ring.
17. Start table, set table speed at 20 RPM.
18. Turn cross feed wheel to feed borozon wheel into sealing ring slowly and continue to feed in until wheel starts to move. Note reading on handwheel, back off and repeat to verify touch off point and recheck number on feed dial. Then back off approximately .010.
19. Set vertical slide control valve to read A-2 which equals 5 ¼ in./M.
20. Set automatic feed at .0005.
21. Turn on coolant.

---

**Figure 7.1** A broad-based methods sheet.

fore, it is broad-based only in comparison to the planning details and time standard documentation. Conversely, note that when work is to be performed, either setup or production, the operator is not told specifically how to do it. For example, in Fig. 7.1, step 11, "mount borozon wheel to spindle if required," the methods engineer is not specifying how that is to be done, but merely states when in the chronological order of events it should be done and assumes that the operator knows how to do it. In a detailed planning sheet the actual sequence of mounting the borozon wheel with the time allowed to accomplish it would be adequately described.

## The Workplace Layout

The methods sheet is a key output of the methods engineer, but not the only output. In addition, a competent methods engineer also produces a workplace layout, as shown in Fig. 7.2. This layout differs from one produced by advanced

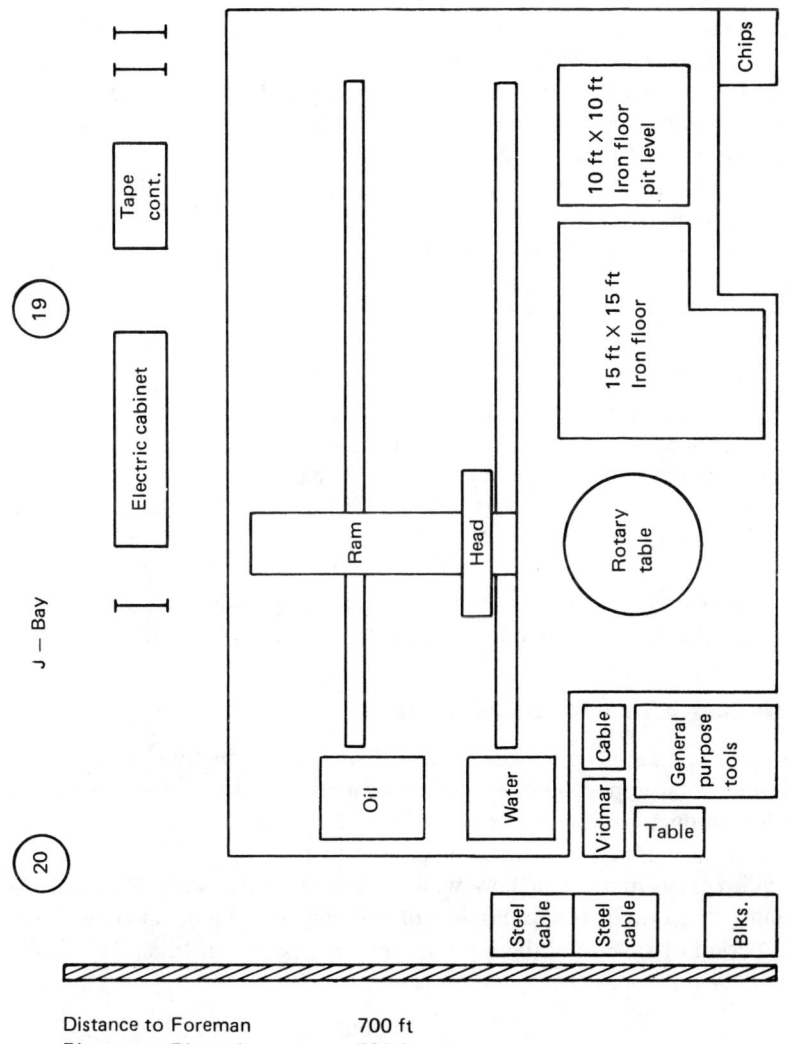

| Distance to Foreman | 700 ft |
| Distance to Dispatch | 700 ft |
| Distance to Tool Crib | 700 ft |
| Distance to Crane Call Board | 320 ft |

**Figure 7.2** Workplace layout.

175

manufacturing engineering in that it contains data pertinent to operating the workstation at an optimal productivity level. The workplace layout is a schematic of the workstation showing the primary and secondary equipment at the particular location.

Primary equipment is the main producing equipment. In Fig. 7.2 the primary would be the rotary table, $15 \times 15$ ft iron floor, $10 \times 10$ ft iron floor pit level, tape control, ram, and head. Secondary equipment is the auxiliary equipment. In Fig. 7.2 it would be the vidmar cabinet, oil and water storage tanks, tables, electric cabinets, chip collector dumpsters, and tool cabinets. Auxiliaries are usually items not directly related to working on the part.

Other interesting items are found on the workplace layout, including (see Fig. 7.2) information such as how far the dispatch cage is from the workstation and the distance to go to sign up for a crane lift.

The workplace layout is oriented in the factory by designating its location with an address. The circled 19 and 20 in Fig. 7.2 indicate the location in J-Bay where this workplace is, for example, its address. This information is vital to the time standard analyst, who needs it to determine the optimal time to perform work at the workstation. The time it will take to move around the workstation and the time to obtain services are included in allowance factors, which are added to the actual performance times to obtain total optimal times.

With the machine tool characteristics supplied by advanced manufacturing engineering, the methods sheet, and the workplace layout, the methods engineer has compiled a complete description of the workstation and its capacity or capability. From this broad-based but detailed data base, specific plans for producing the product and measuring the efficiency of the work in process are generated.

## Productivity and Efficiency Measurements

Another task of the methods engineer is to evaluate the productivity level of the workplace. Since a large portion of manufacturing costs is due to the labor input required during production, it is important to know how well the labor force works in order to know how much to charge for the product.

The simplest measurement of how well the factory is working is output divided by input. The output is the number of product or component parts made over a finite period of time, and the input is the finite period of time. This is the basic productivity measurement:

$$\text{Productivity} = \frac{\text{output}}{\text{input}}$$

Usually the denominator of this equation is kept constant and is taken as 8 hours, the standard time for one shift of work in a factory. Therefore, if the output of the workstation is 8 units per shift, the rate of production is 1 unit per hour:

$$P = \frac{8 \text{ units}}{8 \text{ hours}} = 1 \text{ unit/hour}$$

This system is useful for measuring trends in factory performance, with results recorded on a productivity chart. The input value is almost always 8 hours; however, this may be adjusted to account for unusual circumstances. For example, if a machine breakdown caused the productive time available to be reduced to 6.5 hours, the denominator in the productivity equation would be 6.5. Or if a shift was extended or overtime applied, the denominator could be as high as 12 hours.

Methods engineers use productivity charts to judge whether the workstations are producing at acceptable rates and to evaluate the worth of overtime. In Fig. 7.3 the notation O/T for days 5 and 6 means that overtime was used to meet the daily piece count requirement. It is apparent that on these days the rate per hour decreased to the lowest level of the measurement period. For both O/T days the workstation produced $12 \times 3 = 36$ units per day. If the piece count requirement was less than 36 units per day, the overtime could be justified. Otherwise, it was an unjustified expenditure of labor hours. For comparison, on day 9 the workstation produced $8 \times 6 = 48$ units.

Productivity charts may also be used to investigate method adherence by the operator if the rate was less than anticipated. In the case of the overtime low rates in Fig. 7.3, the methods engineer would investigate adherence to the prescribed method and also whether the extended shift introduced an unfavorable fatigue factor.

Productivity rate measurements are simple and easy to apply if discrete work can be measured at the workstation. They are useful for relatively long production

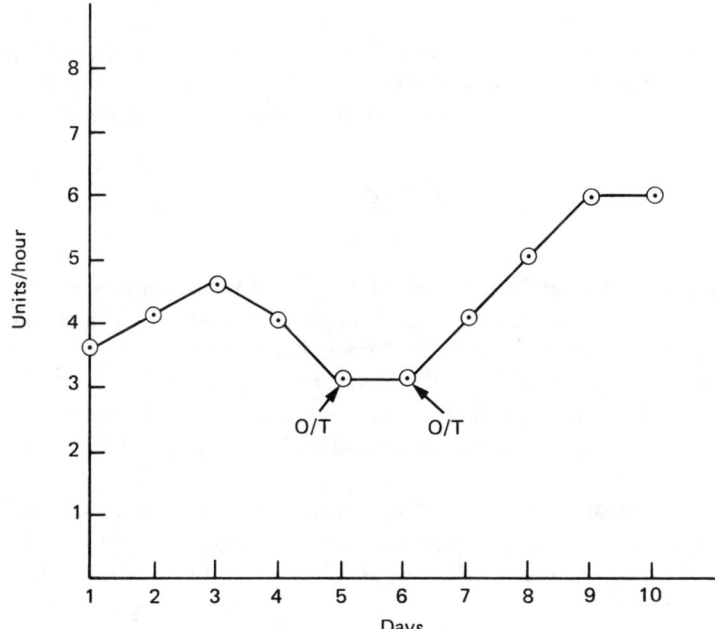

**Figure 7.3** Productivity chart: Workstation 932.

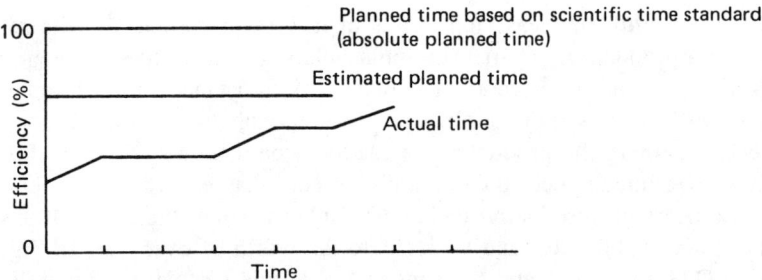

**Figure 7.4** Efficiency measurement.

runs where a specific type of part will be produced repeatedly. For example, such measurements can be employed in the production of electric motor coils or stampings or in drop forging manufacturing. They would not work in a machine shop where there are daily or weekly variations in the size of the parts going across the workstations and the length of time it takes to perform the work. For this type of work the efficiency measurement is used.

Efficiency measurements depend on predetermination of the time it should take to do a specific segment of work. Therefore, rather than specifying how much product will be produced in a finite period of time, the efficiency method of measurement specifies how much time a specific segment of work should take and compares this with the actual time taken.

The mechanics of the efficiency measurement is as simple as that of the productivity rate measurement. The measurement is planned time divided by actual time, giving a numerical value that is almost always less than unity. The reason is that the planned time is the ideal time corresponding to everything being done optimally with no interferences or aberrations occurring during the performance of the work.

$$\text{Efficiency} = \frac{\text{planned time}}{\text{actual time}}$$

Efficiency measurements can be either absolute or relative measurements, depending on how the planned time is arrived at. The planned time may be a simple estimate made just prior to the assignment of the work, or it may be the result of a scientific document taking into account the physics of human body movements and placement of the tools and fixtures at the workstation. The latter type of planned time is called a scientific time standard and will be discussed later in this chapter.

There has been a debate within manufacturing management as to whether scientific time standards to determine planned times are worth the effort. The scientific time standard gives management an absolute value to measure performance against; the question is whether it is really worth knowing how far from the absolute optimum an operation is.

Figure 7.4 shows three efficiency lines. The top one is the planned time based on a scientific time standard. This is the absolute value. It represents the least possible time to do a job because it is the sum of the times it should take optimally to do subsets of the job.

The second line is the estimated planned time to do the work at the workstation. This may be derived by calculating the feeds and speeds to determine how much time the actual machining will take, then estimating how long the setup and teardown times will be. By adding these together, the estimated planned time to do the job is obtained.

The third line, the irregular one, is the actual time to do the work plotted on the chart. Notice that the efficiency value of the actual time is approaching that of the estimated planned time. Also notice that the efficiency value of the estimated planned time is significantly lower than that of the absolute planned time, which is the 100% efficiency value. The question then is whether management should be satisfied with asymptotically approaching the relative measurement or should be appalled that the actual measurement is so far below the absolute planned time.

The answer to this question has to do with how much money is involved in minimizing the spread between the lines. The solution lies in the return on investment. If the estimated standards yield the targeted profit and market share, then lowering costs to gain more profit and a larger market share may not be desirable. Gaining a larger market share may be a poorer return on investment than other opportunities available to the company. On the other hand, lowering operating costs may be essential just to stay in business. This would be particularly true if significant pressure on sales is being felt from foreign competitors who start out with a lower labor cost base. Therefore, the question of whether to use scientific time standards must be evaluated in the same way as any proposal to invest funds to improve productivity.

Regardless of the choice eventually made, an absolute standard focuses attention on minimizing costs to an absolute value, while a relative standard focuses attention on minimizing costs to a relative goal. The relative goal is almost always easier and cheaper to achieve than the absolute one but is never worth as much.

## Methods Analysis

Now that we have surveyed the activities of methods engineers, we will put the key parts together, using an example to demonstrate how methods engineers solve manufacturing problems. This example is a demonstration of the systematic technique called methods analysis, used to produce a manufactured component.

Let us assume we are asked to make a simple screwdriver in lot sizes of 1000 pieces and the factory available for use has all the necessary equipment to produce hand-held tools. The methods engineer must:

1. Determine how to make the screwdriver.
2. Develop tooling and fixtures to optimize production.

3. Develop a useful measurement scheme.
4. Estimate the time it will take to produce the required amount of product.

As we will see, a great deal of the work can be classified as subjective, but it is subjective only in the sense that the approaches used by the methods engineer are based on previous similar experiences. Let us now take each step separately, even though they may develop concurrently.

**Determine How to Make the Screwdriver.** The methods engineer will consult the drawing for the design, which is assumed to be a workable design, and extract all the physical factors such as dimensions, shapes, materials selected, assembly requirements, and any other special requirements specified by the designer. For this example we are producing steel-shanked screwdrivers with wooden handles. Therefore, the manufacture can be divided into three areas: making the shank, making the handle, and assembling the handle to the shank.

For the shank the methods engineers can envision the following sequence:

1. Cut steel rod, $\frac{3}{8}$ in. in diameter, to length on a saw.
2. Rough turn the entire length and thread opposite to the blade end on a lathe; class 2, UNC* coarse thread 8 threads/inch.
3. Form the blade on a grinder to a preset template.
4. Grind the entire shank to a commercial finish or, if specified, to the finish required on the drawing.

For the handle, the following sequence could be proposed:

1. Cut $1\frac{1}{2}$-in. wood dowels to drawing length.
2. Produce the exterior shape on a wood lathe.
3. Produce the internal thread for mounting the shank on a wood lathe; class 2, UNC coarse thread 8 threads/inch.
4. Paint exterior surfaces in accordance with wood finish requirements specified on the drawing.

Next, it is necessary to create the assembly method. A suitable sequence could be:

1. Put glue on threaded portion of the shank.
2. Hold shank in a viselike device.
3. Thread handles onto shanks.
4. Let dry.
5. Package.

*UNC refers to a specific thread design standard found in most reputable machine design reference books.

We have now created one acceptable way to produce screwdrivers in the desired 1000-piece lot size. The methods engineer must now evaluate this method to see whether there is a better, that is, cheaper, way to do the job.

**Example:** Could the shanks be drop forged instead of doing the grinding and lathe operations? For a lot size of only 1000, the answer would probably be no. But the question illustrates the type of thinking the methods engineer must employ.

**Develop Tooling and Fixtures to Optimize Production.** The method described above shows the basic steps to be taken to produce the screwdriver. Obviously, if left to their own devices, the workstation operators would, over time, develop fixtures and tools to help them do the job. Unfortunately, this would involve considerable trial and error and would be very inefficient. For this reason, the methods engineer must devise new or specify existing fixtures and tooling before the start of production. For the major components, the types of fixtures and tools that a methods engineer would propose are outlined below.

For the shank:

1. A device that would limit the feed of the bar stock to the proper length, eliminating time to measure the length before cutting.
2. For the grinder, an angle fixture to hold the shank so that the proper blade angle is ground.
3. A CNC chucking lathe to rough turn, thread, and finish turn the shank. Once a CNC program is produced and debugged, all shanks will be produced identically.

For the handle:

1. A device for measuring the handle length for the saw cut, similar to the shank device.
2. Templates for the outside shape of the handle, and a plunge depth stop device to regulate the depth of internal thread cut in the handle. (Since CNC controls for wood lathes are rare and expensive, the idea of utilizing a computer-controlled machine to form the handle is rejected.)

For assembly:

1. A special viselike device to hold multiple shanks vertical, with blades down.
2. An overhead or suspended glue applicator with an easy to use trigger system to dispense a predetermined amount of glue on each shank threaded portion.
3. A device to hold the handles in a proper attitude to engulf the shanks and to spin the handles down onto the shanks.
4. A speed drying device similar to a hair dryer to shorten the glue setting time.

5. A fixture to hold shipping cartons so that finished screwdrivers can be packaged rapidly.

As we can see from the above description, manufacturing something as simple as a screwdriver requires considerable creative thought on the part of the methods engineer to optimize production. The methods engineer will always conceive of a total system, theoretically considering all possible uses of labor-saving devices and items that improve cycle time. Whether the devices are actually designed and built depends on the economic justification evaluations. Just as a drop forge method would probably not be used, an automated spinning device would probably be eliminated, because for a lot size of 1000, complex, relatively expensive fixtures would not pay for themselves over the length of the contract. If they do not pay for themselves, then they are not economical and would cost the company more than the possible savings to be generated.

To summarize, the methods engineer initially considers all sorts of fixtures, then hones down the actual plan for implementation of tools and fixtures to those that are essential to the job plus those that yield good savings over the duration of the order.

**Develop a Useful Measurement System.** The methods engineer first determines whether the product lends itself to either the productivity rate measurement system or the efficiency measurement system. In either case, the technique for measuring must be one familiar to the personnel involved, or provisions must be made to train personnel. For the screwdriver example, it is evident that the productivity rate measurement system will apply. It is now a matter of determining how many shanks, handles, and assemblies should be produced per hour. This leads to the fourth and final task to be performed by the methods engineer.

**Estimate Time It Will Take to Produce the Required Amount of Product.** The only accurate way to know how long it should take to produce a finished product would be to do a scientific time standard analysis. Usually, for products that will be made over a long period of time such a study would be commissioned. For a long production run, saving 1 or 2 minutes per screwdriver by studying the method for optimization would result in significant cost savings in the direct labor and work in process inventory areas. Saving over $50,000 per year for a product like a screwdriver is a reasonable expectation.

For a production run of only 1000 screwdrivers before the model is changed, scientific time standard analysis is probably not justified. Therefore, the methods engineer will have to approximate. By knowing the feeds and speeds of the machine-dominated portions of the work, the engineer can determine how long it should take to complete the work. For the assembly portion, which is manual work, the engineer will observe the work being performed on similar types of products, time these operations, and then estimate how much time it should take to assemble the screwdriver. If the factory is one in which timing of workers is a sensitive issue, the methods engineer may have to duplicate the assembly format in a laboratory and time a colleague doing the work. Once an estimated elapsed time is calculated for each segment of the work, the values are summed and the time to complete the screwdriver is determined.

Knowing the direct labor time value for the entire order, the methods engineer can now give management the information needed to adequately price the product. Material is known, labor content is known, and tooling required is known. Therefore, price can be determined by simply adding on the appropriate overhead percentage and required profit margin.

It should be realized that the time to produce the product determined from the observations and machine sequence times does not represent the most efficient method of producing the product. The procedure described above does not judge the validity of the production method, and it is reasonable to assume that there could be a better method. Only a scientific time standard analysis could determine whether the method is optimal and, if not, how it could be optimized.

It should also be kept in mind that quotes usually must be prepared prior to the making of a prototype. Therefore, the observation method cannot be used. In its place, the methods engineer will estimate the amount of nonmachine time in the production cycle on the basis of similar experiences. This leads to further inaccuracies. Again, the only way to eliminate such inaccuracies would be to commission a scientific time standard analysis. This would be done at the quote stage for a large enough project, where for marketing purposes the true costs must be known beforehand.

## SCIENTIFIC TIME STANDARD ANALYSIS

Many consider this aspect of methods engineering to be the most abstract and theoretical. Of all the work performed within the MP&WM unit, it requires the greatest attention to detail. The results usually justify the significant work effort required.

Scientific time standard analysis, the product of a desire to use logic and reason to improve the output of manufacturing activities, began in the earliest phases of the industrial revolution. Such names as Frederick W. Taylor, Frank B. Gillbreth, and Henry L. Gantt are associated with its development. It is not appropriate in this text to present the specific details of methods developed for the actual writing of such standards; rather, we will discuss the underlying philosophy of such techniques. In this section we will present the managerial techniques used in organizing for doing the work of a scientific time standard analysis. If projects are organized correctly, all the data necessary to produce a good standard become available. A standard can then be written using the precise techniques taught in typical work measurement courses and documented in the many good texts on work measurements.

Five tasks must be carried out to produce a scientific time standard, and it is management's responsibility to see that these tasks are accomplished in a useful way. The standards written by time standard analysts reflect the data presented to them, and the quality of these data is determined by how well the five tasks are performed. These tasks comprise:

1. Facility information ⎫
2. Workstation information ⎬ description of the workplace
3. Tool lists information ⎭
4. Methods instructions—description of the work to be performed
5. Application data—code used to translate environment and work to be done into time elements

The first four types of information listed above are derived and supplied by others, not the time standard analyst. The data in item 5 are the analyst's input, derived by using the specific motion standards tables, or their computer program derivative, that tell the analyst how much time it takes to accomplish increments of the work. It is difficult to understand this last task without the background of the first four tasks. Therefore, we will review the tasks in order in the following sections.

## Facility Information

Tasks 1, 2, and 3 describe the physical entity called the workstation. This becomes the description of the environment within which the work will be performed. The time standard developed will only be applicable to this specific location with its unique geometry. That is, there will be variations among workstations in how much time it will take to do the entire job. Although feeds and speeds may be the same at different workstations, the supporting services are not and cannot be. For example, a facility 50 feet away from the inspection station is bound to receive quicker inspection service than a similar workstation 500 feet away. This is vital information for the time standard analyst, who now knows that the unavoidable delays incurred in waiting for inspection will be ten times greater for the distant station than the nearer station. Therefore, the time to accomplish work will almost certainly vary at the two workstations.

In addition to providing the geometric location of the facility within the factory, the methods engineer will obtain a detailed description of the primary components of the facility itself. This will include the specification data for the machine or process from the advanced manufacturing engineer—that is, how the machine or process works, the range of capabilities, and so forth. In essence, this is the type of information placed on the bidding specification when the machine was purchased or upgraded.

## Workstation Information

This information usually consists of the workplace layout (see Fig. 7.2) enhanced with descriptions as required. Such descriptions would include comparisons with similar workplace layouts or peculiarities of services. For example, a peculiarity that does not show up on a workplace layout but could affect the time standard would be an overhead crane causing machine vibrations when it passes, so that finish fine tolerance cuts must be interrupted. A time standard that did not take such an interruption into account would be in error and would result in unrealistic data for shop loading calculations. If the interruption caused a loss of 3 minutes every 15 minutes, failure to include it would lead to an error of 20%.

The example above may be improbable, but it points out the need for the methods engineer to supply complete information to the time standard analyst.

## Tool Lists Information

The time standard analyst will be interested in the performance characteristics of the tools available to the operator, since the time to complete a task can vary considerably with the choice of tooling. Also, the life cycle of tools will vary with the material used.

In addition, the analyst will want to know the physical characteristics of the tools. Such properties as size and weight will affect the time it takes to change tooling. A bulky and heavy tool may have to be moved with a jib crane and sling. Therefore, it would require a considerably longer tool change time than a relatively light tool that could be held in one hand.

If the tools were not adequately described, or if the analyst assumed what the tools would be but was wrong, the resulting time standard could be in serious error. The time allowed to change tooling could be totally inadequate or significantly too long. In both cases the time standard would not represent the true optimum time to perform the task. Again, this points out the need for the good communications between the methods engineer and the time standard analyst.

## Methods Instructions

This information is the chronological sequence the methods engineer designed to produce the part at the workstation. If several products are to be produced at the workstation, a method sheet for each product will be produced and given to the analyst. Obviously, the more products going across the workstation, the lengthier and more complex the time standard will become.

## Application Data

At this point the time standard analyst applies the reference library of motion times to the various segments of the chronological sequence. Although this is the least well understood part of the production of scientific time standards, it is quite simple in concept. The first four tasks outlined above present the time standard analyst with a dossier on the workstation for which the standard is to be developed. The analyst envisions what happens at the workstation, what the parts are like, where the operator stands, what tools are available for use, and where the tools are stored, as well as the requirements for quality and the services available to the operator in making the parts. In summary, with the inputs of information from tasks 1 through 4, the analyst is in the position of an operator being given a job to do and having to determine how it should be done. The analyst then mentally walks through the process, determining what kinds of physical movements must be made to accomplish the task. These movements are the keys to the method, because the analyst must determine how much time it should take to do them and whether they can be further reduced if physical changes can be made at the workstation, or if their sequence is altered. The analyst is choreographing movements in accordance with defined requirements and assessing which movements are superfluous and should be discarded.

Once the choreography is complete, the analyst can use the application data. These data come in many forms and many levels of detail, but essentially they comprise a catalog of times for making various body movements. The information can be as simple and specific as how many milliseconds it takes to rotate an arm, or as complex and general as how long it takes to disassemble an automobile automatic transmission. Incidentally, most automobile repair standard hours are combinations of many body movement times summed together to give a time to perform a generic task.

We can see from the above descriptions that task 5 adds the human element to the definition of the job to be done. We start with hardware information on the facility, the workstation, and the tools; progress through a plan to produce the part, the methods sheet; and finally add the human element, the application data. In summary, the time standard analyst is required to complete a very thorough study of the method sequence and the workplace environment, then apply times to the individual incremental steps necessary to do the specific task. These times are obtained from the feeds and speeds of the machine tool, from human motion timetables, and from geometric locations for delay times. The times are tabulated as fixed or variable. Machine times are variable, as they depend on the size of the part, while installing a tool into a tool holder would be fixed for each specific tool.

If a scientific time standard is being developed for only one type of product to be produced at the workstation, the analyst will construct a table in which the only variable is the size of the part, hence the machine cycle time. If the time standard is for various parts, a variable table for different operator requirements as well as tables for the machine cycle times will be designed. The finished time standard then comprises tables or computer programs that allow the planner to extract 100% optimum times for a wide variety of applications at the workstation.

The scientific time standard differs from the estimated time standard in that all the nonmachining or nonprocessing time is specifically accounted for. Hence it is more accurate than the estimated time standard.

> **Example:** If the process requires an inspection, the estimated time standard could not single out this requirement, but rather would tally it along with all other nonperformance times. The scientific time standard would have cal-

culated the time to call for the inspector and the time to perform the inspection. All this would be incorporated in the portal-to-portal time to do the work at the workstation.

A scientific time standard also acts as a check on the method and quite often leads to an improved method. By looking at the motions involved and placement of the tools, the analyst has a unique opportunity to make improvements in the method sequences and workplace layouts.

Clearly, a scientific time standard has beneficial results. But are the results worth the effort? This is a difficult question to answer. If a job is to be repeated over and over again and taking out minutes is important for overall productivity, or if the job is very complex and must be meticulously designed and competitively priced for bidding purposes, then the effort is probably justified. On the other hand, if the job is simple or nonrepeatable, then an estimated time standard is probably sufficient. It is the gray areas between these two extremes that require careful consideration. Often a good scientific time standard can be used for many other products and workstations as an enhanced or upgraded estimated time standard. One may reason that the new area is not very different from one for which a scientific time standard has been developed. In this case, the estimated time standard may be enhanced by basing it on a scientific time standard, thus upgrading the overall planning activity. In such cases the benefit of the scientific time standard somewhat transcends the original purpose for its development.

## Use of Computers in Creating Scientific Time Standards

As with most engineering applications, there is a role for computers in assisting in the development of scientific time standards. As previously discussed, producing time standards is a process involving meticulous attention to details. It is the accumulation of many pieces of data merged with the energy of the human body movements, expressed in time, to give repeatable calculations of how long it should take to accomplish a task. Computers are most helpful when applied to the task of keeping track of large quantities of data and being able to gather, or classify them, in accordance with a certain set of logic instructions. This definitely fits a need in scientific time standards development. So we can see that it would be a great assistance to the time standards development. It would be a great assistance to the time standards engineer to have a tool available to assist in this task.

The manual way to creating scientific time standards is to analyze the method, then determine the motions the operator has to go through to conform with the method. The analyst must decide whether the method can be modified to require less human motion or motions requiring less energy to accomplish. This is done by looking up time and energy contents of the motions the method requires, doing summations, and finally examining alternative scenarios to see if time and energy can be minimized. The ultimate goal is to use the least amount of time, thus energy, to accomplish the prescribed method. Of course, as well as the human

Here, the analyst is simply merging the machine time with the operator times to perform the tasks.

Let's look at where computers can be of assistance in the above process.

**Analyze the Method.**   We can certainly expect that the method transcribed to the time standard analyst will be in a data base format available for electronic interaction between functions. At the least, this will mean the analyst should always have access to the most current version of the method, reflecting any workstation or product changes. This communication enhancement alone ought to prevent wasted effort on the part of the analyst.

More important, we can use the time standard program to evaluate the method broadly to see if it is feasible, particularly in terms of what is or is not physically possible. For example, the program could easily see if we are asking an operator to do something that is beyond a person's physical capabilities to do. Usually this would involve lifting and placing of parts and calibration of settings beyond certain limitations—for example, lifting more than 20 lb and adjusting the position of a part to better than within $\frac{1}{4}$ in. without precise measurement tools. These may seem to be trivial examples. Unfortunately, they do arise quite often in the real world of manufacturing through simple oversight and end up being the cause of emotional labor-management grievances.

**Determine Motions to Comply with the Method.**   This is historically done by looking up tables of motions and movements and then compiling data on incremental motions with their associated times to make up the whole representation of the method. It involves meticulously looking up data, entering on tabulation sheets, and finally summing the data to get a time to accomplish for the total method. Obviously, this is a detailed library-research-like activity. The more complex the method, the more complex the task of defining the time elements will be. It is also quite apparent that the possibility for error is very high. In fact, recognizing this, most MP&WM units divide time standard work into analyst and checker activities. The analyst determines the motions to be employed and does the initial looking up to get the time values and create the summations. A category of analyst called a checker then reviews the analyst's work to minimize errors before the time standard is issued.

With available computer programs, the analyst enters the motions into the program and the program then does the lookups to get the incremental times involved in performing the method step. This has been made possible by the input of the manual data books of incremental motions and times into a computer data base. Also, in most cases, programmers working with expert time standard analysts have combined many incremental motions into commonly occurring macro motions, or sets of motions. This makes the lookups by the computer program even simpler.

**Evaluate Accuracy and Alternative Methods.**   Using these types of programs makes the task of developing time standards from methods inputs easier. In fact, the need for a checker function has virtually disappeared. Most advanced time standards programs also have built into them what we call impossibility factor eliminations warnings. The programs actively search for combinations of motions or sequences of motions that are either impossible or out of sequence. For example, if by error the analyst specified getting a tool and simultaneously fitting the workpiece into the machine chuck, the program would signal error and perhaps even suggest a proper sequence. Similarly, alternative combinations of motions can be suggested by the program, just as word processing programs can check for grammar and spelling. Also, advanced programs of this genre are often asked to find the least-time combinations to do a certain task. Obviously, human evaluation is still required, but the clerical portion of providing alternative suggestions is simplified.

Use of these types of programs has caused one additional thing to occur. They have lowered the barrier to use of scientific time standards. No longer are scientific time standards relegated to products with high costs or large production runs. It is now feasible to use scientific time standards in limited-volume job shops as well as for marketing quotation purposes. This is all due to the fact that the engineering time and cost to produce accurate scientific time standards have been greatly reduced.

There is no doubt that computer programs in the field of scientific time standard development have given new life to this phase of methods engineering. Before these developments it was thought that using scientific time standards was too much of an academic exercise to have much practical use, other than for very large production runs. Now, with these programs available, expert analysts can apply these techniques economically to virtually all planning and quotation activities. We are at the point where not advocating their use would be virtually surrendering obtainable profits.

## PLANNING OF OPERATIONS

This phase of MP&WM operations constitutes the presentation of information to shop operations so that operators know the exact sequence of the work. The planning operation is the part of the routine manufacturing production sequence assigned to manufacturing engineering and is the only production sequence manufacturing engineering is involved in. Virtually all other manufacturing engineering work is out of the direct line of producing the product. The planning operation, similar to the making of engineering drawings by drafting, is a scheduled sequence that is monitored as part of the routine production control activities.

Planning is the most visible of all manufacturing engineering activities in that failures in planning are vividly demonstrated by lack of capability to produce the required products. Therefore, to ensure high levels of competence in the planning activity, MP&WM management has rigidly structured this activity. That is, the formats to be followed for planning paperwork or computer output are specifically set with little room for variance in procedures and technique. In this way, planning can be reviewed quickly and efficiently before it is issued, and errors of omission and other types of errors are easy to detect.

The planning specialist is responsible for producing complete packets of information for work to be done at the various workstations. This activity involves considerable amounts of paperwork and, more recently, large computer data bases. Planning has four major responsibilities in information dissemination and storage:

1. Review drawings produced by design engineering.
2. Define procedures and processes to be followed by shop operations.
3. Produce operator instruction documentation.
4. Integrate functional operating systems.

These four responsibilities are discussed below.

## Review Drawings Produced by Design Engineering

It would not be possible for producibility engineers to review all the drawings produced by design engineering prior to submittal to manufacturing. In fact, producibility engineers work with a higher level of drawing, the engineering layout. This layout, after being approved by the producibility engineer as a producible design, becomes the data base from which a multitude of detailed drawings are produced. It is these detailed drawings that the planning specialists review for errors, adherence to producibility requirements, and materials usage and quantity. The planning specialist plans how the part described on the drawing will be produced, essentially by combining the methods instruction and time standard for the workstation with the drawing and the required materials. The result is a step-by-step detailed sequence to be followed by the operator.

## Define Procedures and Processes to Be Followed by Shop Operations

Using the detailed drawing and the parts list, the planning specialist identifies the kind of material to be used in producing the part. In some MP&WM organizations this includes determining whether the material is stock material or must be purchased; in other organizations this decision is made by the materials function. Regardless of who specifies whether the material is stock or purchase, the planner specifies the size, thus determining how much work will be required to transform the raw material into a finished part.

Once the material size is selected, the planner describes the specific production sequence to be followed for a specific part at the workstation, based on the overall generic method prescribed by the methods engineer. Usually workstations are set up to produce what is called a family of parts—for instance, parts that are similar in shape but have different dimensions and perhaps different auxiliary components. The planning specialist modifies the generic method to produce a customized method for the specific part.

Finally, the planner matches the customized method with the time standards to obtain the item-by-item time for each detailed step. This is done by using the tables, graphs, computer programs, and so forth that constitute the scientific time standard. Knowing what workstation will do the work, the planner uses the matching time standard to calculate the time to complete each operation. This is different from the work of the time standard analyst in much the same way as the planner's work is different from that of the methods engineer. The time standard analyst produces a generic scientific time standard that contains algorithms which, when entered with specific parameter inputs, yield the time to perform unique operations. Determining and entering the specific parameters are the responsibility of the planning specialist.

To summarize, the planning consists of specifying the material, the size to be used, the detailed step-by-step procedure to be used, the tooling required, and stating how much time it should take to complete the work.

## Produce Operator Instruction Documentation

The procedure just described ultimately leads to the package of instructions given to the operator. This package contains the detailed drawing, the instruction sequence sheet, the tool list, the materials list, and some form of document for recording progress or completion. No other information should be necessary for the operator to understand what has to be done and how to do it.

Figure 7.5 is an example of an instruction sequence sheet. It is similar to the methods sheet shown in Fig. 7.1; the major difference is that incremental times are shown for each step. The instruction sequence sheet is the key document contained in the package sent to the workstation, and it must be correct. If it is in error, operators and foremen will lose faith in the planning and productivity will deteriorate. Inaccurate planning in effect gives shop operations a license to set its own standards. The foreman no longer has a persuasive model for the operator to match performance against, and the tendency is to work at an unsatisfactory pace. The shop will then tend to require higher and higher levels of overtime to compensate. To prevent this from happening, the planning must be accurate, verifiable, and believable by all facets of the manufacturing organization.

## Integrate Functional Operating Systems

Planning is not an independent system. The plans constitute the basic data used by the master scheduler to schedule the factory for extended periods of time, the basic data used by production control to calculate the in-process time at each workstation, and the data used by marketing in quotes to customers for similar products. Therefore, the plans become a very important part of the overall data base.

Most manufacturing organizations are striving to create computer data bases in which the plans are a primary component and the basic production data are entered along with the design engineering data. The importance of such data will

| | | Operation of Element Description | Bearing Caps Fabrication | |
|---|---|---|---|---|
| | | | Variable Hour/Inch | Constant Hour/Each |
| 1. | SU400 | set-up | | .14 |
| 2. | 405 | lay out, assemble, burn, grind, and tack weld | | 3.99 |
| 3. | 423 | assemble and tack weld temporary braces (2 braces) | | .04 |
| 4. | 401 | burn openings in wrapper | .0036 | .021 |
| 5. | 402 | burn crop ends of bearing wrapper | .0111 | .02 |
| 6. | 420 | get ring, deslag, check size, position to front vertical flange or thrust bearing, and tack weld | | .18 |
| 7. | 420 | get cover flange, deslag, check size, position to wrapper, and tack weld | | .25 |
| 8. | 420 | get pipe, check size, position to wrapper, and tack weld | | .05 |
| 9. | 420 | get pad, check size, position to wrapper, and tack weld | | .07 |
| 10. | 420 | get brace, check size, position to wrapper, and tack weld | | .06 |
| 11. | 420 | get rib, deslag, check size, position to wrapper, and tack weld | | .30 |
| 12. | 420 | get trunnion, check size, position to wrapper, and tack weld | | .06 |
| 13. | 420 | get boss, check size, position to wrapper, and tack weld | | .06 |
| 14. | 405 | remove clamps and position cap assembly for weld | | .49 |
| 15. | 418 | lay out, assemble, burn, grind, and tack weld | | 4.03 |
| 16. | 424 | Assemble and tack weld back plate to wrapper and horizontal flanges | | .39 |
| 17. | 421 | get front gib supports, deslag, check size, position to wrapper, and tack weld | | .40 |
| 18. | 421 | get front gib block, deslag, check size, position to gib supports and front vertical supports, and tack weld | | .21 |
| 19. | 421 | get boss, check size, position to wrapper, and tack weld | | .06 |

**Figure 7.5** Instruction sequence sheet.

be described in Chapter 11. As the creators of this data base, the planners are also the inputters into the data base storage system.

Computer data bases are becoming more and more important to MP&WM because they allow standardization of planning, which improves its acceptance by shop operations. By being able to recall existing satisfactory plans and simply modifying them to suit slightly changed conditions, the productivity of the planners is significantly enhanced.

## CONCEPTS OF WORK MEASUREMENT

Now that we have reviewed the basic responsibilities of the MP&WM unit, it is of interest to place these responsibilities within the framework of measurement. The methods engineer, time standard analyst, and planning specialist are all engaged in documenting work to be performed within the factory. They specify the way in which work is to be performed and how it is to be measured. However, how this function is carried out depends on how motivated the operators are to do their jobs effectively. The area known as concepts of work measurement deals with how the hourly work force is perceived and led. This perception will set the stage of how the MP&WM unit is to be managed.

There are two basic systems for motivating the hourly work force: an incentive system and a day work system. In the incentive system the operator is paid for each incremental amount of work performed. In the day work system the reward is continued employment. Both systems have their merits and drawbacks and both affect the way in which MP&WM units approach their work.

The incentive system in many ways is self-regulating. In essence the operators are in business for themselves. They earn their living by producing parts and being paid for each one they produce. Therefore, the operators are management's allies in ensuring that support services are adequate, since lack of such services could cause lack of ability to produce the product. In an incentive system the main thrust of MP&WM work must be in the time standard and planning areas, which determine how much product an operator can produce during the work shift and ultimately how much the operator earns. Methods sheets are less important; the operator will tend to improvise if necessary in order to meet or be under the planned time. Incentive systems result in MP&WM units heavily staffed with planning specialists and time standard analysts, with less emphasis on methods engineering activities.

The day work system can be an authoritative system. The operators know they will be paid for hours worked at the workstation and that they must follow the methods sheet in producing the product. They realize that the only way they can lose their job is to constantly abandon the process they are told to follow; they have only a secondary interest in the number of parts they produce. In this system the methods engineer is most important. The methods engineer's creativity will lead to improved tooling, fixtures, and sequences in how the product is to be produced. The day work system puts a premium on creative engineering to produce optimal results and considers operators to be extensions of the workstation where they work. Day work systems typically lead to MP&WM units that emphasize intensive methods engineering activities at the expense of time standard analysis. These units stress the importance of proper layout of the workstation and making sure the operator follows the prescribed method.

Today there are few pure incentive or day work systems. Most organizations employ combinations of both but favor one. It is common in day work systems to use suggestion plans to elicit ideas from the operators for improving quality and productivity, and to pay the operators for all suggestions adopted. Similarly, incentive systems tend more and more to require good methods engineers to create uniformity of approach to product manufacturing.

The idea of using the experience of operators in developing methods has steadily gained favor. Especially in a pure day work system, not using the accumulated practical experience of workers in optimizing workplace methods would be a waste of that experience. Because they live with the products and their workstations daily, operators come in closer contact with production problems than any other employees of the company. They have had to solve many practical, if small, problems in order to produce their required quotas. Therefore, to encourage operator participation in establishing workplace methods, many companies have established formal programs, which are classified under the general title of "participatory programs."

Participatory programs try to improve factory productivity by having the operators participate in setting goals, or establishing methods, or both. One of their purposes is to establish a sense of proprietorship for how the product will be made and how much will be made in a set period of time. In the day work system this raises the interest level of the operators so that error levels decrease and productivity improves. In the incentive system, the participatory approach leads to modified time standards, which should allow the operator to earn more by being able to produce more, again resulting in improved productivity.

Two common participatory programs are called team production and quality circles. Both of these programs have been applied with success in a wide variety of manufacturing operations. However, they do not replace the traditional responsibilities of MP&WM; they simply enhance the activities of that unit to gain further productivity improvements.

## Team Production

In team production a group of hourly operators are assigned the task of producing a set number of units of a product, usually one requiring major assembly operations. The team is given a specification (similar to an acceptance standard given to a vendor asked to produce a given product), told the maximum time allowable to produce the product, and then left to plan how to do it. Of course, the team members are familiar with the product and can draw on methods and planning documentation already created by the MP&WM unit. This is essentially a group incentive system that replaces incentive systems designed for piece part production. Team production allows short-time-interval assembly operations to be put on a meaningful reward pay system.

The team has full access to all the support functions that would normally be available in a factory, including maintenance, quality assurance, methods, and so forth. While MP&WM does not develop detailed methods instructions, it still has the responsibility of developing general instructions pointing out the optimal ways of achieving specific tasks. However, this is an advisory role. The teams have full control of how they do their tasks and are accountable for meeting minimum production goals within cost constraints.

Many conservative manufacturing managers are uneasy with such an approach to factory management. They are concerned that day-to-day decisions on how the factory is run are being made by less qualified or unqualified people and that the results will be unsatisfactory. They say that they lack control to make required changes, which could be a valid criticism. Team production is not for all factories; it requires a skilled work force, people who are more like technicians than production line workers. The team members must be very knowledgeable about their product, its design, and how it is made. Therefore, this approach is usually found in job shops making complex mechanical, electrical, or electronic assemblies, where the operators are indeed experts. It is also often found in service and repair shops.

## Quality Circles

A much more frequently used participatory program is the quality circle, sometimes also known as the productivity circle or methods improvement team. Like the team production concept, its purpose is to elicit ideas from the people directly involved in making the product and to give them a sense of proprietorship. There are many ways of organizing quality circles. One frequently used approach is described below.

A group of workers from a common manufacturing area invited to join an improvement committee. The purpose of the committee is to suggest ways to improve their work situation and the quality of their product, and to reduce the time and cost to make the product. This group consists of the actual machine/process or assembly workers, support people such as inspectors and maintenance personnel, material movers, technical support people such as manufacturing engineers, designers, and management representatives, usually the foreman. Office clerical staff may also be assigned. The management representative is usually the facilitator, that is, the discussion leader. The leader's purpose is to get the circle to put forth ideas and to discuss frankly whether their implementation will improve working conditions, improve quality, or lower costs. All three of these categories will ultimately result in improved productivity. This is not a brainstorming session, although some concepts of that technique may be used. It is a working session where the circle members draft recommendations to management. These recommendations are recorded in the minutes of the circle's meeting and actions are assigned to various members to be achieved, if possible, before the next meeting. These actions are usually in response to requests for information such as clarifications of design requirements, company operating policies, and business outlooks.

The circle's recommendations are reviewed by the manager of manufacturing for possible implementation. What happens at this step determines whether or not the quality circle activity will be successful. As with its antecedent, the suggestion system, a suitable number of the quality circle's ideas must be accepted or the system will collapse. By not accepting the ideas, management is saying that only it can be creative and innovative. If this happens, the purpose of the participatory program has been aborted. To avoid this, management must accept some of the recommendations. In this way employee interest in the circle is maintained and chances are enhanced that a truly significant idea will emerge.

Even with full and active management support of quality circles, they are rarely maintained for over a year. Once they become routine, their freshness and member enthusiasm begin to diminish, until they are no longer useful. It is important for management to recognize this and plan for a replacement activity. This can take the form of reshuffling people into new circles, creating competition between circles for the most successful ideas, with corresponding rewards, or any other idea that stimulates operator participation in productivity improvement.

Quality circles do improve work methods, often through a summation of small items that the methods engineer was unaware of but that make it easier for the operator to do the job. The ideas usually cost relatively little to implement and improve morale if they are implemented quickly. They create an awareness on the part of the operators that may lead to significant ideas and paybacks, and this is what makes them attractive to management.

MP&WM should never consider participatory programs as substitutes for or threats to the unit and its ongoing responsibilities. No sensible management team would consider an idea-generating scheme such as a quality circle to be a replacement for MP&WM. The MP&WM unit is a professional engineering-based organization with highly trained people dedicated to manufacturing optimization. Participatory programs are supplements to the normal creative and innovative tasks required of a progressive engineering unit. Therefore, they are tools to be used by MP&WM to improve their ability to achieve their stated goals.

## SUMMARY

This chapter has described the activities and responsibilities of the MP&WM unit. The methods engineer, time standards analyst, and planning specialist are intimately involved in the daily workings of the factory, and MP&WM is the direct technical contact with shop operations. Whenever it must be decided how a part will be made, whether the part is being processed correctly, or whether the output level is sufficient, manufacturing engineering via the MP&WM unit will be a key contributor.

The difference between a methods sheet and a planning sheet has been explained; that is, the methods sheet precedes the application of the time standard, which, when completed and merged with the methods sheet, becomes the planning sheet. Scientific and estimated time standards have been defined and their uses illustrated.

Productivity measurements and the concept of work measurements, including participatory programs, have also been discussed. Both piece count and efficiency measurements have their place; the overriding concern is being able to determine the success or failure of an activity. This led us to the concepts of work measurements, the differences in how an MP&WM unit will function depending on the type of manufacturing operation it is supporting, that is, incentive or day work or a combination with some type of participatory program.

The concept of the scientific method was introduced early in this chapter, although it could equally well have been discussed with any of the manufacturing engineering units. The scientific method is a very important concept for success in any engineering organization and is one of the basic techniques applied in manufacturing engineering.


## REVIEW QUESTIONS

1. In general, MP&WM work consists of developing specific instructions for the factory from broad directives. Describe how an engineering drawing can be considered a broad directive in reference to instruction details required by the factory.

2. Use the six-step scientific method to explain how a paper clip manufacturing plan could be devised.

3. Referring to the methods sheet, Fig. 7.1, how many steps provide information only and require no action by the operator?

4. The following are items found at a workstation. Which are associated with the primary equipment and which with the secondary equipment? (a) Tool changer, (b) operator platform, (c) compressed air in reserve tank, (d) hand tool rack, (e) tool spindle motor, (f) inspection gauges.

5. Given the following data, which workstation would have a shorter time assigned to it to produce a given product?

|  | A | B | C |
| --- | --- | --- | --- |
| Distance to inspection station, feet | 500 | 250 | 35 |
| Distance to foreman's station, feet | 30 | 120 | 100 |
| Surface feet per minute capability | 85 | 42 | 30 |

6. What type of manufactured product would make satisfactory use of a productivity rate measurement and what types would not?

7. Describe the differences between the productivity rate measurement and the efficiency measurement.

8. Discuss the advantages and disadvantages of adopting an absolute planned time system versus an estimated planned time system.

9. A company has a contract to produce 16,000 steel claw hammers with steel and taped handles. As the methods engineer in a factory equipped to make such hand tools, outline how the following problems will be solved.

(a) Determine how to make the hammer.
(b) Develop tooling and fixtures to optimize production.
(c) Develop a useful measurement scheme.
(d) Estimate the time it will take to produce the required amount of product.

10. A scientific time standard is needed for the steel claw hammer of question 9. Using the five tasks discussed in the text, explain how the work of the methods engineer will be used by the time standard analyst to produce a scientific time standard.

11. Figure 7.5 represents an instruction sequence sheet produced by a planning specialist. Discuss the difference between this and the methods sheet.

12. Discuss the differences that could exist in instruction sequence sheets for a day work factory and a piecework factory. What factory would require the more detailed planning and why?

13. Explain how scientific time standard computer programs can be used to enhance the accuracy of an instruction sequence sheet and how shop operations would effectively employ this information.

14. Using the data of Fig. 7.5, explain how a scientific time standard computer program would review the method sequence to determine human feasibility to accomplish. List the basic conditions you think the program would search for and compare within its programmed logic steps.

15. Using the data of Fig. 7.5, give your opinion as to what steps, or combination of steps, could form the basis for scientific time standards computer program motion macros. Include your reasons.

# EIGHT

# PROCESS CONTROL ENGINEERING IN JOB SHOPS

Job shops do not conform to the usual concepts of statistical quality control. Since every item made may be distinctly different from the one before or the one following it, typical quality control techniques often fail to achieve the goal of ensuring that the products meet the design objectives. Therefore, individualized process control is essential for success. In job shop process control, analytical evaluations are made for adequacy of design, producibility, shop performance, minimization of costs, and so forth. A fully implemented process control system affects almost all phases of engineering and manufacturing operations and is a vital link in the design engineering/manufacturing interface. This chapter explains how a successful job shop process control system works. We will consider the dynamic nature of the system, how manufacturing losses are controlled, how goals are measured, how some statistical concepts can be utilized, how Total Quality Management (TQM) and continuous improvement concepts are utilized, and, most important, why it works.

## THE PROCESS CONTROL SYSTEM

A process control system is a series of coordinated events in a manufacturing cycle that allows achievement of stated quality and production goals.

Coordinated events can be thought of as process steps or sequenced manufacturing operations, or a chronological listing of things to be accomplished that combine the quality requirements with the shop operations requirements. Essentially, the process control system is a method of specifying who does what, when, and with what required quality measurements or checks. Note that it does not specify how a function is to be accomplished. However, by setting the quality measurement parameters with some forethought, it indirectly specifies, or at least strongly influences, the choice of tool or process.

> **Example:** Assume that we are machining surface plates for heavy machine mountings. The design calls for a 2 × 2 ft plate 2 in. thick made of carbon steel with the surface finished to a 32 RMS finish. This is normally considered a moderately fine finish, and in many cases a variation of 10% would mean little. The process control system can dictate how the surface is obtained. If it specifies that a simple comparison gauge be used to measure the operation, then a simple, single-point tool on a vertical boring mill may be used to perform the manufacturing function. However, if an elaborate electronic surface profilometer is to be employed, something better than a VBM would probably be needed; this would most likely be a surface grinder.

The example above shows how a choice of measuring tools can affect the ultimate quality and manufacturing choice without mentioning tolerances. In a real case, a tolerance would be specified on the engineering drawing and a measuring tool capable of differentiating between the tolerance band would be selected. But would the correct tool have been selected and the requirements stated on the engineering drawing have been met without the pressure of the process control system? This is always a question in job shop activities, where there is no production run of any consequence to fine-tune quality methods. It is evident that with short production runs and individualized products, mistakes cannot be easily absorbed. Therefore, process control in a job shop is vital. Discrepancies must be found and corrected very quickly.

The process control system represented by Fig. 8.1 is a closed-loop feedback system in which the results are reduced manufacturing losses, minimized production delays, and improved product quality. The interesting thing about the system is that it never becomes static. In manufacturing there is no reason why the three outputs should ever reach a plateau or maximum value, because the jobs are constantly changing. Thus more opportunities to improve on past performances continually become available.

The system allows extremely rapid corrective action and long-term solutions for similar individualized products. In a sense, since job shops do not have long-term identical product runs, the process control system has a memory loop that allows corrective action to be taken on future products made by similar techniques. Experience gained on one job can be related to others, and these data contribute to an accurate planning and estimating activity as an offshoot of the process control system. In a high-technology job shop, this planning ability often means the difference between success and failure.

The process control system serves another vital function. In many ways it acts as an avenue of communications between manufacturing and design engineering, between different manufacturing units, and between manufacturing and finance. The closed-loop feedback system virtually guarantees this communication with little or no further attention needed from management.

The six blocks shown in Fig. 8.1 constitute the process control system. Because they interact continuously, it is difficult to isolate one block of activity from

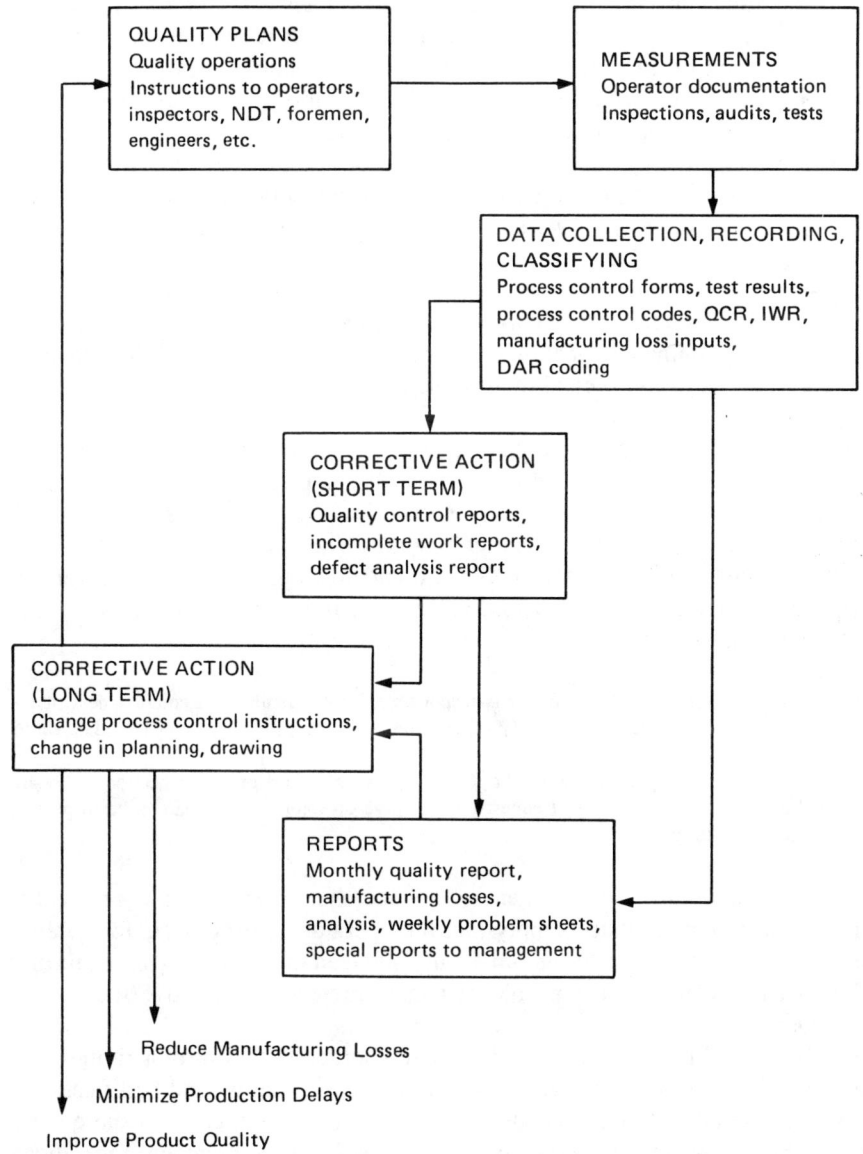

**Figure 8.1** Process control system.

the rest. Therefore, to explain the system, let us assume that we can freeze every-thing in place and examine one block at a time.

## QUALITY PLANS

Who does what, when, and with what quality measurements is spelled out in considerable detail in good quality plans. Quality plans can take many different forms; however, every good quality plan must spell out requirements, assign re-sponsibilities, and be capable of having its specified actions measured. Because job shops handle a great variety of work, their quality plans must be concise, easily understood by all, and complete in themselves.

For job shop quality planning, the "cookbook approach," in which infor-mation is presented with very little room for interpretation or misunderstanding, is excellent. In this approach we spell out exactly when to measure something, who does it, with what facilities, and what he or she does with the information. This type of quality plan may even be completely obvious. The quality plan is made up of three sections: (1) purpose, (2) responsibility for quality, and (3) procedure. An examination of this format will lead to a thorough understanding of the components of the typical quality plan and demonstrate the philosophy of good process control.

*Purpose*: This section of the plan states what the quality objectives are and how they will be achieved. It can be thought of as a preamble or introduction to the quality plan.

*Responsibility for quality*: This section states who is responsible for the qual-ity of the product to be inspected, reviewed, and so forth. It also gives a general description of what has to be reported and how it is reported. For example, this section might state:

> For each operation listed, the operator is responsible for the quality and completeness of the work. The operator is obligated to notify supervision and inspection of incomplete and defective work whenever it is discovered, whether generated at his workstation or elsewhere.
>
> All requests for inspection are to be made by signing a sheet on the inspection call board located at the inspection station. Requests should give operator's name, part to be inspected, and location of the part.

There is no mention of the foreman or any member of supervision or process control management. It states simply that the operator is responsible for quality; therefore supervision must be automatically pledged to quality. This is further reinforced by having the responsible managers review and approve the quality plan by signing it.

*Procedure*: This section states where and when process control activities will be defined. It also states who will make these checks, audits, and certifications, and to a considerable extent how they will be made. This section of the quality plan must be completely synchronized with the workstation planning. Operations are listed in chronological order with the quality requirements for those operations and how the requirements are to be satisfied. It is important that each operation be explicitly defined along with the process control requirements.

**Example (a):**

3.6    *Operation*: Finish machine valves on vertical boring mill.

3.6.1   *Operator*: Complete form PC-210, listing actual dimensions, and return to the inspector at the J-15 inspection station. Hold piece on machine for process control release.

3.6.2   *Inspector*: Audit operator certification form PC-210 that actual dimensions are within drawing tolerances. Sign off PC-210 and send to A-8 inspection station. Notify operator for release of part from machine if dimensions are within tolerance. Otherwise, report all out-of-tolerance dimensions on quality control report and list QCR number on bottom of PC-210.

The operator and inspector are directed to perform particular process control actions. Other indirect work-performing personnel can also be instructed by the quality plan.

**Example (b):**

3.14    *Operation*: X-ray welded connections and other casting areas. Ref. Process Instruction P6G-AL-1223.

3.14.1   *X-Ray Technician*: Completely fill out x-ray folder (TG-1238) noting x-ray numbers and types of defects, if any.

3.14.2   *Materials Engineer*: Review x-rays; determine whether or not x-rays meet standard. P6G-AL-1223. Disposition on face of x-ray folder (TG-1238).

3.14.3   *Foreman*: Have x-ray defect areas laid out on casting for removal. After defect removal sign x-ray folder (TG-1238).

The two examples show the brevity and conciseness of this type of quality plan. It is ideal for a job shop where the size, shape, and design of parts vary considerably but the basic processes vary comparatively little. Many different critical paths may be laid out with this type of quality plan, while only the path employed for the particular operation is implemented.

The numbering system has a distinct meaning. In example (a), 3 indicates that this is the procedure section of the quality plan; 3.6 signifies that this is the sixth sequenced operation of that particular manufacturing area dealing with valves listed under the procedure section; 3.6.1 references the operator's quality action requirements for the sixth sequenced operation; and 3.6.2 specifies the inspector's duties.

The control documents and information needed are also spelled out. In example (a), the operator is instructed to enter data on process control form 210.

In example (b), the technician, foreman, and engineer are informed that the x-ray standard is contained in process instruction P6G-AL-1223. By referring to other documents for specific needs, it is possible to have the information on the quality plan and to keep the size of the quality plan manageable. Also note that the language is simple and direct; abbreviations are used sparingly and only when they are easily definable. The object is to avoid confusion and to tell only what needs telling. Obvious things are normally left out. In example (a), there is no reference to who receives the Quality Control Report (QCR), because it is known throughout the factory that QCRs are sent to design engineering for action and dispositions and that an overall QCR system is in place. In example (b), there is no doubt that the materials engineer makes dispositions regarding acceptance or rejection of the x-ray of the casting. Instructions are given for recording the engineer's actions, because it is not obvious what the disposition method should be.

The cookbook-type quality plan puts a great deal of information on paper, but leaves out the obvious for the sake of brevity and clarity. Therefore, some simple rules should be followed when deciding who should be listed for inputs under each operation and how much information should be given.

1. List everyone who has a routine input for the operation.
2. List all the information necessary to do the job, that is, process instructions, material specifications, process control forms, and so forth.
3. Do not list any information that would be classified as background material.
4. List all operations regardless of whether there is a corresponding process control requirement. For example, a queuing operation is vital to the process, but rarely to the quality of the product. By listing all operations, continuity is maintained.

Quality plans are essentially simple things. Unfortunately, many writers of quality plans lose track of what they are trying to do. To summarize, a quality plan should include the purpose for its existence, a designation of responsibility for quality, and step-by-step procedures that are easy to follow.

## MEASUREMENTS

The type of measurements used, accuracy required, and redundancy depend primarily on the product being produced. In a job shop it is common to assign like work to like areas, thereby minimizing the need for quality plans and the types of measurements to be made in one area. An exception would be job shops that employ group technology layouts. Even here, commonality of parts being processed also minimizes requirements for different quality plans and measurements. For instance, in heavy fabrication and machining businesses, it is common to have all welding assigned to one area; therefore all the measurements, regardless of the products involved, are basically similar. This permits skill development by operators and inspectors and cuts down considerably on errors of omission. Since all the welding is done in one area, there is no need to achieve proficiency in weld auditing in different areas, and it is not necessary in this area to teach inspectors to audit machining techniques. This segregation technique facilitates the development of specific quality plans even though the product mix varies, sometimes almost daily.

In a job shop where like work is dispatched to the same general area, the quality plan is generated to handle a similar product line. For example, heavy-duty induction motor frame welding and high-volume, low-velocity pump convolutes fabrication welding could take place in the same location. The parts are considerably different and may even differ among themselves (e.g., there may be many different styles of motors). However, they are all welded fabrications to be produced in the same shop, so that the weld measurements can be similar, if not identical. In this case, the primary quality plan would be written for welded fabrications, treating all of them in the same way as far as this is practical. Only when a specific requirement for one does not agree with that for the others will there be a need for a subsection of the quality plan. This subsection will be entered in the regular area quality plan and will be very clear about what it is and why it differs from normal. In this section, and nowhere else, specific measurements will be called for. The measurements called for in the general section apply everywhere, while the specific section is narrowly defined.

Measurements must be carefully planned for maximum effect. In job shop environments the planning is such that general techniques must be effective for many specific types of products. Therefore, as many measurements as possible should be made by the operator [see example (a)], inspections should be basically audits, and true inspector-conducted measurements should be of the tollgate type. Hence the basic body of documentation should be the operator's certification (measurements) backed up by in-process audits (sampling) conducted by inspection personnel. This documentation is then backed up by the tollgate inspections to ensure the needed degree of quality.

Tollgate inspections are the control points for job shop process control. Through these inspections the process control unit can assure management that the product is being made to design specifications, that everything up to the point of the inspection is correct, and that the process control system is working adequately. If something goes wrong in the future, it is possible to go back to the tollgate status and either reconstruct the cause of the failure and correct it, or scrap the procedure and start over.

Tollgate inspections should normally be conducted by an inspector and perhaps involve supervisory process control personnel. The procedure involves check-off sheets and verifications that required operations have been completed. Occasionally a physical measurement or test may be called for, but usually this is not required.

**Example:** A tollgate may be required before steam turbine high-pressure shells are furnace heat-treated after welding. The tollgate inspection verifies that all required welding and all required nondestructive testing have been completed and, most important, that the requirements have been completed successfully, that is, have met design engineering requirements.

Another feature of a tollgate inspection is that the authorized performer must sign for the work. In the process control system, measurements are the basis for all long- and short-term corrective actions. It is necessary to make sure that measurements are made correctly, that the work is done correctly, and that incomplete and incorrect work is reported immediately. By making people accountable, there is a much higher probability that the work and the measurements will be done correctly. This is the theory of accountability.

The Theory of Accountability states that people will perform at a conscious level consistent with high quality work only when they can be held accountable for that work.

To put this accountability theory into practice, everything a worker is asked to do must be measurable on a quality basis. The quality plan must state what kind of measurement is required, how it is to be made, and how the action is to be recorded. There must also be a way to record the worker's performance or percent acceptable, and a reward for doing the job correctly. The reward can be positive or neutral. For instance, a welder may be hired to do x-ray quality work. This is a basis for employment. The reward for doing the job to the quality standard is continuance of employment, hence a neutral reward.

We cannot leave this section on quality measurements without mentioning tests. Tests are everybody's idea of process control. There seems to be a feeling that testing is always good and that more testing is better than none. What is often not realized is that testing, unless it is designed for a specific purpose and well planned, can be useless and sometimes harmful. Testing in a manufacturing environment is not like testing in a laboratory. It is difficult to control the variables. Facilities for fine testing normally are not rugged enough to survive the factory atmosphere, and properly trained technicians either are not available or are working in an unfamiliar and often hostile personnel environment. If a test is needed and can overcome these difficulties, it should be successful. But those requesting tests in a factory must plan the exercise to the smallest detail. From experience, a set of rules for achieving good tests in a job shop have evolved:

1. Design the test to fit into the production schedule.
2. Make the test as simple and straightforward as possible.
3. Make the physical part of the test accomplishable by the factory personnel if it is a process requirement, continuous-production type. To do this, supply clear instructions with no ambiguities in writing, specify easily used equipment, and require little interpretation by the operators.
4. Where possible, make the interpretation outside the factory.
5. For one-of-a-kind or seldom used tests, have a knowledgeable person available at all times during the test.

# DATA COLLECTION, RECORDING, AND CLASSIFICATION

The quality plan states what information is to be recorded and measured. The data collection, recording, and classification phase sorts the inputs, categorizes them, and eventually becomes the basis for reports and corrective actions. The quality plan does not specify what reports will be issued; however, when a quality plan is developed, some thought must be given to the types of reports desired. Otherwise, the wrong information may be collected and an erroneous picture of quality presented.

When analyzing data, it is necessary to know beforehand what to look for. The real world of manufacturing is far removed from the scientifically ideal one. We must predetermine the kind of information we wish to have, and then decide whether the available data are sufficient to tell us what we need to know. Another truth about factory-derived data is that they do not generally result in precise curves or fit nicely into a mathematical model. We are dealing with the ability of many people to produce data; they all have their own tolerance bands, so that irregularity is the norm.

In analyzing process control data, we must be willing to make judgments and take actions based on incomplete observations or short spans of data because of the compressed time scale involved in manufacturing. Every decision based on data collected must be backed up with a plan for what to do if it is all wrong. The engineer involved in process control work must be extremely familiar with the manufacturing processes, know the products, and be willing to make decisions based on the data available at that time. This does not mean that the engineer stops analyzing the data because the decision is made. The decision may be modified at a later point. The important thing is that a decision is made and the manufacturing process continues.

I do not wish to give the impression that process control data collected by inspectors and taken by operators is suspect. In fact, it is usually of high quality and taken with diligent care. However, unlike a laboratory experiment, verification is usually not possible and the conditions required to take the readings are not always under tight control. The chance for error is naturally higher than under controlled laboratory conditions. Since this is the case, the process control engineer must know the expected scatter band for any process in order to correctly interpret the data. If there are enough data points, the engineer uses statistical methods. Often, however, especially in job shop process control, it is necessary to make assumptions because of a lack of statistically significant data. Therefore, in order to minimize the risk factor, the engineer is continually testing the assumptions against the incoming data. This leads us to how the data are recorded and analyzed.

The first and most important input is found on the quality control report, which is sometimes called a deviation report. This document is an exception report listing discrepancies from required values. It is most important from a control standpoint because the first order of business is to correct mistakes. Therefore, the process control engineer can monitor the temper of the manufacturing operation by keeping a close watch on QCRs by number and defect category. This may be done manually or with computer programs. It is normally possible to tell by the frequency of occurrence over a period of time whether the quality of the area is increasing, decreasing, or remaining stable. This is done simply by relating the number of QCRs to the number of man-hours worked. If the engineer sees an increasing QCR frequency, it is an indication to look for problems. The simplest way to do this is through a QCR analysis called a defect analysis.

The defect analysis is usually a computerized report with the QCR as its basic input. It is simple to code all QCRs by operation number or description, by responsibility for causing the discrepancy, and by part name involved. The QCRs may then be sorted by these three groupings—operation, responsibility, or part— and much meaningful information obtained. For instance, if a certain part consistently turns up on the report, one would further sort by responsibility to find the primary cause of the defect and by operation to determine where the defective work is occurring. With this information, the process control engineer can now make a detailed investigation and take corrective action. Another important feature of the defect analysis report is that it can be keyed to the costs of labor and material expended in correcting mistakes. If the costs of correcting mistakes (manufacturing losses) are keyed to the QCR, it will be possible through the use of the defect analysis report to associate true costs with individual problems, allowing the business to decide where it is best to expend its resources for corrective actions.

The other major source of data is the in-process data sheets known as Process Control (PC) forms, which contain the primary information obtained from the manufacturing operations as required by the quality plans. In many cases this type of data is used primarily for the specialized reports needed for short- or long-term managerial information and only secondarily as an input for corrective action. The PC form data are preplanned data that measure specific operations. They are often valuable in comparing similar products from time to time to obtain trend information. The process control engineer uses PC form information as a check on various assumptions and decisions that have been made.

With the various inputs—QCRs, the defect analysis report, and PC forms— the process control engineer is capable of monitoring the manufacturing operation, reacting to situations that require corrective actions in a logical manner, and judging the results of the actions taken. Without the process control system inputs, the engineer would be using a scattergun approach to problem solving. The detailed information that can be obtained through the process control system enables manufacturing management to act in a precise way, eliminating the problems without sacrificing any of the good elements of the system.

## CORRECTIVE ACTION (SHORT TERM)

Immediate response to problems is essential for success in business. The process control system, through the use of QCRs and PC forms, offers the ability to respond to problems quickly and intelligently. This "fire fighting" capability is the hallmark of the successful process control organization.

In process control work, one always plans to eliminate or at least minimize the impact of "fire fighting," and the fact that such problems occur should not always be construed as a failure of the process control system. If the problems are due to poor planning or anything within the effective domain of the process control organization, then the system has a flaw and any corrective action must have a dual purpose: to correct the problem and to correct the system. However, if the problem is a result of outside influences, for instance, deterioration of vendor quality, then corrective action can be directed to the deficiency and concern about the system eliminated. Rules for "fire fighting" can be stated as follows:

1. Define the problem as quickly as possible and in as much detail as possible.
2. Determine whether it is a process control system problem as well as a quality problem.
3. Take action to solve the problem, first addressing the immediate quality problem and then, if required, the process control system problem.
4. Make an intensive effort to solve the quality problem and gain control of the situation. Corrections to the process control system can be made at an appropriately controlled pace.

When engaged in "fire fighting" activities, the process control unit is asked to take a leadership role in the problem solution. This means effectively coordinating the interactions between manufacturing, design engineering, marketing, and occasionally finance. To do this successfully, the process control unit must command the technical and managerial respect of the other functions involved, or its leadership role will fall vacant. Cultivating technical and managerial leadership is an independent personal obligation for all members of the process control unit. Since process control is often the art of gentle and not so gentle persuasion, the people involved must be technically proficient beyond reproach and possess good common sense. The successful process control engineer must be both an idealist concerning the quality of the system and a pragmatist about what is achievable with respect to cost in time and money and the prospective payouts.

With this attitude, process control can often temper too much conservation on the part of design engineering and too much radicalism on the part of production control. This balance is essential in successful "fire fighting," for without it, erroneous decisions can be made. Process control often finds itself playing the role of advocate for the missing party when conversing with either design engineering or manufacturing and occasionally marketing or finance. Besides maintaining the balance, process control must decide what the correct course of action is and have it agreed to by all parties. The ability to sift through the data and arguments presented by all parties takes considerable technical engineering skills and rapport with those involved. By virtue of dissecting and presenting the facts about the problem and knowing where it fits into the manufacturing time frame, the process control engineer is in the best position to solve these problems.

The main indicator of trouble requiring short-term corrective action is the quality control report, which details the nature of a specific problem. While all QCRs indicate a problem of some type, not all require short-term corrective action. Most QCRs are routine and the procedure for QCR disposition should be sufficient for solution. It is the QCRs that involve a high cost in time and money that require a high-level, immediate response.

It is the process control unit's responsibility to determine whether a QCR needs short-term corrective action or whether the standard routines should be used. This judgment can be made more objectively by establishing basic working guidelines.

**Example:** A QCR involving an error that costs approximately $100 to correct would not normally justify short-term corrective action. But if the same deficiency was discovered on the day the product was to be shipped, the time frame would dictate that it be handled as a short-term corrective action problem. In some cases a shipping problem, regardless of cost, can require immediate correction.

The question of how to handle a QCR leads to the subject of documentation. In job shops, since every job may be different from its predecessor, the problem, its definition, and its solution constitute meaningful information in planning for similar or even dissimilar jobs. For instance, certain contour welds that failed on one job might be inadvertently tried on a future job if proper documentation on the original problem did not exist. In effect, the organization must keep records so that it does not have to "reinvent the wheel."

With the proper documentation, the next step in the short-term corrective action format can take place. This is establishing a temporary preventive maintenance routine, which is akin to an exploratory solution. Major changes in the quality plan to permanently resolve the problem should not be made until the temporary solution has been properly evaluated. The preventive maintenance routine may be a specially designed PC form to be filled out by the operators, an audit to be conducted periodically by the inspectors, special engineering drawings to correct a specific item, or some other extra effort. Whatever it is, it must be implemented with the complete understanding of all those involved. The temporary preventive maintenance routine is more than a patch to stop a leak; it is a patch that may or may not have to be redesigned to make it permanent. Once the temporary routine has proved effective, it is incorporated into the permanent process control system, usually through the quality plan but just as correctly, although less frequently, through the appropriate engineering drawing.

In summary, short-term corrective action is the part of process control work aimed at correcting important quality problems that require more attention than the normal discrepant quality routines. It involves using the data collection and analysis phase to identify problems, mainly through QCRs. An immediate, vigorous response action may need to be taken. After the immediate problem is resolved, a temporary preventive maintenance routine is established, and finally a long-term solution is put into effect. Long-term corrective action will be discussed in detail in a later section of this chapter.

## REPORTS

Process control system reports display the current status of the manufacturing operation. Their main function is to give an accurate picture of the manufacturing operation at a particular point in time. The reports must be constructed so that they do not give a false sense of complacency to the reviewer or be overly alarming. They must be factual and clearly state how conclusions are reached. They must not contain any form of hidden editorial.

A secondary function of the reports is to aid in trend tracking. Therefore, wherever possible, they should show comparison results. For example, a report stating that there were 42 operator-responsible machining errors in March does not mean very much. For the reported 42 discrepancies to be meaningful, they must be compared with like data. The usual way to do this is through comparisons with similar previous time periods, taking care that the work was similar and of the same relative frequency. The goal in trend tracking is to have the value reported on be the only variable with everything else held constant. This gives the reviewer the ability to judge accurately whether the situation is remaining the same or changing.

In job shop operations, it is not easy to find commonality of data, so that before a measurement is developed and reported it is essential that it be reduced to the lowest common denominator.

**Example:** For welding measurements it is essential that measurements be for like processes with like equipment. Comparing automated welding mixed with manual welding will always give a false picture. Each type of welding should be reported separately.

This leads to another requirement for reporting data. The method by which the data are obtained and what the data mean should be readily identifiable by the reviewer. There should be no secret formulas or codes; the report must be simple and understandable by its least familiar reader. A minimum of four report categories may be recommended to management:

1. Monthly quality reports
2. Manufacturing losses analysis reports
3. Weekly problem sheets
4. Special management reports

The first level of management would receive all four reports; the next higher level would receive reports 1, 2, and 4; and the highest level of management would receive only report 4.

Monthly quality reports explain the status of quality for that time frame. They are heavy on tollgate inspections and compliance with quality plans. One of their functions, in addition to showing the quality status of the products, is to show the status of the quality plan and whether it is functioning as it is supposed to.

Manufacturing losses analysis reports are detailed examinations and explanations of deficiency correction costs. Here manufacturing losses are broken down by part, responsibility, and operation in a format useful for planning corrective action. This is one of the basic reports used for long-term planned improvement programs. Since it is a report on money spent, it identifies the large-payoff items that should be given close attention.

Weekly problem sheets are the most basic of all reports, where the process control engineer lists the problems of the week. The engineer records specifically what the problem was, on what workstation, very briefly what cause the investigation showed, and what corrective action should be taken or already has been taken to correct the deficiency.

Special management reports are either condensations of the three types of reports mentioned above or reports dealing with a particularly difficult problem needing the direct attention of higher management. These reports are prepared quarterly or semiannually for higher management. They are more general and brief than the other reports, as the recipient must be aware of trends, not the specific details involved.

These are the primary process control reports; all other reports are variations on these types. The important thing in reports is to make sure that all vital items are accounted for in a very readable manner, and that the data are presented against a measurable base for comparison. Good reports lead to effective long- and short-term corrective action programs.

## CORRECTIVE ACTION (LONG TERM)

This is the area that makes or breaks a good process control system. The system's ability to evolve and improve through the use of the various feedbacks and its ability to function competently in a dynamic situation are the prerequisites for its success.

The method by which the process control system corrects itself is found within the corrective action (long term) block of Fig. 8.1. Included here are improvements to quality plans, changes in methods and planning, and future problem forecasting. Also included would be the objectives and goals with supporting projects, as discussed in Chapter 2. Long-term corrective actions are taken only when their worth and cost have been carefully calculated and evaluated by the responsible process control personnel. The short-term corrective action is reviewed and steps are taken to make it permanent by issuing revised quality plans, or the short-term action is modified and subjected to more field trials.

In this phase, the results of temporary preventive maintenance routines, short-term corrective actions, and trends shown on reports are evaluated and tested to determine the need for permanent system improvement. While it is recognized that a short-term action would not have been taken without some justification, it must be determined whether the original set of conditions will still prevail over a longer period of time.

> **Example:** If a temporary preventive maintenance routine was set up to control a particular weld variable, and the welded fabrication was later replaced by an easy-to-make forging, there would be no need to incorporate the temporary preventive maintenance routine. A record of the routine and how it worked would be in a reference file. Therefore, the short-term corrective action would be terminated, and no corrective action on a long-term basis would be taken.

Now let us assume that in the above example a new size assembly is required and it cannot be produced by the forging vendor at a reasonable price. Hence, the old welded fabrication design must be employed. Going through the experience file on this operation turns up the record of the temporary preventive maintenance routine used previously. At this point, it would be wise to incorporate this routine in the quality plan to prevent the problem from recurring. Thus the temporary preventive maintenance routine becomes a permanent part of the quality plan.

The objective of the process control organization is to use all the feedbacks available to develop evolutionary changes in the process control system. The organization should employ the trends in anticipating future needs so that it can supply management with information to be used in decision making. Solving small problems revealed by trends before they become big problems is the most cost-effective way to function.

Process control uses trend information to guide manufacturing engineering actions concerning producibility and equipment operation. It informs shop operations about potential operator performance and manufacturing problems. It gives design engineering the quality results related to their calculations. Process control tends to control the manufacturing process by being a vast information storehouse and by having the responsibility for quality plan initiation. Therefore, any long-term changes to the plans require considerable evaluation before they are made. But changes must be made, and the "red tape" procedures common in many well-meaning organizations cannot be allowed to prevent change. Therefore, another factor in long-term corrective action is that indecision and procrastination must not be allowed to gain a foothold. Process control has to be a pragmatic action center or it will fail.

The process control unit must always be aware that it is through long-term corrective action that the three continuously generated benefits of the process control system are gained. These benefits are reduced manufacturing losses, minimized production delays, and improved quality.

## RELATIONS WITH OTHER MANUFACTURING ENGINEERING UNITS

Process control is the newest addition to the manufacturing engineering subfunction, and it is not derived from the traditional base of manufacturing engineering, that is, industrial engineering, tool and dye engineering, methods engineering, and design engineering. It was split off from quality control when quality control became a nonmanufacturing entity called quality assurance. Process control is an amalgam of applied engineering that derives its strength from statistical concepts considerably modified by pragmatic approaches to manufacturing and design engineering. In some ways it is like producibility engineering, except that it deals in after-the-fact rather than before-the-fact evaluations of products. Similarly, it resembles methods engineering, but again in after-the-fact aspects of production, not before. Unlike any other manufacturing engineering functions, process control must react to and correct immediate production problems, while the other manufacturing engineering functions can deliberate for considerable periods of time in solving problems. Process control represents manufacturing engineering to shop operations for virtually all product-related technical questions.* Thus it is the counselor, confidant, and sometimes disciplinarian of shop operations. However, in order to do its job effectively, it must draw on all the resources of the other manufacturing engineering units as well as support those other units. Let us now look at the relations between process control and the other manufacturing engineering units.

It is the responsibility of the manager of manufacturing engineering to ensure that all of the manufacturing engineering units function as a coordinated team. Since process control is a little different from the traditional manufacturing engineering units, it is important to understand what it does and what its relations with the other units should be. We will start with advanced manufacturing engineering.

The Advanced Manufacturing Engineering (AME) unit deals with the mechanical systems that the factory needs to produce its products. Thus it needs to know whether the equipment is working adequately to produce the parts in accordance with engineering drawings. This leads to a natural communication between the area planner and the process control engineer. The area planner is interested in knowing whether the plans for the manufacturing layout and the equipment are working well. The process control engineer cannot answer directly for the performance of the equipment, but can account for the human-machine interface and the results of putting it to work. The PC engineer supplies performance information related to failures and why they occur. In contrast, the performance information supplied by the methods engineer is related to the efficiency of operating the equipment. The process control engineer's information deals with the reasons for failure—whether it is operator-caused or machine-caused. In both cases

---

*In contrast, methods engineering is the shop operations contact in all technical matters pertaining to the operation of the machine tools and process equipment.

the area planner will be interested in learning whether or how the equipment contributed to the failure, and in discussing possible corrective actions with the process control engineer. Knowing the nature of the failures, the process control engineer is in a position to review the equipment and offer avenues of exploration for correction to the area planner.

Producibility engineering is responsible for ensuring that only producible designs are released for manufacturing. Process control engineering is at the other end of the spectrum and evaluates the results of the work of producibility engineering by determining how producible the design really was. Its measurements of percentages of failure and costs of failure show how well producibility engineering did its job. A good deal of the success of a producible design on the factory floor must be credited to the methods engineer, but without a producible design to start with, the efforts of the methods engineer would be futile. Hence the failure rate of a producible design is basically a measure of the effectiveness of the producibility engineering function. This helps us to understand the relationship that exists between the process control engineering and producibility engineering units. The producibility engineer can ask the process control engineer about the probable effects of allowing specific designs to reach the factory floor. The process control engineer can provide needed information for producibility engineering to represent manufacturing in design evaluation activities.

The Methods, Planning, and Work Measurement (MP&WM) activities are close to the evaluation activities of process control. Both are interested in measuring the output of the workstations—process control for product conformance, and MP&WM for efficient production of a conforming product. To be efficient, both types of measurements should complement each other and, whenever possible, be derived from the same data. Methods engineering measures the rate of productivity, while process control measures the quality of productivity. Both must be optimized for a manufacturing operation to be competitive, and one cannot be optimized at the expense of the other, hence the need for cooperation. This cooperation usually takes the form of designing the quality plan to conform with the prescribed method for producing the product, so that measurements required by process control will also be useful in measuring productivity. A typical example would be a process control measurement requiring the number of items produced over a shift to be recorded along with the necessary quality characteristics. This gives a measurement of productivity rate as well as a measurement of quality.

Finally, process control provides a statistical service to maintenance engineering by categorizing and quantifying machine-caused quality problems. This is the type of information maintenance needs for planning corrective actions. Defective products may or may not be due to a complete machine failure but they can show equipment that is not functioning as it was specified to. Therefore, the process control data point out symptoms of equipment in the process of failing and enable maintenance engineering to take action while the deficiencies are still relatively inexpensive to fix. For this reason, the maintenance engineering unit must have a close relationship with the process control unit. It is not unusual for maintenance engineering to request specific statistical comparisons of equipment from one time period to another on machine-caused quality failures. The ability to detect trends of machine performance and plan future maintenance activities helps the maintenance engineering unit to maximize its utilization of repair funds. Thus process control data represent not only quality trend data but also, in this case, machine deterioration data.

The examples presented above show that process control engineering is one of the technical support functions of manufacturing and explain why it is now considered part of manufacturing engineering.

## USE OF CONTROL CHARTS

Control charts are sometimes used in job shop process control work. They represent a statistical concept that has limited applications but can be very effective in monitoring machine tool quality. Control charts are based on the concept of the normal distribution, in which the values for like things, such as values for the diameter of roller bearings, cluster about an average measurement. This average measurement, called the mean, would be the diameter specified on the engineering drawing, and the plus and minus tolerances would be the allowable deviation from the mean.

The normal distribution is illustrated in Fig. 8.2. For process control applications we identify the mean, $U$, as the drawing dimension and either a two-sigma, $|2S|$, or three-sigma, $|3S|$, spread as the plus and minus tolerances. If we have a $|2S|$ spread we are stating that 95.5% of all parts made will be acceptable if the process is under control. This is derived from probability theory and corresponds to the area under the normal distribution curve, where the area from $-2S$ to $+2S$ includes 95.5% of the total area. This means that in a process under control 95.5% of all parts produced will be acceptable, that is, within the tolerance band—provided, of course, that the selected tolerances fall within the capability of the human-machine system. Selection of tolerances to fit a $|2S|$ spread is an iterative process based on design requirements and machine tool capabilities. If a $|3S|$ spread is used, the probability of good parts being made increases to 99.7%.

> **Example (a):** Suppose we are making roller bearings on a cylindrical grinder. The drawing calls for a diameter of 0.500 ± 0.002 in. Therefore the mean $U = 0.500$ in. and the two standard deviations are $|2S| = \pm 0.002$ in., or $-2S$

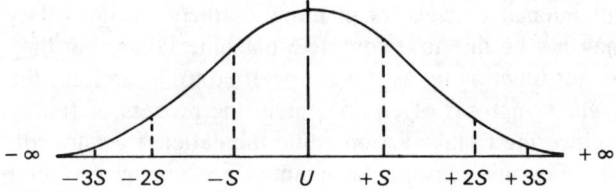

**Figure 8.2** The normal distribution.

$= -0.002$ in. and $+2S = +0.002$ in. This means that if the process is under control, 95.5% of the sample population would fall within the $0.500 \pm 0.002$ in. tolerance band. Now we collect data to see whether we are performing satisfactorily. We take 10 samples of roller bearings at random as they are produced, measure the diameters, and calculate the mean:

| Sample | Diameter |
|--------|----------|
| 1 | 0.499 |
| 2 | 0.500 |
| 3 | 0.501 |
| 4 | 0.501 |
| 5 | 0.502 |
| 6 | 0.500 |
| 7 | 0.498 |
| 8 | 0.500 |
| 9 | 0.500 |
| 10 | 0.499 |
| | $\Sigma X_i = 5.000$ |

$$U = \frac{\Sigma X_i}{N} = \frac{5.000}{10} = 0.500$$

and the standard deviation:

| Sample | Diameter | Difference from mean | Square of difference from mean |
|--------|----------|----------------------|--------------------------------|
| 1 | 0.499 | 0.001 | 0.000001 |
| 2 | 0.500 | 0 | 0 |
| 3 | 0.501 | 0.001 | 0.000001 |
| 4 | 0.501 | 0.001 | 0.000001 |
| 5 | 0.502 | 0.002 | 0.000004 |
| 6 | 0.500 | 0 | 0 |
| 7 | 0.498 | 0.001 | 0.000004 |
| 8 | 0.500 | 0 | 0 |
| 9 | 0.500 | 0 | 0 |
| 10 | 0.499 | 0.001 | 0.000001 |
| | | | $\Sigma(X_i - U)^2 = 0.000012$ |

$$S = \sqrt{\frac{\Sigma(X_i - U)^2}{N}} = \sqrt{\frac{0.000012}{10}} = 0.001095 \approx 0.0011$$

Therefore, the standard deviation of this sample would be acceptable, since we are well within the selected tolerance band of $|2S| = \pm0.002$ in.

The last step is to place the data on a trend chart called a control chart. Since we cannot discern trends from the above calculation, we must collect observations over a period of time, arrange them in a table or graph, and then decide whether the data show any trends. A control chart for periodic average values is a plot of the means found in periodic chronological sampling checks and is illustrated in Fig. 8.3.

If the time scale selected is days, then we are recording daily sampling averages. In Fig. 8.3 we can see that the mean steadily climbs toward the upper tolerance limit and on day 7 exceeds it. This would indicate that the process needs to be brought back under control and would lead management to take action to investigate for cause and then take the necessary steps to regain control. In practice, this action would probably have been started as soon as a trend was discerned, possibly after day 3 but certainly after day 4. Days 8 through 10 show the effect of corrective action taken to regain control.*

The control chart technique is useful, but only if we can relate actual tolerances to the basic ability to produce. If the tolerance band is too tight for the human-machine process, then no amount of statistical maneuvering will be of any use. Therefore, two things are necessary precursors for successful use of control charts: (1) tolerances must be chosen within the capabilities of the processes, and (2) the design and the manufacturing process must be correctly matched.

Let us look at another interesting bit of information we can obtain about the roller bearing manufacturing. We would like to know, based on actual measurements of several samples, how many or what percentage of the roller bearings are acceptable, that is, within the tolerance limits. This can be done by using probability theory and data derived from the normal distribution of events.

**Example (b):** In example (a) we measured 10 roller bearings from the cylindrical grinder and found a maximum diameter of 0.502 in. and a minimum diameter of 0.498 in. We calculated the mean and the standard deviation as $U = 0.500$ in. and $S = 0.0011$ in. Now, using probability theory, we want to find out what percentage of the total population would yield the same re-

---

*Figure 8.3 shows a nonrandom distribution. If the distribution were random, a trend line would not be discernible. The significance of a nonrandom distribution is that the process is changing, and it will continue to change until a new steady-state level is reached. At that point randomness will reoccur.

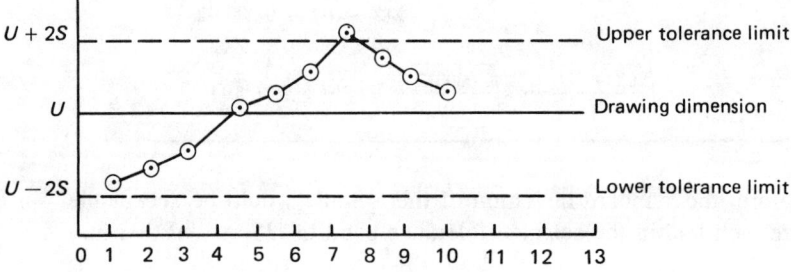

**Figure 8.3** Control chart for periodic average values.

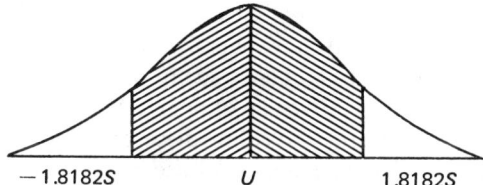

−1.8182S          U          1.8182S

**Figure 8.4** Normal curve; the area of interest in example (b) is hatched.

sults (i.e., be acceptable) if the processing characteristics remained the same. To do this, we must determine the number of standard deviations between the maximum and minimum diameters and then determine the area under the normal curve that this spread represents. Since the area under the normal curve represents the entire universe of like parts, the percentage of the area that we calculate represents the probability that the future production will fall within these tolerance bands (which are tighter than the ±2S spread).

We will use the formula to find the number of standard deviations from the mean, then enter the normal curve areas chart (Table 6.1) to find the probability.

$$Z = \frac{X - U}{S}$$

where $Z$ is the number of standard deviations, $X$ the variable away from $U$, $U$ the mean, and $S$ the standard deviation. Using the data of example (a):

$$Z_r = \frac{\text{maximum diameter} - U}{S} = \frac{0.502 - 0.500}{0.0011} = 1.8182$$

$$Z_1 = \frac{\text{minimum diameter} - U}{S} = \frac{0.498 - 0.500}{0.0011} = -1.8182$$

If we enter these data on the normal curve, the area we are interested in will be the hatched part shown in Fig. 8.4. Thus, $Z_r = 1.8182$, which, when entered in Table 6.1, is interpolated as 0.4657. Remember that the maximum value for the right-hand portion of the curve is 0.5000. Similarly, $Z_1 = -1.8182$, corresponding to 0.4657. Since we are looking for a total area we add the two values and obtain 0.9314 or 93.14%. Therefore, we would expect 93.14% of all roller bearings made by this process to fall within the tolerance range of the sample.

Probabilities are used in process control activities in other ways, as in determining the percentages of defectives in a population of parts to see whether a satisfactory population is being maintained. In this case the satisfactory population includes a percentage defective agreed to by a customer and a vendor. There are also procedures for calculating the probability of accepting a bad lot of goods. These are not useful in job shops, which normally have small lot sizes where virtually every part must be acceptable. Also, the parts are usually complex and do not lend themselves to representation by large universes of simple but similar parts where acceptance criteria are simply measured. Within these constraints, this section shows the present practical, statistical techniques for use in job shops. Even the roller bearing example is not a very good example of job shop use, but it does illustrate the techniques.

These techniques can be used to monitor the performance of the human-machine system. For example, we may not have identical parts in great quantity, but we have similar parts. If we make shafts of 2.000 ± 0.002 in. diameter for week 1 and switch to shafts of 2.350 ± 0.002 in. diameter for week 2, and so forth, we can still calculate the mean and standard deviations of the sample and use control charts to monitor the process, provided that the process remains the same for the different-size shafts. However, our use of probability theory to determine the percentage of good shafts to expect would have to be revised constantly as the mix changes, so we would not use this tool very often. The results would be interesting but of little use for future manufacturing planning.

From our examples, we can see that the statistical concepts used for control charts are most useful for continuous manufacturing, but can be partially adapted for job shop use. We should also be aware that we are measuring an entire human-machine process, not just the human or the machine. This, of course, introduces subjective decision making into something that might at first glance be considered entirely objective. When a control chart indicates an out-of-tolerance situation, we must subjectively determine whether it is due to an operator deficiency, a machine malfunction, or a combination of both before we can determine how to correct the situation.

Areas of sampling for control charts must be carefully selected. The more complicated the part, the less likely it is that we can measure one or two key items that will accurately indicate overall quality, and the more difficult it will be to calibrate all the individuals who will be taking measurements. This implies correctly that there is a measurement accuracy problem in sampling work. Variations in taking measurements from individual to individual can lead to misleading interpretations. In many process control units, only a few specified individuals will make sampling measurements. Ultimately, if only one person takes measurements, the only concern is how that individual will perform from sample to sample. Obtaining consistency in sampling measurements is a difficult task.

In theory the use of control charts is appealing, but in practice it can be difficult, especially in a less than antiseptic factory. The control chart is a powerful tool, but the user must always be aware that it is only a tool and must be used correctly and interpreted wisely.

## TOTAL QUALITY MANAGEMENT

The topic of process control would not be complete without a discussion of Total Quality Management (TQM). In recent years TQM has been touted as one of the few new management practices that will make companies competitive, particularly against the onslaught of Japanese competition. The shorthand TQM has come to mean a powerful solution to all that ails modern North American industry. It implies that if we will only pay supreme attention to the needs and desires of our customers, and we deliver on those aspirations, we are bound to be successful. Obviously, no such nirvana will occur. There is no one magic cure. There is no doubt, however, that the philosophy and techniques attributed to TQM will enhance a company's competitive position if they apply the principles properly.

TQM focuses on awareness techniques for making products to the best of the organization's abilities. Let's take a look at these principles and see how they lead to a set of objectives compatible with the process control philosophy.

TQM theory can be described via the TQM triangle, as depicted in Fig. 8.5. We see at the points of this equilateral triangle the labels Customer, Process, and Data. The focus is on the customer's requirements at the apex of the triangle. This on turn generates a plan (the process) for achieving the requirements. The plan is implemented and data are generated, from which the effectiveness of the plan is evaluated. The results are then compared with the customer's requirements and the process is modified in order to get improved results. Notice that the statement is very direct and forceful. Modifications will be made, even if the initial results are satisfactory. This implies a very strong objective of forcing continuous improvement at all times. The modifications are implemented, and then the results (data) are analyzed again to see if they are in compliance with the customer's needs. Around and around the cycle goes, each iteration (hopefully) resulting in further improvement until eventually total compliance with customer need is achieved and we have reached a state of zero defects. This is the theory: continue improving until perfection is achieved.

There are a few important concepts at work here: continuous improvement and the striving for zero defects. These are what I call awareness factors. No new techniques for achieving improvement are employed. In fact, the process control techniques described in previous sections of this chapter are probably optimal in attaining the TQM objectives. We are simply focusing on using what we currently

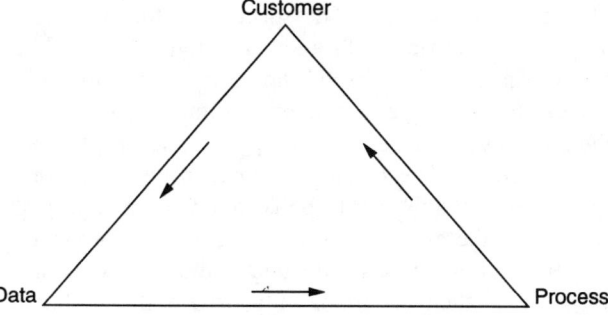

**Figure 8.5** The relationship of Customer, Process, and Data for continuous improvement as shown via the "Total Quality Management" triangle.

employ but not being satisfied with status quo results. We are focusing the entire organization's efforts on finding ways to improve the process so that the next time we will get even better results, and the cycle of improvement never ends until the system is perfect.

Perfection, of course, is like the mathematical concept of infinity—approachable but never achieved. This is also true in manufacturing via TQM. However, we can see that by focusing on perfection we will easily surpass any strategy that assumes that some percentage of manufactured output will be substandard.

Example (b) in the section on the use of control charts is an example of the old versus the new concept of management control of operations. In example (b) we see that the process is capable of producing 93.14% acceptable roller bearings or, conversely, 6.86% defective. We then observe that this is a satisfactory population of percentage acceptable versus defective, probably agreed to by the customer and the producer. Agreeing to buy a percentage of parts that are defective is, unfortunately, still a common practice in too many companies, but it no longer need be. With TQM, the data of example (b) would not be accepted as is. They would become the starting point for further pursuits of improvement along the road to the ultimate goal of 100% acceptable parts. With processes and procedures invented decades ago, we find that if we actively and resolutely pursue perfection the results are astounding. Now, the concept of Acceptable Quality Levels (AQLs) is no longer valid in commercial considerations between buyer and seller. Instead, the concept of parts-per-million defectives is the standard between buyer and seller and can be approached only via the concepts of continuous improvement and zero defects philosophies. Process control techniques approached as the *modi operandi* within the TQM frame will definitely generate superior results. TQM awareness concepts merged with process control techniques should be adopted as the overriding policy in applying quality control concepts within an industrial organization.

The question then becomes, how do we apply this awareness factor required by TQM? The answer is through identification of the customer on a generic and a specific level.

We all know the definition of a generic customer. Our organization performs its macro functions to satisfy the requirements placed on us by the ultimate customer, the buyer. This is the generic customer. The specific customer is an individual with a receiver-supplier relationship with another individual. Precisely, this means that the specific customer is the person who receives the immediate output of another person's work. For example, a machinist's immediate customer would be the individual who receives his or her work output. Therefore, the need to satisfy the customer is very focused. The machinist's part in his or her organization's desire to satisfy customer requirements boils down to satisfying the demands of the next workstation in the entire chain of producing the product. If the localized provider-customer relationship is satisfactory and all subsequent relationships are likewise satisfactory, then the probability of overall satisfactory results is high.

In essence, then, organizations practicing TQM espouse the following theme:

Know your customer and do everything feasible to satisfy your customer.

This is done on the individual, mainly internal, level and also on the macro level as a provider to the client. This principle of customer identification is the key to successful process control practice in a TQM environment.

## SUMMARY

This chapter has presented the process control system for use in job shops. Most textbooks on quality control concentrate on continuous-flow manufacturing and emphasize statistical analysis. Continuous manufacture is used for commodity-type products, so that an individual failed part can be disposed of and be replaced with one of many more satisfactory parts. However, 70% of all manufacturing in North America involves discrete parts and is job shop–oriented. Hence there is a need for relevant information on job shop process control.

Process control in job shops requires attention to detail for each part, whereas in continuous or mass production it would be impossible to check each part after each operation. Each part in a job shop usually carries an identification, that is, a part number and a shop order number. This means that each part has a distinct value for a specific project. In continuous-flow manufacturing, individual parts have no identity; therefore, techniques based on statistics and probability theory can be effectively applied. If a part fails in manufacturing in a job shop, an entire project may be delayed. This difference sets the tone for process control in job shops. The specific part is important and must be carefully evaluated throughout the manufacturing cycle. Statistical concepts can be used, but the importance of the individual part must be maintained. The checks and balances must be aimed at thorough control of processes to ensure that individual parts are indeed of adequate quality to achieve their design purpose. Finally, TQM theory espousing continuous improvement leading to zero defects in manufacturing applies to job shop process control.

## REVIEW QUESTIONS

1. Refer to the methods sheet in Chapter 7, Fig. 7.1. For each step shown, list possible quality actions that the operator and/or inspector will perform.

2. Describe how the choice of a measuring tool can dictate the process by which a part will be made.

3. Discuss how process control engineering acts as the communications link between the manufacturing and engineering functions. Use Fig. 8.1 to illustrate two key points of the discussion.

4. Write a cookbook-type quality plan for the methods sheet in question 1.

5. Discuss why job shop quality plans are normally written around generic processes rather than parts.

6. Explain why operator certification is a better form of process control than inspector certification.

7. We wish to test frame support welds, by ultrasonic techniques, for railroad car undercarriages. The production schedule requires two undercarriages to be completed per 8-hour shift. No defects in the welds can be tolerated and an angle beam test ultrasonic equipment is available. Only trained ultrasonic technicians are allowed to use this equipment, of which there is one unit available per shift. It is estimated that one frame support weld test can take 15 to 30 minutes to perform. Using the five rules for testing in a factory, prepare a test plan outline for the frame support welds.

8. For question 7, discuss the type of data that can be gathered and how the data can be presented to give management an accurate evaluation of the quality of the work being performed.

9. Evaluate the quality trend of a shaft axial manufacturing area. The following data have been collected.

| Week | 1 | 2 | 3 | 4 | 5 | 6 |
|---|---|---|---|---|---|---|
| Operator errors | 6 | 8 | 13 | 5 | 9 | 12 |
| Machine-caused errors | 2 | 6 | 6 | 8 | 5 | 3 |
| Defective material | 2 | 1 | 2 | 3 | 2 | 4 |
| Shaft axials produced | 30 | 33 | 29 | 36 | 32 | 30 |

10. When is "fire fighting" a valid process control engineering techuique? Discuss it in comparison to long-range corrective action programs.

11. With respect to process control techniques, state what a temporary preventive maintenance routine is.

12. For the data in question 9, discuss how the data and/or the interpretation would be reported to the following levels of management: (a) manager of process control engineering, (b) manager of manufacturing engineering, and (c) manager of manufacturing.

13. Explain how a temporary corrective action program becomes a permanent addition to the quality plan.

14. Using the data of question 9, construct a scenario showing how process control engineering would interact with each of the other manufacturing engineering units. Show how these data could affect the activities of the other units.

15. Valve cover joint studs of 3 in. diameter are produced on a CNC chucking lathe. The data below are from a sample audit taken to verify that the process is under control. The engineering drawing requires a stud of 3.000 ± 0.015 in. From the data, determine whether the process is under control by determining the mean and standard deviation.

| Sample | Diameter |
|--------|----------|
| 1      | 3.005    |
| 2      | 3.007    |
| 3      | 2.987    |
| 4      | 3.006    |
| 5      | 3.008    |
| 6      | 3.009    |
| 7      | 2.995    |
| 8      | 3.010    |
| 9      | 3.003    |
| 10     | 3.002    |

16. Sample means for the valve cover studs in question 15 are: 3.010, 3.009, 3.007, 3.005, 3.006, 3.008, 3.004, 3.001, 3.000, 2.998, 2.995, 2.992, 2.990, 2.991, and 2.989 for days 1 through 15, respectively. Construct a control chart for periodic average values and interpret the data.

17. Based on the data of question 15, use the probability technique of the standard deviation of a normal distribution to predict the percentage of valve cover studs expected to fall within this tolerance band of better than two standard deviations if the process remains the same in the future.

18. Discuss the issues related to converting a process control function's methods of operation in a company entering into contracts with suppliers based on AQLs to one requiring parts to be acceptable in the parts-per-million range.

19. Define the term "zero defects" in the context of a manufacturing organization: continuous flow and job shop. Is it possible to achieve zero-defect manufacturing? Explain your answer.

20. Why do the process capabilities of job shops increase many times over when continuous-improvement awareness factors are introduced?

# NINE

## MAINTENANCE ENGINEERING

It is essential for a factory to maintain its equipment in order to produce a product within an acceptable cost and time period. If the equipment in the factory is erratic in performance, only an underoptimal result can be obtained. How erratic the equipment performance is will determine the degree of underoptimal operation. The mission of maintenance engineering is to severely depress the occurrences of erratic equipment performance, thereby minimizing a negative productivity factor.

The basic task of maintenance engineering is to develop preventive routines for equipment so that it is capable of running to specific design requirements. This means that maintenance engineering is interested in determining what checks, inspections, and replacement of parts should be performed during scheduled downtimes so that the machines run as designed and unscheduled downtimes are prevented. However when breakdowns occur the unit must react quickly to return the affected equipment to service. With these objectives in mind, let us examine the techniques used by the maintenance engineering unit.

## TASKS ASSIGNED TO MAINTENANCE ENGINEERING

Four basic tasks are assigned to maintenance engineering:

1. Preventive maintenance
2. Rapid emergency repair

3. Toolroom activities
4. Buildings and grounds maintenance

All four activities can require a wide variety of engineering disciplines to achieve successful results. When making repairs to machine tools, the skills and knowledge of electronic engineering may be required to diagnose a Computer Numerical Control (CNC) problem, and those of mechanical engineering may be necessary to determine the risks of bearing instability if a substitute lubricating oil is used.

In emergency repair work it is common to make replacement parts in the toolroom without having the vendor's detailed drawings, so that the maintenance engineer becomes a design engineer, producibility engineer, and methods engineer all at the same time. The maintenance engineer must have more than passing competence in these areas if the repair parts are to work and the equipment run again. We can see that maintenance engineering is complex and challenging. I often think of maintenance engineers as the land-based equivalent of shipboard marine engineers who can keep a ship afloat and operational in what appear to be impossible circumstances. The true maintenance engineer is of the same mold: creative, resourceful, and possessing a positive attitude toward keeping the equipment operational. Management of maintenance engineering then involves creating a system of controls and reporting within which these engineers can operate effectively. Before examining maintenance engineering management techniques, let us review the four basic tasks of the maintenance engineering unit.

## Preventive Maintenance

The basic goal of maintenance engineering is to prevent unscheduled shutdown of equipment due to failure. Therefore, preventive maintenance is a primary activity of the unit.

Successful preventive maintenance requires a precise and thorough understanding of the equipment used in making the company's products. This includes understanding the design philosophy of the equipment, including what makes it work and what must be considered the weak links of the system, and understanding how the equipment is to be used in the factory. This dual view of design use versus actual use sets the tone for how preventive maintenance will be applied.

**Example (a):** Suppose that a lathe designed to carry a load of up to 600 lb is to be used frequently at or slightly above this load. What would the preventive maintenance scheme be like? Obviously, operating above recommended load levels is not the best idea, but occasionally a firm considers the risk of overloading preferable to an expenditure of capital funds to purchase different equipment. Management in this case is deciding to cut into the equipment's factor of safety, recognizing that the load rate was set conservatively low by the vendor to protect against warranty failure. Therefore, the maintenance engineer would look for the weak links in the lathe design and schedule them for frequent examinations and possibly replacement of parts before a failure occurs. The design's weakest link would probably be the bearings of the machine, because they are being asked to carry a continuous overload. Therefore, after a certain period of operating time the bearings would be inspected, and after a longer period the next weakest link along with the bearings would be inspected, and so forth, until a time interval was reached where the bearings would be replaced before a failure occurred. The replacement time would be determined by calculating the expected bearing life with the sustained overload.

In the above example, the weakest link was identified and a preventive maintenance program was designed to fit the user's specific circumstances.

**Example (b):** If the lathe was to be used commonly for far lighter pieces, then a preventive maintenance program might have been focused on other things—perhaps the lubrication systems to make sure oil lines are not plugged— and the bearings looked at only rarely. Here the usage dictates examining not load-carrying components but systems components.

These examples show that maintenance engineering cannot simply follow vendor-suggested preventive maintenance techniques and ignore the specific utilization of the equipment. Preventive maintenance techniques suggested by vendors are based on a generalized concept of the use of the equipment and a desire to protect themselves from negligence lawsuits. Therefore, maintenance engineering should use the vendor's techniques as guidelines but not as rigid requirements.

In these examples operational parts are sometimes replaced before they fail. This is done to control downtimes and not let failure dictate when machines are not usable. Many parts that appear to be usable have reached the end of their statistically usable lives and are in imminent danger of failing. Good preventive maintenance is based on knowing the useful life of components and replacing them prior to failure.

The techniques for determining the effective life of component parts, such as roller bearings, are well known and are covered extensively in textbooks on machine design. Maintenance engineers are competent in these techniques. However, the body of knowledge applied by maintenance engineers encompasses virtually the entire spectrum of engineering. What is necessary in a particular company depends on the type of product being manufactured and the type of equipment used. In an electronic assembly operation we would expect to find automated equipment and the equipment containing electronic circuits, especially for testing. Therefore, the basic skills employed by the maintenance engineer would be those of an electronics engineer. Similarly, a factory producing artificial fabrics would be basically a chemical plant, and the maintenance engineer would be skilled in chemical engineering techniques and perhaps the fluid flow techniques of mechanical engineering.

## Rapid Emergency Repair

Most people think of emergency repair as the focus of maintenance engineering. However, like the emergency room of a hospital, it does not represent all the techniques and services represented and is not the preferred method of treatment.

Rapid emergency repair is the ultimate fallback position of the maintenance engineering unit. If programs of preventive maintenance fail, or an accident occurs, or an operator error results in machine failure, the only practical recourse is repair in the most expeditious and economic manner. The only alternative to repair is to scrap the equipment and replace it. If this is the case, the maintenance engineering unit's involvement comes to an end and the project is assigned to advanced manufacturing engineering. Since the latter is a significantly rarer occurrence than making the required repair, let us look at the philosophy of emergency repair and see how it affects the actions of the maintenance engineering unit. Remember that this is emergency repair, not repairs made as part of a planned downtime where the activity has been extensively planned.

The first aspect of emergency repair is that there can be no specific preplanning. The nature of the repairs can vary widely, and the exact requirements are never known until the need for an emergency repair becomes evident. This means that the maintenance engineering unit must be competent to handle a large variety of repairs and be stocked in a general sense with a wide variety of repair materials. This broadness of application is often a problem in determining stocking levels for spare parts. How many sizes of pressure regulators or capacitors can the maintenance engineering unit afford to stock? Think of how often home repairs are thwarted because the jar of screws and nuts does not contain one of the proper length, diameter, or thread pitch. The same is often true in factory repair activities. To minimize the need to stock a variety of sizes, maintenance engineering makes sure that AME knows what components for all equipment should be standardized, down to the level of thread sizes and electrical hardware. One of the critical factors in being able to effectively cope with emergency repair is to have a well-stocked inventory of generalized and standardized spare parts.

The second aspect of successful emergency repair is the ability to diagnose the problem quickly and correctly. This has a large impact on the economics of the factory. A machine that is down costs the company productive capacity, hence lost profitability. Machines usually cost in the neighborhood of $100 per hour to run, so that every extended hour of lost production will cost $100 plus the cost of coping with the problem that one of a series of machines needed to produce the product is down. Therefore, there is a need for the maintenance engineer to respond quickly and make an accurate assessment of the damage and how to fix it.

The third requirement in successful emergency repair is that the equipment be fixed quickly after it is determined what has to be done. This quite often requires a balancing of resources by the manager of maintenance engineering. There are only a finite number of people available to be dispatched to any number of jobs, so it is necessary to set priorities. The manager of maintenance must keep informed about the situational needs of production in order to correctly assign priorities to the repair jobs. The need to perform preventive maintenance economically must be balanced against the need to accomplish rapid emergency repairs. This may sound like a contradiction; however, not all breakdowns of equipment dictate a need to repair the equipment rapidly.

**Example:** If a company has a half-dozen engine lathes engaged in making similar parts, it may not be necessary to fix one lathe when it breaks down. It may even be possible to schedule the repair at the convenience of the maintenance engineering unit. This would be the case when all six engine lathes are not needed to make the production quota at that time. If only four lathes are necessary, then two would be idle before the breakdown and work could be shifted to one of the idle machines without production output suffering. If the breakdown occurs at a critical machine, then maintenance engineering has no choice but to react rapidly to repair it.

The setting of priorities must have a rational basis. Critical machines must always be serviced first. Other equipment will receive immediate attention only if personnel are available. Preventive maintenance work is abandoned only to fix critical machines; otherwise it is carried out on schedule. Whether a machine is considered to be critical can change with production circumstances, and the manager of maintenance engineering must be aware of present conditions.

Preventive maintenance work can sometimes turn into rapid emergency repair. Suppose that a critical machine tool has been scheduled for a preventive maintenance check and minor overhaul. Since it is a critical piece of equipment, the maintenance engineering unit is given a specific time when it must be available for shop operations. If a serious problem is discovered during the work and the return to service by the due date is jeopardized, then a scheduled downtime must be treated as an emergency repair to meet the commitment. In this situation other critical machine breakdowns will not be allowed to supersede this preventive maintenance work.

Computer programs have been developed to help management decide how to set priorities for repairs and preventive maintenance. In most cases these programs are based on a hierarchical branching network decision-making algorithm. The answers to a series of questions are rated to indicate how critical an incident is to current production needs and therefore how high its priority should be. The problem is usually how to decide whether one product component manufacturing area is more vital than another. No matter what kind of algorithm is designed, this is still basically a subjective decision. All the programs can be expected to do is perform the much needed clerical function associated with job dispatching. To illustrate how transient priorities can be, suppose that the manager of manufacturing gets a call from an important customer who needs replacement bearings immediately. Since the customer is important, the manager will expedite the manufacture of the bearings, and any breakdown in equipment becomes a critical machine failure. Now suppose that the same customer calls later and needs electrical circuit boxes before the bearings. The machines that cannot be allowed to stay down now become the sheet metal forming and assembly equipment. The bearing machines are replaced by the electrical circuit box machines at the top of the critical list. With such a dynamically changing situation, it is difficult to design computer programs that are flexible enough to set priorities for maintenance responses.

We have also seen the introduction of Artificial Intelligence (AI), particularly expert systems, computer systems for use in repair strategy diagnostics. AI programs are discussed in Chapter 11.

## Toolroom Activities

The toolroom is a support function for preventive maintenance and repair activities. It is set up to supply "homemade" parts to the maintenance crews, and is also a primary source of manufacture of tools, jigs, fixtures, gauges, and other miscellaneous items for production needs specified by the methods engineer. In addition, the toolroom is a place where prototypes of new products can be built and evaluated.

Another important feature of toolrooms is that they are staffed by the most skilled machinists and craftsmen on the company's payroll. They must often make precise parts to meet urgent needs without definitive engineering drawings, commonly by using the failed parts as models. To accomplish this type of task, the toolmaker must possess not only a high level of competence in operating a variety of machine tools, but also an excellent knowledge of machine component parts in order to reconstruct the original from the failed reference part. The maintenance engineer aids in this process by explaining what the part does and how it works, but the major reconstruction task still falls to the toolmaker.

Toolrooms are characterized by having many precise but manually controlled machine tools. All are capable of very fine tolerances but usually have a limited output rate. Toolrooms are measured not by output volume but by the quality of the output and the ability to make single parts as required. Therefore, the key measurement associated with toolrooms is their readiness to serve. This is a highly valued attribute and most manufacturing companies are willing to fund activities on this basis in preference to most other measurement concerns.

Material stocks for toolrooms are usually kept separate from other company stocks. One reason for this is that the toolroom material stocks usually represent a cross section of materials that match the basic components of the firm's production equipment. Unless the company is a machine builder, it would be unusual for the tool steels and gear blanks needed for repairs to match the types of materials the company uses for its sales products. Another reason is that it is risky to keep the material in a common location with production stock. Its volume is always minuscule compared to that of production material, and it could be used in error by shop operations or could be lost. Finally, if the toolroom material is segregated it can be found and sent to the proper machine tool rapidly, which is often important.

Making repair parts for machines and equipment is the primary purpose of the toolroom and always receives top priority. However, the bulk of its activities revolves about the methods engineering task of producing jigs, fixtures, and gauges for production. Making production fixtures usually carries the major financial write-off and covers part of the cost for maintaining the readiness to serve capability. The toolroom will prepare quotes for making the jigs and fixtures, carrying the proper overhead rates to a designated portion of the costs for readiness to serve.

The maintenance engineering manager must be conscious of outside costs and must make sure the toolroom does not become a profit center at the expense of the shop operations units. This would contribute to raising the selling cost of the company's products and would ultimately lead to pressures to minimize the extent of toolroom activities. A proper balance must be struck in pricing toolroom work so that the toolroom costs are covered and costs to overall operations do not have a negative impact on competitiveness.

Maintenance engineers use the expertise of toolroom personnel in determining how repair parts can be made and estimating whether the repair part can be considered a permanent or a temporary fix. Often the materials stocked by the toolroom are insufficient for permanent repair but satisfactory for temporary repair. Since the toolroom does not have access to the vendor's design calculations, it is not surprising that a materials match is not made. Usually the toolmaker will test hardness to determine whether the available material is sufficiently strong to withstand the loads that caused the original part to fail. With this information the toolmaker can discuss the risks involved in using the homemade part with the maintenance engineer, who can then determine whether it will work and for how long. In evaluating the risks to be faced in getting the down machine on line again, the maintenance engineer benefits greatly from the master craftsmen expertise of the toolmaker. Therefore, one of the ongoing tasks of the maintenance engineering manager is to foster this relationship.

## Buildings and Grounds Maintenance

The fourth and final responsibility of maintenance engineering is the "all other" category called buildings and grounds maintenance. This is a broad responsibility covering activities that range from good housekeeping, such as sweeping up metal chips in the factory aisles, to complex engineering activities such as managing energy conservation efforts. Virtually anything deemed to be technical but not specifically assigned to other manufacturing engineering units is assigned to the maintenance engineering unit via the buildings and grounds category.

No specialized organization is set up to manage buildings and grounds maintenance (i.e., nothing similar or parallel to a toolroom). Typically the work is dispatched in the same way as any preventive maintenance or repair activity. Buildings and grounds maintenance activities are usually assigned in an integrated dispatching system used for all maintenance work. For example, in making work assignments to the plumbers, no distinction would be made between serving the cooling unit of a machine tool or the hot water system in the plant cafeteria.

Minor exceptions to this commonality policy exist, but they always occur within very specialized areas of activities, such as the operation of the cleaning crews. Cleaning of the offices and workplaces is a specialized activity within the maintenance engineering unit because there is no other place to assign the responsibility.

Buildings and grounds maintenance covers so much that it is difficult to adequately describe the breadth and scope of the activity. The ability to respond creatively is of paramount importance. The maintenance engineering unit may consider this activity to be secondary to keeping the production machinery operational. However, the success of this activity usually has a positive effect on morale. Keeping the workplace environment pleasant tends to improve morale, hence indirectly helps to improve productivity.

One buildings and grounds maintenance activity, that of energy management, does have a direct influence on the company's profit and loss statement. Since the oil embargo of the early to middle 1970s and the subsequent rise in the price of energy, management of energy usage has become a vital aspect of business control. The maintenance engineering unit has the responsibility of monitoring energy costs and planning for energy cost savings. This is done through a multifaceted program. First, energy-producing and delivery equipment is monitored for efficient operations.

**Example:** Steam traps should be inspected at frequent intervals to ensure that they are functioning properly. Steam traps that allow steam to pass directly to the condensate lines obviously cost the company money because of the inefficient use of energy.

Second, the maintenance engineering unit is involved in conducting surveys and programs to reduce energy costs.

**Example (a):** A program to replace broken window frames, sashes, and glass with high-impact plastic and insulated substructures.

**Example (b):** Measurement of lighting levels throughout the offices and factory area. If lighting levels are found to be significantly above those recommended for particular types of work, energy savings can be realized by removing the excess lighting fixtures.

The third facet of the energy cost reduction activities would be conservation through awareness programs and the establishment of policies aimed at conservation.

**Example:** Appointing area energy monitors with responsibility for turning out lights at the end of the workday.

Most companies are making strong efforts to reduce energy costs, and the significant progress needed will depend on a sound engineering approach. Maintenance engineering is usually assigned that task.

Now that we have reviewed the four basic tasks of the maintenance engineering unit, let us see how the unit is organized.

## ORGANIZATION FOR EFFECTIVE MANAGEMENT OF THE MAINTENANCE ENGINEERING UNIT

This unit consists of engineers, specialists, and all the skilled tradesmen necessary to maintain the company's physical equipment. Physical equipment means virtually everything the company owns, from buildings to typewriters to machine tools. In order to organize this vast territory, the unit is usually broken down into the categories shown in Fig. 9.1, with the leader of each reporting directly to the manager of maintenance engineering.

### Preventive Maintenance and Repair

This is usually headed by the maintenance foremen, where the plural is used because the responsibilities are usually divided by shift and perhaps by plant geography if the plant is large. The foremen direct the activities of skilled tradesmen such as machine repairmen, electricians, plumbers and pipe fitters, structural steel workers, and carpenters. In addition, a number of laborers are assigned to do the nonskilled portion of the work. No engineers are assigned to this subunit.

### Toolroom

The toolroom consists of the skilled toolmakers, senior machinists, and other equipment-related tradesmen. In addition, most toolrooms have draftsmen and perhaps a CNC programmer assigned to the staff. Since this activity tends to be

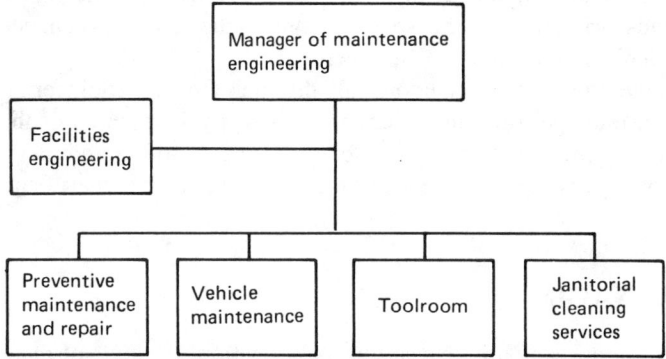

**Figure 9.1** Categories of activities in the maintenance engineering unit.

large, with a staff of 40 or more, a supervisor reporting to the manager of maintenance engineering usually administers the toolroom. Under the supervisor there would be one or more foremen and a purchasing clerk. The toolroom is set up in virtually the same way as shop operations units. No engineers are assigned to this subunit.

## Janitorial and Cleaning Services

This is the cleanup crew for the company and usually consists of low-skilled, entry-level personnel who all have one job description or perhaps two (e.g., cleaners, and floor polishers). The staff varies in size depending on the extent of the company's facility and is managed by a foreman who reports directly to the manager of maintenance engineering. If the activity is large enough, the staff will also include a purchasing clerk. The activity operates semiautonomously, and many of its people operate on the night shift. No engineers are assigned to this subunit.

## Vehicle Maintenance

This activity exists only if the company has an extensive fleet of vehicles, for both outside use and internal material handling. If the vehicle population is small, vehicle maintenance is either handled by the preventive maintenance and repair subunit or contracted to an outside firm. If the maintenance engineering unit does have a vehicle maintenance subunit, it is headed by a foreman who reports directly to the manager of maintenance engineering. Under the foreman are a purchasing clerk and a staff of auto mechanics who are experienced in lift truck and other specialty vehicles as well as ordinary trucks and automobiles. No engineers are assigned to this subunit.

## Facilities Engineering

All the engineers assigned to the maintenance engineering unit are in this subunit. The organization may be of two types. For a large organization, a supervisor of facilities engineering is appointed. For a smaller organization, each facilities engineer would report directly to the manager of maintenance engineering. The choice of organization depends primarily on the span of control that the maintenance engineering manager feels able to handle effectively.

The facilities engineering subunit supports all the maintenance engineering subunits. In addition, it must perform all the engineering analysis required of the maintenance engineering unit. Therefore, this subunit serves in both an advisory and a direct-action capacity, hence its special place in the organization chart (Fig. 9.1).

## METHODS OF MANAGING

Since the maintenance engineering unit is a service organization, the style of management must be such that it can respond quickly and be able to change priorities just as quickly. Unlike the other manufacturing engineering units, a significant part of the detailed work content of the unit is essentially unknown. The preventive maintenance and janitorial and cleaning activities can all be scheduled. However, the repair work cannot be scheduled except for short periods, that is, a week to perhaps a month. For this reason, although many of the assigned responsibilities of the maintenance engineering unit are managed in the normal manner, the quick-response portion requires unique management techniques to be performance and cost effective.

Three unique management techniques are used by maintenance engineering:

1. Order entry system
2. Shop maintenance information system
3. Preventive maintenance system

Each of these systems can be either manual or computer assisted. Each affects the others but can be bounded and studied separately.

## Order Entry System

This system prevents chaos in responding to requests to repair product-producing and other equipment. It consists of a method of recording jobs, determining whether the jobs can be worked on, determining the work load for a particular time period, recording results, and projecting the future work load. Figures 9.2 and 9.3 represent the basic documents of the order entry system.

The job log and job effectiveness forms make it possible to plan the work load and measure the effectiveness of the work force. With this system it is possible to determine a manpower level for the short to medium term and schedule effectively for at least 1 month. It is also possible to do partial work scheduling for approximately 3 months.

Let us review the job log form (Fig. 9.2) and discuss the data to be found there. The form provides for recording all unplanned repair jobs given to the maintenance engineering unit by date of entry, job description, and job number. When a job is entered on the form, the responsible facilities engineer must estimate the number of hours it will take to fix the failed item and select the craft or crafts necessary to do the work. Note that the hours are placed under the respective craft columns. If a repair job requires more than one craft to accomplish, the job will be entered as often as there are crafts required to accomplish the repair, with the estimated hours shown for each phase of the work. It is also necessary to estimate whether or not material is available for making the repair. If not, either the material is ordered or a work order is placed in the toolroom to make it. The job log also shows the available for work and unavailable for work backlogs in hours for each individual job and, for statistical information, the number of weeks that the job has been carried on the report. The report is issued weekly, and all jobs that have not been completed as indicated by the last weekly report are carried

Job Log

| Date entered | Job description | Job no. | Machine repair | Carpenter | Electrician | Plumber | Steel work | Other | Material available (yes/no) | Hours backlog | Hours available for work | Weeks on report | Job complete |
|---|---|---|---|---|---|---|---|---|---|---|---|---|---|
| | | | Estimated work hours by craft | | | | | | | | | | |
| 10/23 | Fix roof leak | 83 | | 110 | | | | | yes | | 110 | 8 | complete |
| 10/31 | Repair motor, lathe 2201 | 94 | | | 6 | | | | yes | | 6 | 5 | |
| 11/1 | Repair drill press 9024 | 95 | 16 | | | | | | no | 16 | | 4 | |
| 11/25 | Repair leaking sink, ladies room | 116 | | | | 2 | | | yes | | 2 | 1 | |
| 11/25 | Fix stair well crack | 117 | | | | | 20 | | yes | | 20 | 1 | |
| Total | | 26 | 273 | 178 | 210 | 85 | 176 | 16 | | 288 | 620 | | |
| Work force hours available | | 17 | 240 | 160 | 200 | 80 | 160 | 80 | | | | | |
| Work available | | | 165 | 162 | 142 | 38 | 113 | 0 | | | | | |
| Backlog | | | (75) | 2 | (58) | (42) | (47) | (80) | | | | | |

Comments
Hours available for preventive maintenance
Machine repair = 75
Electrician = 58
Plumber = 42
Steel work = 47

**Figure 9.2** Part of a job log form.

Week 9349

| | Machine repair | Carpenter | Electrician | Plumber | Steel work | Other |
|---|---|---|---|---|---|---|
| Required hours, staff available for repair work | 165 | 160 | 142 | 38 | 113 | 0 |
| Hours worked on repair work | 194 | 156 | 191 | 68 | 143 | 0 |
| Efficiency | 85% | 102.5% | 74.3% | 55.8% | 79% | — |
| Jobs available | 6 | 3 | 5 | 2 | 1 | — |
| Jobs complete | 5 | 3 | 5 | 1 | 1 | — |

Comments: (1) M.R. job no. 85 not started, wrong material received. (2) Plumber job 116, required replacement, sink cracked, pipe thread worn and needed rethreading, original estimate was for washer replacement only.

**Figure 9.3** Job effectiveness form.

forward and entered on the updated version. This ensures that all work not accomplished is carried on the unit's open work status.

The part of the form showing hours available to work and work force hours available establishes the next week's work plan. Here we see that the number of hours available to work represents an accumulation of all jobs that could be worked on that week. This means that all materials and instructions are available and, if manpower is applied, the job can continue on to a successful conclusion. This is an important concept. Maintenance engineering can only work on jobs that are available to work on, and all scheduling is based on being able to apply an effort to successfully complete the job. If this is not possible, for example, if there is no repair material, then the job must be ignored for scheduling purposes and not put back on the schedule until it can be worked on.

The number of hours of work force available is simply the allocation of people per craft converted to hours per week. The bottom line of the form, backlog, is the difference between the work available and the work force available. If the number is negative (indicated by parentheses in Fig. 9.2), there is an excess of work force available. This excess is shown in the comments section as hours available to do preventive maintenance work. If the backlog is a positive number, there is more work to do than work force available to accomplish it. This is an available for work backlog and will be carried over to the next week, usually in the form of a complete job rather than a partial job.

**Example:** The backlog of 2 hours for the carpenter craft, if it is a job estimated to require 2 hours, would be carried as a 2-hour job in the job available column for the next week. If the job is estimated to take 5 hours, it would appear as a 5-hour job for the next week and 3 hours would be subtracted from the work available line of the log.

The job log readily informs management of its capability to do the work that is available to be done. A look at the backlog line shows whether it is possible to accomplish the needed work and whether there is an excess of manpower available. To determine whether excess manpower actually exists, it is necessary to consider the unavailable backlog hours, because they will eventually become available hours. Another factor to be considered is the hours that are delegated to perform preventive maintenance, which are also shown on the form. These are not the total hours available for preventive maintenance, since usually a maintenance engineering unit will reserve certain hours for preventive maintenance work, and the excess hours shown on the job log are supplemental to this reserved time.

The job effectiveness form is used to evaluate the efficiency of the repair operation. The first line of the form is a data input from the job log showing the work available to be performed during the planned week. The second line, hours worked on repair work, indicates the hours during which the scheduled work was actually performed. The efficiency of the operation, shown on line three, is simply the planned hours divided by the actual hours. Here the planned hours are the job estimates made by the facilities engineers and are not as exact as planned hour determinations developed for direct labor activities. It must be recognized that planning how to make a product or part is different from planning how to fix a failed piece of equipment. In the former case all the facts are known, whereas in estimating a repair job the facilities engineer must first diagnose what is wrong and what caused the failure. This diagnosis can be a simple procedure, as implied in the second comment in Fig. 9.3, or it can be very complex, such as evaluating the cause of a malfunction on a CNC unit. Even simple diagnoses can be in error, again as illustrated in the second comment in Fig. 9.3. From these brief illustrations we can see that estimating for repairs is an inexact procedure and cannot be fairly compared with the detailed time standards used for production planning.

Nevertheless, the estimate is useful. Machine tools and equipment tend to fail at the stress points of their design. Therefore, failures tend to be repeatable over a period of time, and a similar failure would result in a similar estimate. In other words, series of symptoms tend to indicate particular causes of failure, which leads to somewhat standardized repair procedures. For this reason, the use of planned hours versus actual hours does give a good measurement of maintenance labor efficiency. The reasons for measurements of the work force are the same as those discussed in Chapter 7.

The last two lines on the job effectiveness form indicate the ability of the maintenance engineering unit to complete jobs according to schedule. The hours of the first three lines make up the jobs in total shown on the jobs available and jobs complete lines. It is necessary to know this to properly evaluate the efficiency measurement. In the machine repair column of Fig. 9.3 we see 194 hours worked versus 165 hours available for work (planned), which gives an efficiency of 85%. This is the efficiency value to be compared with other weekly performances. However, note that only five jobs were completed, while the hours planned were based on six jobs. Hence, the efficiency calculation is in error; the 194 hours were performed against somewhat less than the indicated 165 planned hours. Thus the last two lines of the job effectiveness form show how valid the efficiency measurements are.

## Shop Maintenance Information System

This is an archival record system used by maintenance engineering in order to have data available for decision making. The data are needed in determining whether a machine has reached the point where a complete rebuild or a replacement is necessary, or in determining work assignments based on machine availabilities. Decisions are also required to effectively determine and control operating costs. The shop maintenance information system has three parts: (1) the machine tool capability and replacement log, (2) the daily operating status, and (3) the current period and cumulative cost status.

The titles of these parts may vary, but the content of the SMIS is relatively consistent. This system is usually handled through a computer-based data input program, where the outputs are always the same general triad. Let us consider the philosophy of the system in order to understand the management techniques involved.

**Machine tool capability and replacement log.** The MTCRL is the basic document that shows equipment performance data. It includes such information as the accuracy of current machining compared to original specifications and runoff data, the capability of the machine to operate across the entire range of feeds and speeds, and the number of times the equipment has failed and the reasons for failure. The MTCRL can be called up to give the facilities engineer the history of the specific equipment to aid in diagnostic work.

Assistance in diagnosis is an important reason for having an MTCRL, but an even more vital function is supplying data to the AME and MP&WM units. The AME unit uses the data to determine when equipment is approaching the end of its useful life and must be replaced. This information is essential for planning capital investment spending and is used particularly in the development of Long-Range Forecasts (LRFs). Without this information LRFs would be based on the faulty assumption that all existing equipment is in first-class shape, which would be tantamount to assuming that large expenditures for repairs never occur.

Similarly, planning for product production is based on the ability of equipment to perform in accordance with its design. If planning is based on an original or specification sheet value of machine tool performance, it is probable that the hours calculated will be too optimistic. For this reason it is important for methods engineers to have access to current data pertaining to equipment performance. This MTCRL information allows the MP&WM unit to generate planning that is realistic, hence believable by the operators and foremen.

**Daily operating status.** The daily operating status is a report showing what pieces of equipment are not available for use ("down") and what pieces are available for limited use ("limping"). If a lathe can only turn up to 50 RPM compared to its normal capability to run at 100 RPM, it is considered to be limping. Similarly, if the lathe cannot rotate the workpiece at all it is considered to be down.

The "down and limping" report is very important to shop operations, particularly the production control unit, which must constantly decide what product components should be loaded to what workstation. The daily operating status is an important data input in determining how such loading should be done. Although production control would like all equipment to be operational, it is important to identify the machines that are inoperable so that alternative manufacturing arrangements can be made.

The production control manager also has information to decide whether a limping machine can be used with any degree of effectiveness. By knowing what is nonfunctional and relating that to the operations the machine must perform on the part, the manager can decide whether to assign work to the machine. Another useful piece of information is the percentage of nonworking features, as shown in Fig. 9.4. If a high percentage of the capability is unavailable, as shown for workstation 12C20, it would be unlikely for the production controller to assign work to that workstation.

The daily operating status report shows at a glance the status of the repair business. This enables the maintenance engineering manager to decide whether to divert personnel from preventive maintenance work or other non-production-oriented work to repair work. The three lines at the bottom of the report summarize the magnitude of the current repair work load. Most maintenance engineering managers are keenly aware of the trends of numbers of machines down and limping as well as the number of days a piece of equipment shows up on the report. In many ways this report reflects the performance of the maintenance engineering unit. There are two particular features to evaluate when looking at this report over a period of time.

First is the length of the list. Increasing numbers may mean that preventive maintenance is not being done properly or that there are inexperienced operators at the workstation. Whatever the cause, the maintenance engineering manager would wish to investigate the changes that have caused a change in the length of the list.

The second feature is how many days an item remains on the list. A trend toward longer stays on the list could indicate problems with the performance of the repair personnel or problems with the repair parts supply system. It might also indicate that the toolroom is not keeping up with maintenance demands, perhaps

Date: 11/29/93

| Workstation No. | Description | Down | Limping and non-working feature | Promise repair date | Days on list |
|---|---|---|---|---|---|
| 2J32 | CNC lathe | | 30% Auto feed | 11/29/93 | 2 |
| 17A12 | VBM | √ | | 12/2/93 | 6 |
| 12C20 | CNC HBM | | 70% tool changer | 12/1/93 | 5 |

Number limping: 6
Number down: 4
% Workstation unavailable: 14%

Figure 9.4 Typical daily operating status report.

because it is favoring prototype production. Distinct changes in the average days on the list warrant investigations for cause.

**Current period and cumulative cost status.** This third major output of the shop maintenance information system is a summing of the hours required to repair down or limping equipment converted to dollars, plus the cost of materials and other services required to make the equipment operational again. It is important for any business that must work against a predetermined budget.

The information generated here is useful to AME because it helps to quantify decisions involving appropriations of funds for repairs, rebuilds, or replacements. Without such information, and short of a machine failure, it is difficult to convince management to replace worn-out equipment. The manager of manufacturing engineering usually delegates the collection of maintenance cost data to the manager of maintenance engineering, who is charged with keeping accurate records of costs to repair and notifying AME when the cost to repair any piece of equipment becomes excessive.

The cost-collecting activity is also done to compare actual figures with estimated figures related to repair work. The purpose is to identify variances between actual and forecast dollar amounts so that effective corrective actions can be taken.

## The Preventive Maintenance System

So far in this chapter we have referred to the existence of such a system, and have learned that excess hours available after repair work is assigned are allocated to Preventive Maintenance (PM) and that the daily status reporting may indicate whether PM has been effective. We will now discuss PM in more detail and show how the maintenance engineering unit approaches this responsibility.

In PM we are interested in preventing complete or partial machine failure by inspecting the machine, replacing old parts with new parts, and adding disposables such as oil and grease as required. Therefore, a manufacturing engineering definition of PM would be the replacement of parts or components prior to their actual failure in order to keep the equipment operating as close as practical to an as-new condition.

The primary difference between PM work and ordinary maintenance work is that the latter involves replacing parts that are worn out, while the former consists of replacing parts that soon will become worn out (old parts) or that have reached the end of their useful existence.

A good example of PM is the electric lighting program used for many large office buildings. The calculated life of fixtures is used to establish a schedule for removing old bulbs and replacing them with new ones. Bulbs are replaced even though they are still functioning properly because from the calculation of hours of life and simple probabilities and statistics, it is known that failure is imminent. Replacement is based on the hours of utilization, not on failure. The result is that darkened rooms are a rarity.

In practice, manufacturing PM systems are not as pure as the light bulb system. They do specify changing oil and filters based on hours of usage, but they do not require taking out bearings that have reached a theoretical effective lifetime limit. The cost of doing so would not be justifiable because machine component lifetime depends on many factors that are almost unmeasurable. For bearings some of these factors would be load parameter histories including temperature differentials (i.e., hot or cold starts), the quality of the lubricating oil, and the exact lineup of the bearing components relative to the shaft. With so many variables the life expectancy is difficult to quantify, and a slightly different approach is taken in practical PM.

A practical PM program would require changing the cheaper or simple things based on the life expectancy approach. For instance, filters and oil would be changed on the basis of hours of operation. Next the maintenance operator would be required to check machine components such as bearings and gears on the basis of hours of operation. Here a probabilistic approach is used to determine when there is a likelihood that a component will fail. If the inspection shows indications of minute failure or wear, the maintenance operator will replace the part. Otherwise the part will be returned to service. Machines are inspected on a set schedule of operating hours based on probabilities of failure of the least survivable parts or components.

Depending on the complexity of the equipment, several stages of PM may be carried out. For example diesel engines usually have PM checks at 500, 1000, 1500, and 2000 hr. The 500-hr check is the most basic one. The 1000-hr check would include the items of the 500-hr check plus additional ones, and so forth, until the level is reached where a complete overhaul is required. After the overhaul the cycle begins again.

The managerial decision regarding PMs is to determine what they should be and how to schedule their accomplishment. To determine what the PM items should be, the facilities engineer obtains recommendations from the equipment vendor. These recommendations are evaluated for practicality and risks involved in not accomplishing them, and based on the findings the PM items are developed for the particular equipment.

The question of scheduling is always a sensitive one. Shop operations never wants to shut down a production line for PM, while the manufacturing engineer wants to maintain the schedule to prevent damage to the equipment and to allow for PM manpower planning. The maintenance engineering organization must remain flexible in accomplishing the PM work, and it is often done during the night or early morning shifts and on weekends. This poses a practical scheduling problem, which is partly resolved by a prenotification procedure employed by maintenance engineering. That is, the maintenance engineering unit informs shop operations when a machine tool is due for PM and asks shop operations to report back when the machine will be made available. These notifications, which are often computer generated, become flags for work tickets to be prepared and issued through a dispatch system to the PM and repair subunit. Based on the replies from shop operations on equipment availability, the PM work load is scheduled. This system works fairly well if the factory has a good load-scheduling capability and is operating on or close to schedule, and if there are no emergencies such as machine breakdowns to disrupt the schedule.

Traditionally, PM work fits in where it can even though efforts are made to schedule it. Therefore, most maintenance engineering units keep a skeleton staff permanently assigned to PM work and supplement the staff with other personnel as they become available. This manpower allocation system works if management has a good log system showing when equipment requires PM and if the PM work becomes a scheduled item similar to emergency and routine repair work indicated on the job log.

## SUMMARY

The maintenance engineering unit is a very diverse unit and in many ways a business in itself. In this chapter we have tried to develop an understanding of the management problems involved in maintenance engineering, a subject not usually covered in maintenance engineering texts.

The primary management problem for the maintenance engineering unit is how to control an unknown and constantly varying work load. The job log and the job effectiveness report are used by management to manage this uncertain load. These methods, coupled with the financial control procedures discussed in Chapter 2, give management the ability to adequately handle the maintenance responsibility.

We have also covered the reporting system used to monitor the status of equipment and create a history log. Status reports and documentation are vital in maintaining a well-run factory. The status of equipment reports are short-range data useful for daily shop loading and measuring the effectiveness of maintenance engineering, while the history documents compel management to recognize that no equipment lasts forever and that eventually the company will need to spend money for replacements.

Finally, we have discussed the philosophy of preventive maintenance and seen that it must be tailored for the individual workstations, with the goal of preventing unscheduled outages due to unanticipated malfunctions. The theory of PM is based on calculations of the number of operating cycles to failure, and parts are replaced before the total number of cycles is reached. In practice, modifications must be made to take into account costs of replacement parts and the difficulty in making accurate life-cycle calculations.

## REVIEW QUESTIONS

1. What is the basic mission of maintenance engineering? Show how this mission is integrated with the rest of manufacturing engineering.
2. What is the difference in concepts between preventive maintenance and

rapid emergency repair? How is this difference reconciled under one management team?

3. Describe the basic steps in creating a preventive maintenance plan for a machine tool such as a CNC vertical boring mill used to produce tight-tolerance inside diameters.

4. Describe the constraints placed upon the maintenance unit when it is engaged in emergency repairs and how they affect its methods of operation. In particular, include the lack of preplanning time and adequacy of spare parts availability and selection.

5. Discuss how senior management emphasis can influence the selection of emergency repair priorities.

6. In order of priority, outline the tasks assigned to the toolroom. Give reasons for your priority selection.

7. Outline the basic energy management program. Explain why it is a legitimate assignment for maintenance engineering.

8. The basic purpose of the job log (Fig. 9.2) is to identify all jobs and those that are available for work. How would a maintenance engineer decide whether a job is available to work on, and what factors could change it from unavailable to available and vice versa?

9. The job log tells management whether preventive maintenance time is available to be performed by the respective trades. Discuss how an emergency or breakdown repair scheduling system can assign time for nonemergency repair activities.

10. Why is a job effectiveness form (Fig. 9.3) a necessary part of the job log system?

11. The shop maintenance information system exists to provide management with an up-to-date evaluation of the factory's equipment. Describe how the three parts of this system are related and the data that each contains.

12. In reference to the data contained in Fig. 9.4, interpret the meaning of each of the three lines of data shown from the viewpoint of the production control manager and the manager of manufacturing engineering.

13. Discuss the basic tenets of practical PM systems. In particular, what pragmatic compromises must be made, and how does manufacturing engineering strive to comply with the needs of shop operations?

# TEN

## COMPUTER NUMERICAL CONTROL OF MACHINE TOOLS

The introduction of the computer to control a machine tool or process is the most revolutionary and profound occurrence in industry since the invention of the steam engine ushered in the industrial age. Computer Numerical Control (CNC) promises to allow low-skilled operators to produce craftsmanlike products simply by telling the machine to do it. It can permanently capture the craftsman's skills in the operation of the machine or process and duplicate or triplicate those skills throughout the entire factory. It opens up new opportunities for mass distribution of complex parts that were previously the sole domain of the skilled machinist or toolmaker. No longer is it necessary to merge a creative mind with an artisan's skills to produce complex shapes; CNC only requires the creative mind to tell the computer what steps are needed to create the complex part. This is a different type of creativity, one that understands how to mathematically model physical shapes and contours, and then devise a sequence a machine tool will follow to create those shapes.

But CNC is not without its costs. Since its introduction in factories in the 1960s, CNC has led to profound changes in manufacturing, the most important being a shift in required manufacturing skills. The skilled operator population has steadily declined, while the skilled CNC programmer population has grown. Factory maintenance personnel have changed from master electricians to electronics technicians and in some cases electronics engineers. Machines and processes have increased enormously in complexity, while the skill levels necessary to use the machines have decreased. The supporting structure, namely manufacturing engineering, has grown considerably to plan and service the equipment. This chapter concerns the philosophical management problems that must be dealt with to succeed in the CNC revolution.

## DESCRIPTION OF A CNC MACHINE

A machine tool is a device that can do one or more of the basic metalworking operations:

1. Drilling and boring—making holes
2. Turning—making externally round pieces
3. Milling—making flat pieces with a rotary tool
4. Planning, shaping, and broaching—making flat or concave pieces with a stationary tool
5. Grinding—shaping either curved or flat surfaces with an abrasive
6. Forming—bending, forging, drawing, and shearing

Each operation requires a degree of intelligence to guide the action. With an ordinary machine tool this is provided by the operator, who tells the machine what to do typically by setting the feed (depth of cut of the cutting edge, e.g., 0.010 or 0.025 in.) and speed (rate at which the cutting tool traverses the workpiece, e.g., revolutions per minute or surface feet per minute). Some machine tools require more than two settings. For instance, a lathe or a boring mill requires three settings: (1) depth of cut, the amount of material removed per pass; (2) in feed, the translation rate of the tool into a workpiece; and (3) part speed, the rotational speed of the workpiece.

With manual control machine tools the operator physically sets these parameters by controlling knobs, levers, and switches directly connected to transmissions, motors, and other devices for activating the machine. The machine continues to operate as set until the operator physically makes a change in the settings, so it is possible to machine off too little or too much material. The operator changes the settings when he has machined off just enough material, has the proper finish, and has achieved the drawing dimension—that is, the operator is intelligently directing the activities of the machine. This is the way industry has operated since the industrial revolution. It requires intimate knowledge of the machine tool's capabilities, skill in operation, and a relatively long apprenticeship to become a master machinist.

Suppose that the operator is absent. Unless we can find another operator, we cannot get the machine tool to work. But if we put a timer on the controls with a mechanical linkage that turned the dials and pushed the levers a measured amount, we could set the timer so that at period 1 the machine runs at feed 1 and speed 1 for $X$ minutes and at period 2 it runs at feed 2 and speed 2 for $Y$ minutes and then stops. Then all we would have to do is push the start button and walk away.

We would have created an outside source of intelligence, albeit crude, to control the sequence of operation of the machine.

Now if we had a great many of these mechanical trip lever times, each set slightly differently to give different amounts of mechanical dial twisting and lever pushing, and if we set these timers up in the proper sequence, we could conceivably automate the machine to make many different parts without directly controlling the machine. This would be accomplished by rearranging the mechanical trip levers to suit the particular need at the time, that is, by creating a program. All we would have to do to make parts would be to arrange the timers and trip levers in a programmed order and push the start button. This, essentially, is hard automation.

We have created a programmed sequence and mechanically hooked the controls to an assembly of many timers, spring trips, levers, and so forth. If we correctly calculated the time at each step versus the physical movement of the part and tool we should end up with a good part produced through various operations in sequence without operator guidance. This is automated control. If it is easy to rearrange the sequences, then the control is flexible automated control. If we replace the mechanical trip lever and timer apparatus with electrical signals and can vary those signals at will, then we are approaching logic or computer control. Finally, if we can write instructions in plan (programming) language and have a microprocessor interpret this into electrical signals to which the machine's controls can respond, we have computer numerical control.

Thus CNC is a modern version of the many different mechanical timer trip lever assemblies. A computer is used to provide intelligence to control the variable operations of the machine tool. The specific steps that we want the machine to progress through can be changed at will, and we have a form of control intelligence directing the machine's activities.

## CNC PARTS PROGRAMMING

To make a CNC machine tool work, the computer must be programmed, that is, instructed in the sequence of electrical signals to be sent to the controls of the machine in order to activate the transducers, motors, and gears connected to the operating components. This is done by having a programmer create a CNC parts program. The sequence of creating the CNC program is as follows.

1. The programmer studies the workplace drawing looking for a sequence of operations, much as a skilled machinist does. The skilled machinist determines how many cuts or passes of the machine are required and what knobs and dial settings will be used by visualizing the part being machined. The CNC programmer also looks at how many cuts are required and visualizes the part being machined, but lists the machine motions necessary to machine the part in a sequential order. This becomes the manuscript, and it is like a step-by-step instruction sequence. However, unlike a set of instructions, the manuscript is written in mathematical notation using a space coordinate system. In its most basic form it is a

contour tracing system. Imagine visualizing a rectangular block being turned into a circular shaft on a lathe. Figure 10.1 is a simplified before-and-after drawing of a 5 in. long, square-faced block converted on a lathe to a 1 in. diameter round dowel 5 in. long.

We can describe the path the lathe tool will take in cutting the rectangular block, from face A to face B, as the block is rotated. Figure 10.2 shows the trace of the tool as it traverses from right to left and back again and is indexed in along the radius $R_1$ to the position $R_1 = 0.5$ in.

The manuscript is the tool path written in a mathematical form. The basic manuscript associated with Fig. 10.2 becomes:

1. Location 0,0,1; $R_1 = 0.707$ in.: traverse left to location 0,—5,1.
2. Location 0,—5,1; $R_1 = 0.60$ in.: traverse right to location 0,0,1.
3. Location 0,0,1; $R_1 = 0.50$ in.: traverse left to location 0,5,1.

In this case $R_1$ indicates how much the tool must be indexed in along the final part radius from the diagonal of the rectangular block.

This manuscript is not a working manuscript. For one thing, the angle of the tool is not considered. It illustrates the thought process used to prepare manuscripts, not the technically correct procedure. To show the technically correct procedure here would make this example difficult to follow.

2. The manuscript is then converted into something the computer can understand: a punched tape, floppy disk, or cassette, for example. The tape or other medium containing the proper instructions is the work output of the CNC programmer.

3. The information translated to a tape is the CNC program and is now ready to be fed into the machine tool's computer. The trade jargon for this is feeding the Machine Control Unit (MCU). The MCU is a small computer or microprocessor dedicated to a particular machine tool, and the information in it makes the machine tool sequence occur in the desired order. If another sequence is needed for another type of part, another manuscript must be prepared.

Let us now go back and see what is fed into the MCU so that we can better understand what a numerical control program is.

The Society of Manufacturing Engineers defines numerical control as follows:

> Numerical Control is the operation of a machine tool by a series of coded instructions, which are comprised of numbers and other symbols. Webster's definition of a program is "a logical sequence of operations to be performed." Coded commands, gathered together and logically organized so they will direct a machine tool in a specific task, comprise an N/C program.*

Thus, a CNC program describes a tool location relative to the workpiece or the machine tool's own coordinate system. In addition to describing the locations of the tool, the program contains information dealing with rates and types of mo-

---

*Numerical Control Fundamentals,* Society of Manufacturing Engineers, Dearborn, Michigan, p. 56, 1980.

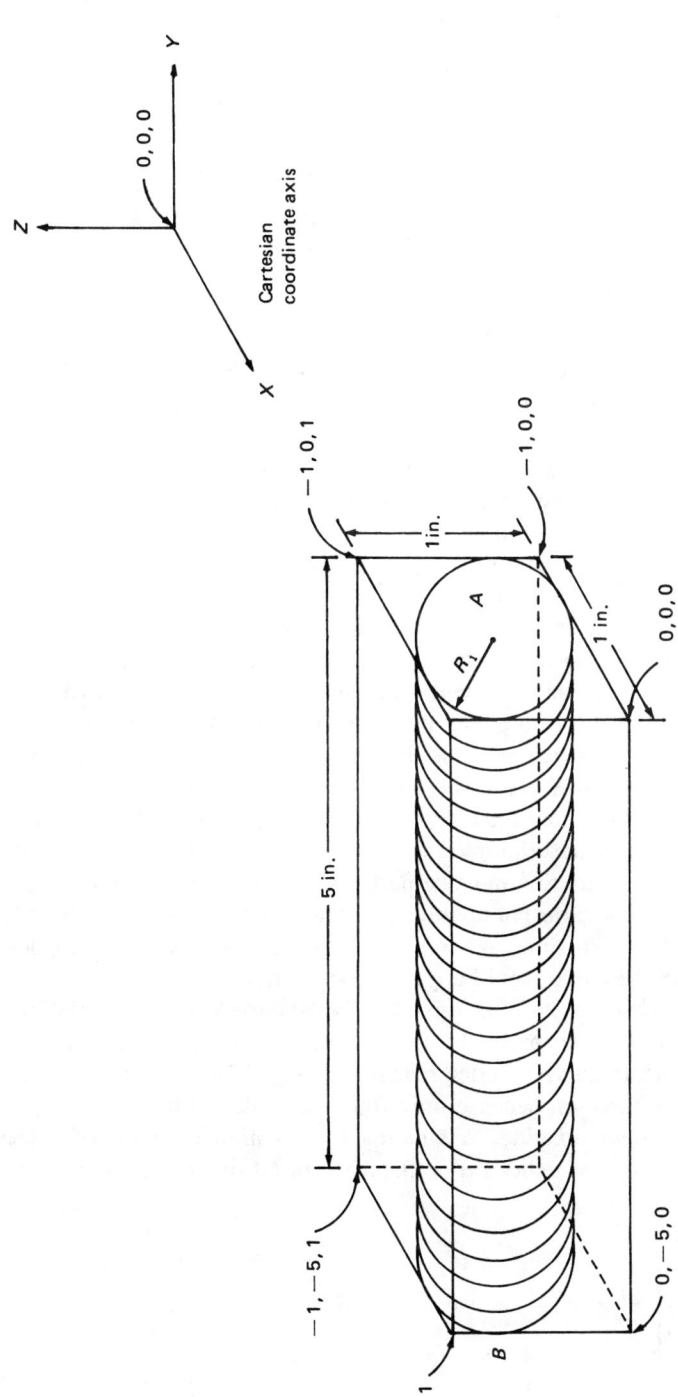

**Figure 10.1** Square-faced block converted on a lathe to a round dowel.

251

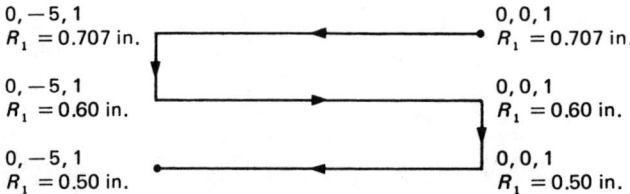

0, −5, 1
$R_1 = 0.707$ in.

0, 0, 1
$R_1 = 0.707$ in.

0, −5, 1
$R_1 = 0.60$ in.

0, 0, 1
$R_1 = 0.60$ in.

0, −5, 1
$R_1 = 0.50$ in.

0, 0, 1
$R_1 = 0.50$ in.

**Figure 10.2** Lathe tool path.

tions and auxiliary operations the machine tool must perform, such as changing tools. Tool locations are usually relative to a cartesian coordinate system, shown in Fig. 10.3.

Cartesian coordinates fit machine tools perfectly, since the tools are constructed on two or three perpendicular planes. All machine tools have a (0,0,0) reference point on the XYZ axes. Depending on the number of degrees of freedom of the machine, the location of the tool is based on an X, Y or X, Y, Z point in the space coordinate system.

This reference point can be at one end of the bed of a lathe, at the tool changer turret location of a horizontal boring mill, at the center of the rotating table of a vertical boring mill, or anywhere the machine tool builder designates. The important thing is that a zero datum point in space is designated so that all positions the cutting tool can reach are described in space in reference to that datum point. The coordinate point X, Y, Z shown in Fig. 10.3 is referred to the 0,0,0 datum point. The cartesian coordinate system is utilized in writing the manuscript and in translating the manuscript into machine language and inputting it into the MCU.

Sometimes contouring must be performed. This can also be programmed into the MCU by using cartesian coordinates, but that requires a considerable number of points to define a curve. As shown in Fig. 10.4, the fewer points defined, the more steplike the curve would appear to be. Fortunately, a cutting tool is a finite mass and by its size alone can smooth out most curves as it is translating in space. Therefore curve B can probably be machined adequately well with eight points defining it, while curve A after machining may look like an unsmooth or step function approximation because only four points define the curve.

As one would imagine, calculating the location in space of seven closely adjacent points can be tedious. For this reason CNC programmers define the curve

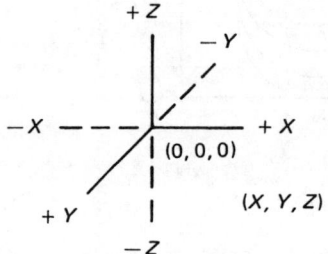

**Figure 10.3** Cartesian coordinate system.

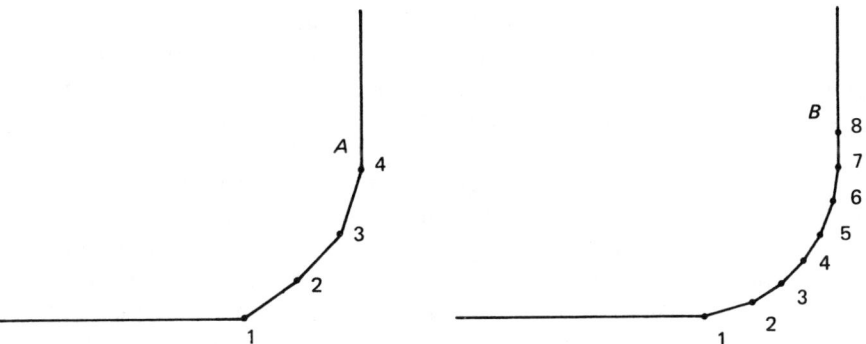

**Figure 10.4** Shape of contour as a function of the number of points used to define the curve.

mathematically, as illustrated in Fig. 10.5. Figure 10.5 is a geometric representation of the curved tool path shown in Fig. 10.4B. Note that in this case we only show three points on the curve plus the point designated as $R$. Point $R$ is located in space and described in the manuscript as any other point would be. However, $R$ is the center of a circle of radius $r$. When the cutting tool reaches point 2, it is instructed to sweep a circle of radius $r$; the arc of the circle to be swept will begin at point 2, which is at an angle $\alpha$ away from the vertical perpendicular, and be complete at point 3, which is at an angle $\beta$ above the horizontal perpendicular. Thus, to make programming easier, arcs of parabolas, circles, and so forth are employed. This is done by locating the center of the arc desired and instructing the translating device of the machine tool to position the cutter there and then trace the arc. Although this is simpler than defining all the points shown in Fig. 10.4B, even it can be a relatively slow process to program. Therefore, most MCUs are designed so that the description of the desired curve, e.g., $y = ax^2 + bx + c$, can be input directly into the manuscript and none of the points $R$ must be located. The CNC programmer describes a best-fit equation for a curve

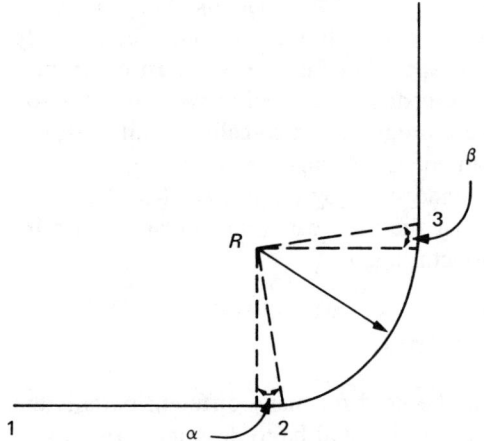

**Figure 10.5** Defining a curved tool path.

and the MCU causes the machine tool to perform the necessary translations. By using mathematical descriptions of curves, an infinite number of points can be employed and an exact curve as defined by the design engineer can be created. This was not possible before because a manual machine must be stepped through a curve in a manner similar to that shown in Fig. 10.4.

We can see that the process described above offers many opportunities for error. Therefore, the skill of the CNC programmer, while different from that of the master machinist, is vital to the success of the modern manufacturing concern.

To summarize, we have created a unique set of instructions, called a program, for a specific part. Like any other new program, this program requires debugging, which is very time consuming. Now let us look at ways to reduce the debugging time so that programs can be supplied to the CNC machines at a significantly higher rate.

## COMPUTER-ASSISTED CNC PROGRAMMING

Many programming languages have been developed to help program parts for production on CNC machines. These languages allow the programmer to describe part shape, machine tool motion, and machine functions in English-like commands. The programmer still must prepare a manuscript, but the computer-assisted CNC program does such things as calculation for tool offsets and determination of radius centers. The computer also does all the geometric checking to ensure that a continuous tool path or, if desired, a step function tool path has been input. The program defines the part to be made as a series of points, lines, circles, curves, and planes in three-dimensional space. The output is a series of tool paths that have already been compensated for tool sizes and shapes.

Computer-assisted CNC programming is the only practical way to perform this function. Most companies would find it economically unjustifiable to have a great many programmers on the payroll, if they could be found and recruited, to do manual programming for each new part. This would be slow and error-prone, and would tend not to derive the full benefit of the CNC machine's capability for productivity improvements. These machines are extremely productive, but only when running, and not when waiting for tapes. For this reason, most companies buy "canned" programs from software companies for specific types of common geometries such as shafts and bolts. These programs, often called family-of-parts programs, are normally written in the common CNC languages such as Apt, Adapt, Compact II, and others that are readily understood by the popular MCUs. These are actually variations of Fortran, so that they are easily taught and generally familiar to technical personnel in manufacturing.

## FAMILY-OF-PARTS PROGRAMMING

Creating general CNC programs that can be used for many different parts is the goal of all manufacturing managers. The ideal would be to describe a part to be made, indicate what machine tools should be used, and quickly produce usable CNC tapes. This ideal has not been achieved yet. However, family-of-parts programming begins to approach this ideal. Family-of-parts programming is a subset of computer-assisted programming that is used where parts that are similar in shape will be made, and the variations of instructions to the machine, though large, are not infinite.

For example, Fig. 10.6 shows a generalized outline for producing a rotor shaft. Every dimension and radius is indicated by an appropriate symbol representing a generic variable. A program can be written to perform all radius, length, and diameter machining. That is, all the $R$'s, $L$'s, and $D$'s can be described in general mathematical terms so that tool paths can be described, and substituting specific values for these variables determines the specific geometry for a desired product.

These programs are developed to machine a part of a general shape. It is only necessary to fill in specific values for all the variables, which can be found on the engineering drawing. The manuscript is now prepared and ready to be entered into the MCU, which can be done rapidly. Then, associated with the family-of-parts program, the computer's editing system evaluates the changes (including required omissions for simplified versions of the product) and produces a tape for the specific job. This amounts to matching the generic program to the specific task through the input of specific values for the variables.

## USES OF CNC MACHINE TOOLS

Numerical control technology allows the use of general-purpose machines for large lot sizes, that is, mass production. More than one set of operations can be performed on one machine tool, which means more flexible machines, hence less expense incurred for model changeover. With manually controlled machines, the key to mass production has been special-purpose machines, which are very costly and time-consuming to set up but, once running, very economical for producing large numbers of identical parts. With CNC machines, the setup changes are stored in a memory and are essentially electronic rather than mechanical. Special jigs and fixtures are not needed to guide the machine; instead, the proper manuscript must be fed into the MCU.

Numerical control technology also permits small lot sizes to be produced economically with general-purpose machines. Manual control machines may require a cam-type template for the machine to follow (tracer control), requiring the expensive and time-consuming operation of making a template in the toolroom. Small lot sizes are expensive because the cam costs are written off over a small number of pieces, and they take an extended period of time to produce. With CNC machines many of these problems are eliminated. If the shape can be determined mathematically then CNC machines, especially those with six degrees of freedom, can produce the part. This means that many complex parts can now be made quickly and at considerably less cost.

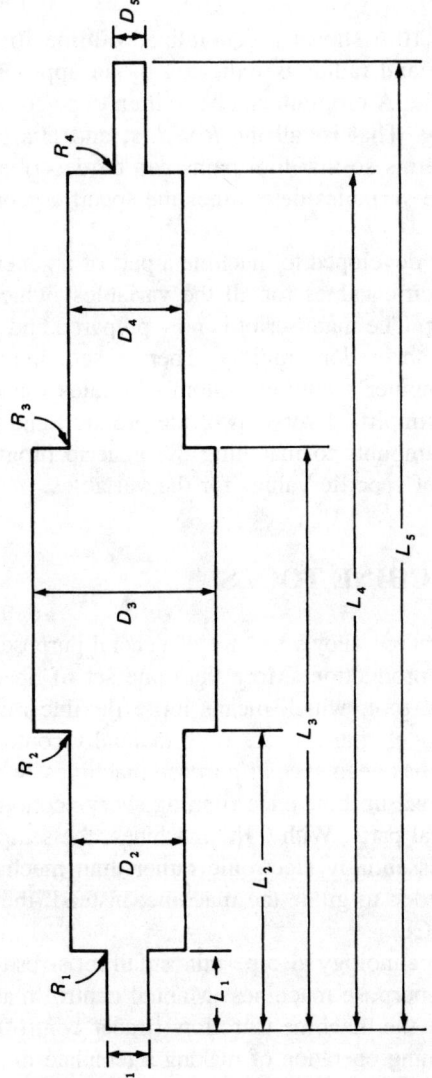

**Figure 10.6** Generalized outline for producing a rotor shaft.

**Table 10.1 Times and costs for production of a motor base by conventional and N/C technology[a]**

|  | Setup | | Cycle | | Tooling | |
|---|---|---|---|---|---|---|
|  | Conv. | N/C | Conv. | N/C | Conv. | N/C |
| Time (hr) | 1.30 | 0.40 | 0.86 | 0.65 | 0.64 | 0.21 |
| Cost ($) | 3140 | 800 | 2350 | 835 | 1890 | 260 |

[a]From E. DeGarmo, *Material and Processes in Manufacturing Engineering*, 3d ed., Macmillan, New York, pp. 916–917, 1969.

It can certainly be said that CNC technology improves the economics of manufacturing. Some data from a study made by DeGarmo are shown in Table 10.1. These data are representative of time savings reported by many industrial users. Typical savings range from 1.7:1 to 3.5:1. Notice that not only is the setup time reduced and practically eliminated, but the cycle time is significantly reduced, because the machine can operate at a higher rate, particularly when changing from one operation to another.

**Example:** In lathe work a manual machine's operator can proceed at a relatively high rate of metal removal while roughing out the length of a shaft. However, in starting to contour a fit, the operator will probably measure the shaft diameter to make sure it is correct, reposition the cutting tool, and then carefully step-index the cutting tool position in the radial and longitudinal directions. In doing the same job, a CNC lathe would do the roughing to size along the length at about the same speed as a skilled operator. It might do it a little faster because the CNC machine would not have to be concerned about machining the length under size, since its program has instructed it to start traversing at a specified point in space and stop at a second specified point in space. The workpiece is perfectly centered in that space and securely fastened in the lathe by the head stock and tail stock, so the exact dimensions will be machined to. After the roughing operation the cutting tool begins to trace the contour tool path, as the CNC program instructed it to. There is no stopping to position the cutting tool as in the manual operation, because the end of the traversing operation, the second point in space, is the beginning of the contouring operation. The contouring operation can be accomplished much more quickly because the radial and longitudinal indexing is preplanned in the program, not done by operator experience. In the manual procedure the operator must take a little metal off in each pass to avoid removing too much, while in the CNC mode the design geometry is input into the MCU by the manuscript and a virtually perfect contour can be achieved rapidly.

With CNC technology the thought process of the design engineer conceiving of the part is virtually duplicated in the MCU of the CNC machine before the part is made. We know exactly the part geometry and the sequenced steps required to achieve the concept within the raw material. Before CNC manufacturing had only a picture, a drawing with dimensions on it. Skilled operators guided basic cutting machines to try to duplicate the necessary shapes. The MP&WM personnel specified the number of traverses required, but they could never detail in minute steps the very fine contouring step functions required. With CNC it is practical to do so. An exact mathematical match to the engineering drawing is possible and can be produced two to three times faster than in the inherently less accurate manual method.

The lathe is probably the simplest machine tool and the most capable of forming relatively complex shapes in a manual mode. The discussion above shows how even this simple machine is greatly enhanced by the application of CNC. One of the benefits not even imagined when CNC machines were first introduced was that of multiple tool use. Before CNC, only special-purpose, complex machines had the capability of doing more than one operation per workpiece setup. A machine of this type has been used to produce internal combustion engine parts for automobiles. In actuality, this is a hard-automation process where parts are fed on a moving rack past milling heads, drills, reamers, boring bars, and so forth. As the workpiece passes each tool, that tool performs its work. These machines have the capability to rotate and translate the workpiece to present the proper attitude toward the specific cutting tool. They are ideally suited for very long production runs because the cost of initial setup can be prorated over all the parts produced. However, when the design engineer changes the design, the machine must be literally rebuilt to accept the new design. The positioning and tool sequencing must be modified, to an extent determined by the degree of the design change. To make only a few parts on such a machine would be out of the question; the cost would be prohibitive. Therefore, smaller lot sizes had to be made on general-purpose machines, one operation at a time, and economies of scale were nonexistent. With the use of CNC it is possible to construct a general-purpose machine that can do much more than one operation per setup. Economies of scale can be approached with much simpler machines than the hard-automation, special-purpose machines. Let us see how this is done.

The CNC machines discussed above are called multitool changer machines or flexible manufacturing machining centers. They are characterized by a track or turret containing different types of cutting tools, such as drills, milling heads, and reamers, that can be called upon and presented to the workpiece. More complex machines of this type also have work setup tables that can rotate, translate, and in general present the workpiece to the tool at a variety of angles and positions. Virtually all six degrees of freedom of translation and rotation are possible. In essence, the workpiece is positionable and the cutting tool is brought to it, all under the control of a computer—the familiar MCU instructed by a properly conceived manuscript. Here we have the capability of planning virtually the entire machining of a part on only one workstation. The program gives an instruction to the machine for what tool should be used for each operation and then an instruction to the machine to position the part for the next operation. The more complex machines are also given instructions for how the workpiece should be rotated on the positionable worktable or holder to present the proper aspect for machining.

**Example:** Suppose we are required to machine a gearbox casing, as shown in Fig. 10.7. The casing is a fabricated structure designed to hold the stationary portion of a gear set, with the rotating component contained within the stationary components. The purpose of the casing is to provide a foundation for the gear set. To manufacture the casing, we could mill the faces, drill the numerous holes for securing the gear set, bore the diameters to the proper size, and finally mill the feet. Before the advent of CNC machining centers we would do all the milling on one machine, the drilling on another, and finally the boring on another. With a multipurpose machining center we can do the entire operations on one machine. Let us construct a manuscript summary to show how this is accomplished.

1. Set up workpiece in position one. This is the position shown in Fig. 10.7 and is probably a manual operation.
2. Mill face 1. A tool path is created for a milling tool to do the milling work. A milling cutter is selected and its location in the tool turret identified.
3. Drill holes on face 1. A tool path is created for drilling locations to do the drilling. A drill bit is selected and its location in the tool turret is identified.
4. Rotate worktable 180° horizontal. CNC program instructions are given to the worktable to rotate after completion of step 3.
5. Mill face 2. Logic similar to step 2 is created.
6. Drill holes on face 2. Logic similar to step 3 is created.

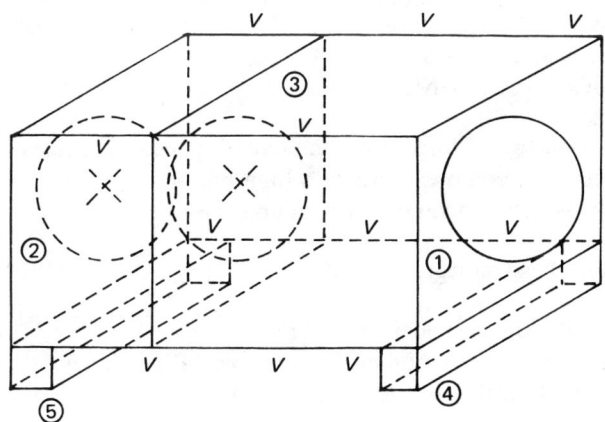

**Figure 10.7** Gearbox casing. The $V$'s denote clamp positions for supporting the workpiece on the worktable.

7. Bore diameters on faces 1, 2, and 3. A tool path is created for a boring tool that has both longitudinal and radial component instructions. A boring tool is selected and its location in the tool turret is identified.
8. Rotate worktable 90° up vertical. Logic similar to step 4 is created.
9. Mill face 4. Logic similar to step 2 is created.
10. Mill face 5. Logic similar to step 2 is created.
11. Drill holes on face 5. Logic similar to step 3 is created.
12. Drill holes on face 4. Logic similar to step 3 is created.
13. Rotate workpiece 90° down vertical. Logic similar to step 4 is created.
14. Rotate workpiece 180° horizontal. Logic similar to step 4 is created.
15. Off-load workpiece.

The list above is a linked series of tool paths and machine instructions designed by the CNC programmer. An entire plan to manufacture a part is created and can be carried out by one machine that is being directed by a computer in accordance with a preconceived logic. Since the computer can only do what is programmed, the sequence will be carried out identically time after time. We have created soft automation and can change it as often as we wish by creating another set of instructions for the computer. Thus CNC creates mass production without the necessity for special templates, cams, or ingenious devices for mechanically instructing a manual machine.

## ADVANTAGES AND DISADVANTAGES OF CNC MACHINES

All new technologies introduced in manufacturing can create new types of management problems that must be overcome. To find acceptable solutions to new problems, it is important to understand all aspects of the problem. In the case of CNC that means understanding the advantages and disadvantages to the entire manufacturing organization brought about by this marriage of the computer with production. We will discuss these advantages and disadvantages to develop an understanding of the management options available with CNC.

There are three basic advantages of CNC:

1. The ability to produce parts at two to three times the rate of manual machines.
2. The ability to produce accurate, complex parts at a high rate.
3. The absence of a need for skilled machine tool operators.

There are also three perceived disadvantages of CNC:

4. A cost roughly twice that of manual machines.
5. A maintenance downtime roughly 50% higher than for manual machines.
6. The need for a new type of skill, that of the CNC programmer.

The six items listed above may be thought of as management opportunities, because any identified situation can be a management opportunity if handled properly. Item 6, the need for CNC programmers, is classified as a disadvantage because it means that a different type of factory skill must be nurtured and people trained in that craft. Scarcity of programmers can create difficulties in using the CNC machines to their maximum capacity. On the other hand, it means that extensive apprenticeship programs to train skilled operators, with their attendant indirect costs, can be pared back, which is a benefit. In addition, a unionized plant may be a disadvantage for management. Skilled operators most often are union members and part of the hourly work force, while CNC programmers are usually part of the technical staff and not union members. Obviously, having the machines directly controlled by the CNC programmer tends to offer management more flexibility. Finally, the effects of the scarcity of CNC programmers can be creatively minimized by the adoption of computer-assisted CNC programs. This, coupled with the fact that CNC programmers can often service several CNC machines while a skilled operator can only operate one machine at a time, makes the advantage/disadvantage trade-off of the need for programmers lean toward being an advantage even though it may originally have been perceived as a disadvantage.

Chapter 9 implied that there are differences in maintenance between CNC and manual machines. At present, CNC machines do require more maintenance than manual machines. However, the two- to threefold increased production rate more than offsets the 50% and higher maintenance time increase. Even with a rate of downtime approximately 1.5 times that of manual machines, CNC machines yield a greater production rate and hence a greater profit.

The same reasoning is involved when a company decides to spend twice as much for a CNC machine as for its manual counterpart. If the machine can produce over twice the volume of a manual machine, that advantage outweighs the disadvantage of higher initial cost and higher maintenance cost. Of course, capital equipment decisions must be based on a case-by-case analysis of the specific situation and not on an industry average. As CNC becomes even more firmly linked to the computer-assisted management techniques the differences in costs between CNC and manual machines will become less and less important. We are rapidly approaching the time when integrated computer control of factories will require only CNC equipment.

One advantage not discussed so far from a management viewpoint is the capability of making complex parts in volume quantities. This enables management to sell tighter quality specifications to its customers, which in many cases is either a marketing advantage or a necessity to stay in business. In addition, by being able to produce complex products in volume and at relatively low cost, it is possible for the company to offer new products or enter businesses that were closed to it before. For instance, consider a company that could only make nuts and bolts on its manual lathes and now can make intricate cams. The company can now enter the machine tool business as a supplier of new equipment. Or it can become competitive in the machine tool rebuild business because it can make complex repair parts quickly and inexpensively.

An analysis of advantages and disadvantages should be looked at as an analysis of different types of opportunities. This is especially true in connection with CNC machines. Here we are dealing with a creative revolution, and all perceived problems associated with that revolution must be looked on as potential opportunities for greater profitability. The examples discussed above show how perceived problems with CNC can be turned into advantages if approached properly. When we speak about the high costs of purchasing CNC equipment, we should think of the even higher costs associated with running uncompetitive factories. Our discussion of advantages and disadvantages should perhaps be rephrased: the advantages of implementing CNC versus the disadvantages of not implementing CNC. Not implementing CNC is stating clearly that a firm intends to withdraw from the marketplace and no longer compete.

## MANAGING THE USE OF CNC MACHINES

Effective management of CNC equipment, like that of any other resource, requires an understanding of the resource. In most industrial organizations the responsibility for understanding the CNC equipment and setting guidelines for its use is delegated to manufacturing engineering. Within manufacturing engineering, it is further delegated to the methods, planning, and work measurement unit. Although advanced manufacturing engineering is the procurer and specifier of CNC machines, it is not the responsible unit, because we are concerned with the use of the CNC machines and not their development and implementation. Therefore, the MP&WM unit, not the AME unit, is responsible for interpreting CNC to shop operations.

The MP&WM unit must plan for the use and provide the day-to-day methods support necessary to operate the CNC machines effectively. Let us look at the major considerations involved in operating with CNC that are different from the traditional MP&WM functions.

1. *Establish planning guidelines for selection of parts to be made on CNC machine.* Since most manufacturing concerns find themselves with a mix of CNC and non-CNC machines, it is necessary to route parts to both types of facilities. The MP&WM unit is charged with establishing guidelines for effectively loading both types of facilities. In doing this, they would want to plan parts to be made on the most efficient machines first and the least efficient as load demands. Since CNC machines tend to be the most efficient, most guidelines require that CNC machines be loaded first, or base loaded, before other equipment is scheduled. Also, since CNC machines are usually more adaptable, they would be favored over manual machines for complex parts manufacturing. This leads us to the second consideration.

2. *Review parts to ensure that CNC application is correct.* We know that CNC machines are ideally suited for making complex parts because all the steps can be preprogrammed. But should this be given preferential scheduling treatment over the ability of CNC machines to make simpler parts significantly faster than manual machines? That is, if CNC machine capacity is limited, should that ca-

pacity be utilized for very small lot sizes of complex parts in preference to large lot sizes of simpler parts? This is a consideration MP&WM must deal with constantly. The planner must ensure that the CNC machine is the proper workstation for the part assignment. The decision is based on whether there are technical and economic advantages of making the part on the CNC machine. The overriding consideration must be technical, since it would not benefit the company to run an economic batch on a CNC machine and then struggle unsuccessfully to make a complex part on a manual machine. If the complex part can only be made on the CNC machine, it must be made there, and the batch-run parts must either wait or be scheduled for a manual facility.

3. *Determine machining method and tooling required.* Before a CNC programmer can instruct the CNC machine in performing the required work, a method and tool selection must be made. Unless the CNC programmer is a qualified planner or methods engineer, he is unlikely to know the proper sequence of making the part. Therefore, the planner usually states the method to be followed, including the tooling required, and the CNC programmer then develops a tool path sequence and a manuscript. The difference between this approach and the initiation of manual planning is that in the CNC case, the planner does not have to detail the exact sequence of cuts to be taken. This evolves automatically once the tool path is described mathematically by the programmer. The planner must only describe the tool to be used, which in turn will dictate depths of cuts and feeds and speeds.

4. *Translate drawing dimension into input format.* Since most CNC work is now carried out with generic family-of-parts programs, planners now find themselves entering the geometric data into generic programs on a day-to-day basis. A family-of-parts program written for a specific CNC machine is entered into by the planner to generate the required output for the CNC machine tool. The planner uses this output, typically in the form of a punched tape or a direct stored instruction, as part of the planning package dispatched to the workstation. If there is no tape, the planning package simply informs the operator what program should be called up to instruct the CNC machine in making the part.

5. *Develop family programs for new part families.* Usually the CNC programmers are part of the MP&WM organization, and one of their responsibilities is recommending when individual programs should be consolidated into family-of-parts programs. It is MP&WM's responsibility to recognize when a family begins to exist and a productivity improvement could be achieved by allowing nonprogrammers to generate specific parts tapes for CNC machines. A cost trade-off is usually evaluated here, weighing the cost of developing a family-of-parts program against that of continuing to program the parts individually. The decision is based primarily on volume.

6. *Provide training for CNC operators.* Unlike the apprenticeship program for training skilled operators, there is no traditional program for training relatively unskilled people to operate CNC equipment. Contrary to the sales brochures of some CNC vendors, CNC equipment still needs operators—perhaps not the skilled craftsmen needed to get the most out of manual machines, but nevertheless people

to monitor and load the CNC machines. These people must instruct the machine, load the raw materials, off-load the finished products, and monitor the machine for proper operation. By instruct and monitor, we mean inputting the properly designed program and determining whether the machine is operating correctly. MP&WM must have the capability of training operators to do this.

Usually the MP&WM unit will have staff responsible for handling the training. These are CNC programmers or methods engineers who are familiar with the machines and usually have the machines as part of their responsibilities for methods and planning. They are the service staff shop operations calls on when there is a need to train new operators.

## ROBOTICS

The idea of machinelike imitations of humans has always caught the interest of the public. Therefore, it is hardly surprising that the public was interested when the term robot was used to describe programmable positioning and loading devices. In fact, even the general news magazines have devoted space to discussing this offspring of the CNC revolution. No discussion of CNC can be considered complete if robotics is not included.

A robot may be defined as a mechanical device capable of being programmed to move or manipulate in three-dimensional space and capable of doing work in that space imitating, in a general way, motions capable of being accomplished by a human being. Let us analyze this definition. A vast number of machines can move in space. Fewer machines can manipulate, that is, rotate, translate, and traverse. The ability to imitate human motions is even more severely restricted within the machine population. However, there is still one more important part of the definition that must be fulfilled: the machine must be programmable; its sequence of movements must be capable of being varied by changing instructions to its controlling computer.

The last sentence states the link between a CNC machine and a robot. Manipulators that mimic human motions have been available for a long time. For instance, manipulators controlled from a distance have been used by atomic scientists to handle radioactive materials behind a radiation barrier. In order to use these manipulators, the scientists had to develop a manual skill, probably by trial and error. These are analogous to manual machines rather than CNC machines. Much greater precision would be possible if these manual manipulators could be driven through three-dimensional space based on a programmable positioning device commanded by a prescribed mathematical sequenced program. This is exactly what a robot is—a manipulator that a CNC programmer has instructed to follow a sequence of commands dictating its spatial motion and when it should grab, start spray painting, start welding, and so forth.

A robot, then, is a specialized CNC machine. The controlling mechanism is the familiar MCU and the intelligence is provided by a manuscript, that is, a mathematical compilation of the specialized tool path. In this case the tool path may be the journey through space to pick up a part and place it somewhere else. The purpose of this tool path is different from that of a CNC lathe's tool path, but the principle is the same. The CNC programmer programs its motions, just as he does for the CNC machine tool's cutting apparatus. The robot's motions usually must be very finely controlled, while the CNC machine can rely on the accuracy of the lead screws and beds. Therefore, the original MCUs are replaced by more powerful mini- and/or microcomputers capable of much finer control.

The rationale for using robots is the same as that for using CNC machines. Robots produce higher levels of quality and work faster than novices and journeymen. They can be programmed to do the work of skilled operators—in this case not machine tool operators but welders, assemblers, and painters. The use of robots to do human manual work is not nearly as advanced as that of CNC machines taking over from manual machines, but the trend is virtually identical. As the MCUs for robots become more powerful and as sensing devices develop from crude pressure sensors to finer sensors for touch, feel, and sight, more human motions, particularly of the fingers, will be mimicked by robots.

Ultimately, this will lead to flexible automated factories where robots do the human work of loading and unloading CNC machines and moving material from workstation to workstation. The technology for control is available now, but the price to accomplish this may still be too high. This is a logical outgrowth of the CNC revolution—the use of a computer to direct a machine based on a program designed by the human controller.

Is this good or bad for society? In my opinion, it is good. It releases humans from performing drudge work on a production line or uncomfortable and possibly dangerous work in foundries and mills. Humans evolve better and better tools so that we can exercise our intellect and have our machines do the drudgery to support our standard of living. This view may seem utopian, but we will never achieve it unless we continue with the CNC revolution. Certainly, there are disruptions in the ability of some to earn a living as these jobs are replaced with robots and CNC machines. We no longer use wheelbarrows and shovels as the main tools for constructing roads; we use machines, and society applauds the improvement. The same should also be true of switching from manual machines and manual work to CNC machines and robots in our factories.

## SUMMARY

We have discussed the concept of CNC machine tools by first explaining the meaning of manual control, then hard automation, and finally electronic control via the computer, in order to demonstrate the logic that led to their development.

We have also covered the philosophy of programming for CNC machines. Manuscript development and the relationship to the cartesian coordinate system of points in space were explained. Further simplifications were discussed through the use of equations representing geometric forms to program curves in space.

Management techniques related to CNC machine use have been described.

The six steps in planning and implementing work on CNC machines show the impact of CNC on the manufacturing engineering organization.

Finally, the concept of robots has been introduced. Robots are a logical extension of numerical control development. Their programming is similar to that of CNC machines, but they are involved in nonmachining activities such as assembly, welding, and painting. From the discussion of robots it became evident that the flexible automated factory is not only possible but inevitable.

## REVIEW QUESTIONS

1. Discuss the differences between the skills required to produce a complex part on a manual lathe and on a CNC lathe.

2. A shaft having several different diameters and a threaded portion is to be machined on a manual lathe. Discuss the judgments the operator must exercise in making the part and what the machine will do in response to these judgments.

3. For the example in question 2, describe how the judgments to be made by the operator can be predetermined so that we have an automated process.

4. Define a programmed sequence with respect to a specific machine tool operation.

5. A manuscript is the CNC version of the detailed sequential planning sheet. Make a manuscript for the machining operations used to manufacture a stud threaded at both ends according to the accompanying sketch. Include a graphical representation of the tool path. Make from carbon steel 1.000-in. diameter rod stock.

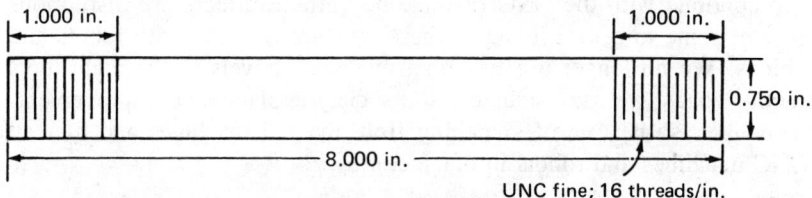

Diameter tolerance = ± 0.001 in.
Length tolerance = ± 0.002 in.

6. What is the zero datum point of a machine tool, and how is it related to manuscript preparation?

7. There are two basic ways to define curves in reference to manuscript preparation: the point-in-space method and the radius method. Compare the two, discussing advantages and disadvantages of each.

8. For the shaft shown below, create a set of values for a family-of-parts program as shown in Fig. 10.6. *Hint*: start from the left side; all $L$, $R$, and $D$ values must be used as inputs.

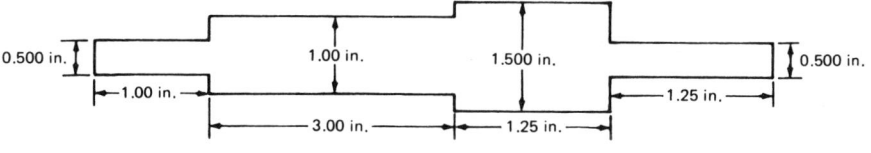

All radii = 0.020 in.

9. Discuss the advantages and disadvantages of special-purpose machine tools and general-purpose CNC machine tools with respect to flow (mass) production of machined components.

10. Why are machine cycle time and setup time significantly reduced with CNC machines versus manual machines?

11. Why can a CNC multi-tool-changer machine (a machining center) accept changes from design engineering more readily and at less expense than a special-purpose multitool machine?

12. Discuss the ways in which CNC machines can open additional marketing opportunities for manufacturing companies.

13. Why are CNC machines always considered to be base-loaded facilities?

14. Under what circumstances should a part not be loaded to a CNC machine?

15. Describe in outline form the similarities between robots and CNC machines.

16. What areas of development are necessary for robots to achieve the same dominance in assembly operations that CNC machines have achieved in machining operations? Give reasons for your answers.

17. Describe the potential benefits of robots over other types of transfer, assembly, and welding methods.

# FUNDAMENTALS OF COMPUTER-INTEGRATED MANUFACTURING

Computerization of the management process is the emerging technology in manufacturing, particularly in the areas of design data manipulation, manufacturing data manipulation, and decision making. In this chapter we will explore the fundamental principles involved in defining, developing, and putting into use computer systems that enhance a company's ability to compete. This theme of unifying via computer is a natural follow-on to the introduction of CNC machines and financial accounting systems. In Chapters 12 and 13 we will explore the immense potential for factory data collection systems and computer-aided process planning through group technology. However, it is first necessary to understand the Computer-Integrated Manufacturing (CIM) fundamentals.

CIM is the current title for this unifying concept. Previously it was widely known as Computer-Aided Design/Computer-Aided Manufacturing (CAD/CAM). But that term fell into disuse because it was felt that mentioning only the engineering (design) and the manufacturing functions implied that other business functions, such as marketing and employee relations, need not be integrated for optimum performance. Of course, this is not the case. So another name evolved: Computer-Integrated Manufacturing. This states integration via computers and focuses on the manufacturing nature of most businesses. Also, some experts felt that the slash (/) between CAD and CAM implied a linear nonconcurrent approach that favored islands of automation instead of integration of efforts.

Efforts are still being made to rename this philosophy. Some say it ought to be called Computer-Integrated Business (CIB) because in reality it is more than manufacturing oriented. They argue that the theory is just as applicable to service businesses as it is to those producing physical products. Some say even that does not go far enough. They contend that any organization serving a defined purpose can use this philosophy. Hence they propose the name Computer-Integrated Enterprise (CIE). We will use the acronym CIM. It is still the most widely used name and it suits our purpose because this book is about manufacturing.

## INTEGRATION OF FUNCTIONS WITH CIM

Continued progress in productivity improvement is essential for a company to remain competitive. This is another way of saying reduce operating costs, which are defined as direct labor costs, overhead costs, and material costs. A properly conceived CIM system affects all three of these costs categories and, by making possible significant reductions, results in productivity improvements (see Fig. 11.1). No longer can a manager rely on all functions to perform their jobs adequately well but in an isolated manner. With the geometrically rising costs of labor and material, what were once acceptable design cycle times, material procurement times, planning times, queue times, and setup times now lead to losses in market share, lower profits, and, in the extreme, business failure. Methods have been found to significantly improve these factors. These methods involve full integration and they are classified under the title CIM.

CIM is the only way to integrate all job shop functions to minimize the total cost of manufacturing. It is done through the use of common computer data bases.

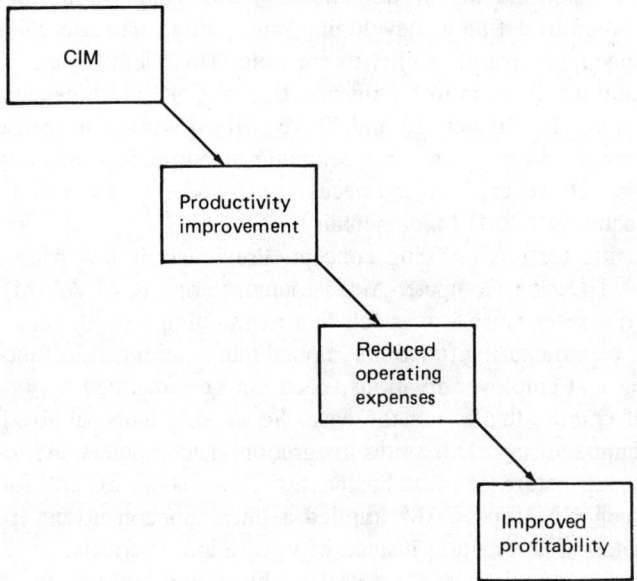

**Figure 11.1** Benefits of CIM.

This means that all functions can access each other's accumulated and stored information, so that the same information can be used in different ways by the different functions. Before common data bases were available, functions created their own data bases from the information they were capable of gathering, the information passed on to them from others, and the data they created in performing their responsibilities. Each had its own exclusive repository of information, and the only data reasonably sure of being defect-free were the data they inputted themselves.

As an example, design engineering uses the outline of a part or assembly to perform a stress analysis to make sure the part geometry is sufficient for the task. The engineering function is reasonably sure that their work is being done correctly because they are responsible for creating the geometry via the engineering drawing. Eventually the part is released for manufacturing. Manufacturing engineering will then create a CNC tool path manuscript and make the part. This sounds simple, but it can be an error-prone activity. In creating the CNC tool path, manufacturing engineering must recreate the geometry. The engineering drawings produced by design engineering are conceived as instructions for operators to follow in producing parts on manual machines. The final shape is shown along with the final dimensions, but these dimensions are not calculated in three-dimensional space, the cartesian coordinate system. This system is necessary for the MCU input to the CNC machine. Therefore, the manufacturing engineer must create a new data base for the MCU input, selecting a zero datum point and then creating points in space relative to the datum point, using the engineering drawing as the source of information. The manufacturing engineer is creating a geometric data base by converting the absolute dimensions on the engineering drawing into relative dimensions via the coordinate system, as illustrated in Fig. 11.2. This translation creates opportunities for making mistakes. Data entered incorrectly and misinterpretations of engineering drawing information are common examples. Errors are not only possible, but probable, which means additional costs and slowdowns in scheduled production to make corrections.

The proper solution to this problem is to have the person who designed the part select the datum point and then show the required dimensions in cartesian coordinates. Design engineering can do this, and does for its own stress analysis needs, for instance, in finite-element mesh generation. However, before the advent of common data base generation, there was no practical way to give manufacturing this necessary information, with the result that it had to be recreated. The common data base, as required for CIM, allows all functions to have access to primary data, which eliminates both translation errors and duplication of effort. In our example, the CNC programmer could simply access the data base and construct the CNC tool path, without any risk of making interpretation errors. Reduced duplication of effort leads to reduced queue times, setup times, and design times and to an overall improvement in productivity. Thus common data bases can optimize the total manufacturing cycle time. We will expand on this theme of the common data base throughout the remainder of this chapter and show why it is important and is the heart of the integrated CIM system.

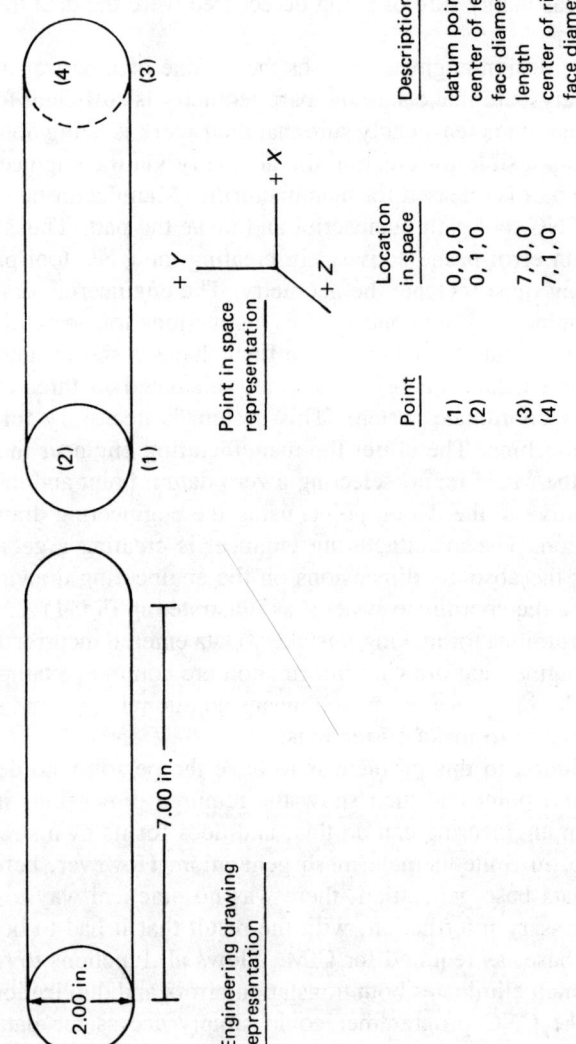

**Figure 11.2** Converting from engineering drawing to data base.

## PRODUCTIVITY VIA CIM

CIM is primarily a productivity enhancement system, which is why it is of interest to business. The first thing the manufacturing engineering manager must ask is, "How do I improve productivity with CIM?" If the manager does not know, then obviously no aspects of CIM will be introduced. Similarly, if so-called productivity improvements do not result in lower costs or improved levels of profitability, CIM will not be introduced. CIM must improve the cost picture, and this must be clear to the decision-making manager.

To answer the rhetorical question above, let us look at how job shops are generally run. An order is received and is put into the manufacturing master schedule by marketing and agreed to by manufacturing. The specific design is evaluated and produced by design engineering, which translates the design into drawings and instructions for manufacturing. Manufacturing, in turn, takes these drawings and instructions, creates specific task instructions for the various workstations, and orders materials to be used during the fabrication of the product. Manufacturing then schedules the job to be made in the factory. Finally, the job is completed and shipped to the customer, whose technical questions, if any, are answered by marketing or design engineering.

The one common factor in this entire sequence is information (Fig. 11.3). The information needed for all facets of operations is essentially the same: what it is, how it is supposed to work, who has to do the work, when the work has to be done, and why it has to be done in a certain manner. These questions are asked by the various functions of the total company organization in relation to their respective charters. For example, the design engineers are interested in how it works so that they can do the necessary scientific analysis to ensure that it will work and has a high enough margin of safety against failure. The manufacturing

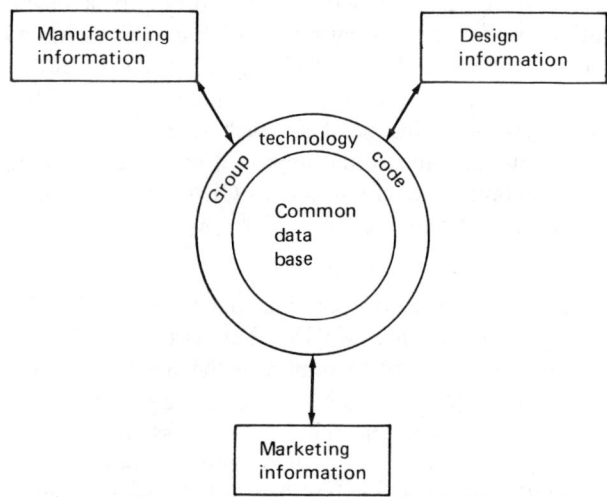

**Figure 11.3** Inputs and outputs of the group technology coded common data base.

engineers are interested in how it works so that they can design the manufacturing process in a compatible manner, that is, not create a situation in which required tolerances cannot be met. Both design engineering and manufacturing engineering are asking the same question in relation to different needs, but based on the same set of facts, the data base.

Without CIM, it is difficult to obtain all the facts to answer the question from the viewpoint of each function, and in many cases each function has to create its own library of facts for each individual order. With a CIM system it is possible to have one centrally located library of facts for each job, with each function having access to that library. If manufacturing engineering wants to know how tolerances for a particular job are to be set, it simply visits the library and looks up the information. The library analogy goes even further. Most good libraries are continually expanding, and the same is true for data bases. As manufacturing engineering determines how to process a part, this information is fed back into the data base. This is important because it allows other functions to know what that specific portion of the plan is so that they do not do further work that may be contradictory.

> **Example:** Suppose that manufacturing engineering has decided on a particular form of automated welding because it is suitable for the design created by design engineering. Further, suppose manufacturing engineering has spent company money to purchase the necessary welding equipment. Then, for some reason, design engineering changes the physical nature of the parts, making it virtually impossible to use the recently purchased welding equipment. This is not an impossible occurrence. In fact, this type of "the left hand not knowing what the right hand is doing" goes on every day. With CIM, a data base accessible to all could be searched by design engineering before it makes any change, and a situation such as that described above could probably be avoided. Design engineering would know that special-purpose welding equipment had been bought to do the job. A more productive welding technique would be used; there would be no need to further redesign after manufacturing engineering protested; and there would be no need for each side to prepare its arguments. Hence, many opportunities for overall improved productivity would result from the existence of a common data base. Thus CIM amounts to providing a good, reliable communications system.

A second answer to the rhetorical question posed at the beginning of this section can be found by looking at another use of CIM. This is the coupling of the common data base with group technology to minimize the need for totally redoing quotes, designs, and plans for each new job offered to the job shop. We may define group technology as a systematic search for sameness in manufacturing techniques, such that larger production runs, commonality of tooling, dedication of machine tools for families of parts, and efficient utilization of designs and planning can be accomplished. The facilitating mechanism is a group technology classification coding system. We will cover group technology in Chapter 13.

The common data base includes large quantities of design, marketing, manufacturing, and financial data pertinent to all contracts worked on by the manufacturing firm. After a relatively short period of time, it is found that specific aspects of new jobs are very similar to aspects of previous jobs for which data are stored. Therefore, a great deal of redundant work for each function can be eliminated if we can recall previous work. The group technology classification and coding scheme, used as an index for the common data base, allows recall of that information on the basis of coded design characteristics, manufacturing characteristics, and other specific information. That is, we use the group technology scheme to code needs and characteristics of new jobs and then, using the code as an index, search the common data base for matches or near matches from previous projects. This recall process is optimized with CIM. For example, when asked to design a new component, the design engineer can search the file quickly and thoroughly for similar designs used previously, and greatly shorten the design cycle by not having to start from scratch. The same is true for manufacturing engineering in determining the planning sequence of the new component.

This shortening of the cycle time means that more useful work can be performed in the same standard work week. It lowers the indirect costs of manufacturing because work is accomplished faster, but just as thoroughly. This is measurable productivity improvement and enhances the company's competitiveness in the marketplace.

A third answer to the rhetorical question concerns the use of CNC machine tools and processes. General-purpose CNC machine tools can be utilized to produce parts at a higher rate than is possible with manually controlled machine tools. The basic reason for improved productivity with CNC technology is that CNC mimics the performance of the most skillful machinists for all of the CNC machines, whereas the best machinists are only occasionally available to run all the manual machines. The characteristics of CNC machines and manual machines are discussed and compared in Chapter 10.

| CNC Machine | Manual Machine |
|---|---|
| • Skills of best machinists programmed | • Rarely best machinists operating |
| • All operations sequenced automatically | • Pauses between operations to review next step |
| • Potential for more parts per shift | • Less parts per shift |
| • Consistent, highly repeatable | • Inconsistent, less repeatable |

Another way in which CIM improves productivity is through the productivity/quality measurement capabilities offered by the use of the computer. Before the advent of computers, the only way to keep track of various parts on various workstations was with manually prepared tally sheets prepared by production expediters and foremen. Since machines were slower then, it was possible to effectively control operations in this way. With the advent of CNC the pace of manufacturing quickened. Queue bottlenecks occurred as work completed by fast CNC machines reached slow transfer systems. The resulting need to move material more rapidly from station to station effectively eliminated time available to produce manual tally sheets. Human input and manipulation of data was not adequate for the task, and initially good interstation control was lost. With the application of the computer to manufacturing control, it again became possible to know where parts were in comparison to where they should be according to the schedule. In fact, the schedule became a computer-integrated activity quickly generated from the planning cycle and the delivery dates promised to the customers. With CIM techniques, it is possible to have real-time information on where parts are, where bottlenecks are, at what workstations quality problems are occurring, what workstations are consistently showing poor performance, and other information needed to manage the operation. It is also possible to react to current information to take corrective action. Before CIM job shops frequently reacted to information that was weeks old, after sizable losses had occurred, and the problem was aggravated by the implementation of fast CNC machines without the rest of CIM. This capability of CIM is illustrated schematically in Fig. 11.4.

## CONTROL OF JOB SHOP OPERATIONS WITH CIM

The use of CIM to control shop operations is becoming one of the most important factors in manufacturing. In its role as the technical resource for the manufacturing function, manufacturing engineering must promote this capability. Let us review this revolution of information management and control made possible by CIM.

The key to success in any endeavor involving many people and many parallel

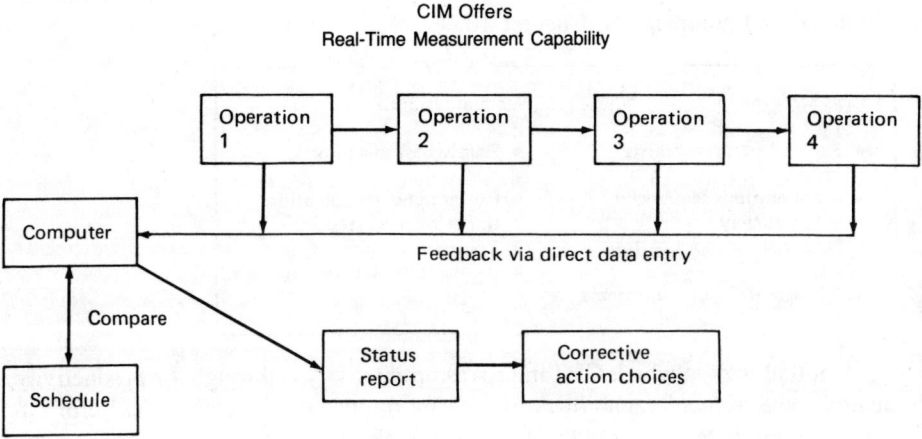

**Figure 11.4** Real-time measurement capability of CIM.

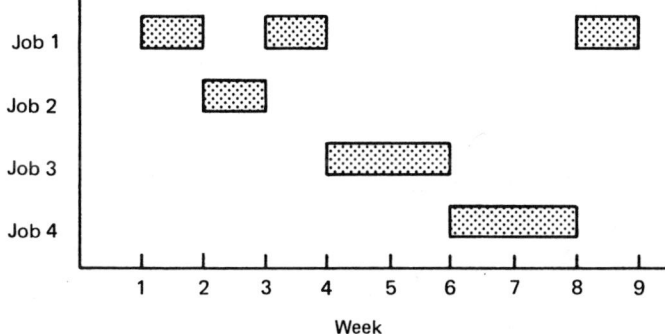

**Figure 11.5** Workstation schedule chart.

and sequential operations is good communications. It is necessary to let all those involved know what they are supposed to do and when they are supposed to do it. All treatises written about effective management of businesses are, in essence, expounding communications philosophies. Therefore, it is reasonable to say that any device capable of improving our ability to communicate is desirable. CIM computer systems significantly enhance human communications related to design and manufacturing. To understand the revolutionary effects of CIM on these aspects of job shop manufacturing, we must first consider the manual control procedures used before computers, and then see how CIM offers significant improvements.

A typical manual control system would be a gantt chart showing work going across a workstation as a function of time. Figure 11.5 is an example of such a chart. Jobs are plotted in sequential time order across the workstation. In effect, this is a reservation system for jobs at that workstation. If a job shop has only a half-dozen or so different workstations, this manual system of reserving space on a machine tool or process workstation is adequate and relatively easy to use. To make sure the jobs are completed on time, the manager will compile all the individual workstation schedules into a master production schedule. In addition to the factory portion, the manager must schedule the design engineering function and possibly the assembly of the product at the customer's site. There is, however, a subtle difference in creating a master production schedule from a series of gantt charts for many workstations. Each of the charts applies to a particular workstation with times reserved for various jobs. The master production schedule deals with the chronological sequence of specific steps required to produce each job. It is a compilation of the gantt charts for the workstations, but with the charts dismembered to show the progress of individual jobs as they make their way from raw material and design concept to finished product. Figure 11.6 illustrates the master production schedule.

This difference between individual workstation gantt charts and master production schedules requires a very important mental shift. Shifting and cross referencing are continually going on in the production scheduling of the factory, from workstation primacy to job primacy, and they must complement each other. We

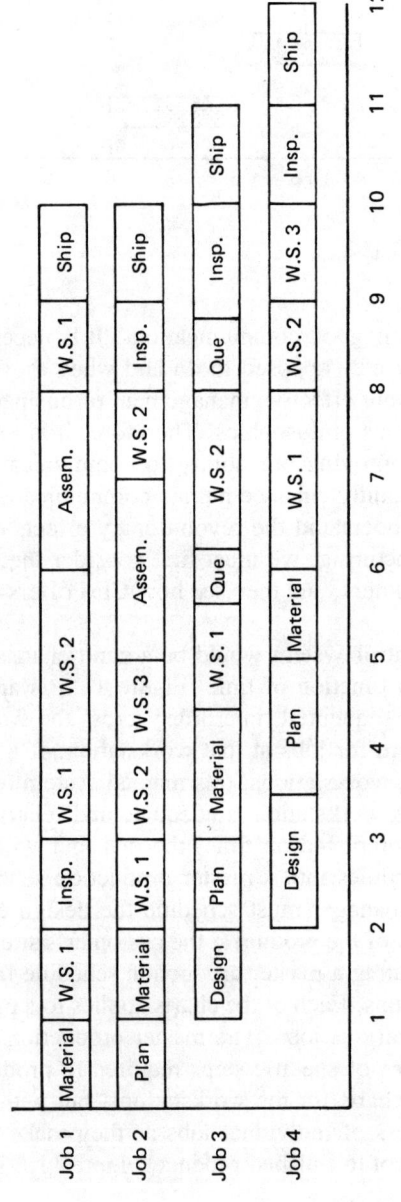

**Figure 11.6** Master production schedule.

cannot have a situation where the master production schedule appears to be balanced but an individual workstation ends up having to accomplish more than one job at the same time.

Let us now consider a typical job shop that has 50 jobs running across 50 different workstations. There could be a matrix of 2500 workstation-job permutations, resulting in an enormous scheduling task. This can be done manually—and was before CIM—but it requires careful attention to detail. A schedule is made out for each workstation based on the planned or estimated time to perform the work. These time increments are then added up for all the workstations to obtain the total cycle time per job. All the jobs are collated to make the master production schedule, and we have what is called a manufacturing plan. This is a static manufacturing plan in that it is difficult to change.

Obviously, the manager tries to schedule enough time at each manufacturing or design stage to take care of unforeseen events. But, as one would expect with 2500 possible permutations, it would require a great many planners and production schedulers to anticipate all that could possibly go wrong and have alternative strategies available. This would be economically unjustifiable. Therefore, at best the manager is compelled to tolerate inefficiencies. Jobs are done well before their scheduled due dates and kept waiting until the next workstation becomes available, which increases work-in-process inventory costs. Or additional equipment is purchased to eliminate scheduling bottlenecks. At worst, problems such as delays in receiving materials or worse than anticipated machine breakdowns stretch out the schedules and delay shipments to customers, with all the cost penalties associated with these failures.

The major problem is the effect of these unknowns on the master production schedule itself. With so many variables to consider, it is very difficult to restructure the schedule when problems occur with one job or a group of jobs. This leads to more than one job being available to work on at a workstation, and requires a decision to favor one of the jobs. The problems multiply as short-range decisions must be made and soon the entire master production schedule is incapable of being followed. The result can be a business out of control, with many emergency production meetings, products scheduled day by day, shipping dates to customers missed, and herculean efforts made at great costs to satisfy preferred customers. When this happens a decision to start all over again is made; the master production schedule is thrown away and the cycle starts once more. The inefficiencies are partially absorbed and partially passed on to the customer, and if neither can be done effectively, the company leaves the business.

This is the classic manufacturing problem: how to create a dynamic schedule, which reacts to change, rather than trying to maintain a static schedule that is too cumbersome to change and hence cannot deal with unforeseen problems. The management of information through CIM is the solution to this problem.

The computer is capable of handling 2500 permutations and many more. Hence, when unexpected problems occur with a job or a group of jobs, it is possible to reschedule the entire job shop and maintain control. The original master production schedule was a controlling document. If the factory followed the schedule, the product would be produced on time and at an acceptable cost. When the schedule becomes void, the manager may be able to control near-term product production, but will probably lose control of start operations, such as purchasing and planning. So much effort will be required to make the necessary corrections to near-term finishes of production that the manager will not be able to devote the necessary resources to revamping all the starts. A decision to leave the starts alone is not satisfactory, since to do this when production is bottled up will result in an inventory excess at best, which saps the economic vitality of the company. What usually happens is that the delays at the back end of operations lead to relaxed efforts at the front end, so that the starts, too, begin to fall behind. Then as the back end problems are cleared up, the starts are late and cannot support the repaired master production schedule. This is the "accordion effect" that managers strive to avoid. Large amounts of overtime to get back on production schedule and expensive purchasing decisions to get materials in house quickly are two of its many costs. This can be avoided if schedules can be revised quickly and thoroughly, as they can with CIM.

The need to reschedule the entire shop when significant problems occur is evident. A master production schedule is a sequence of different parts going across common facilities in a chronologically arranged order. Therefore, all jobs in the factory are interdependent for scheduling purposes. With the scheduling ability of the computer based on the common data base, it is possible to adjust schedules and even make prognoses related to possible manufacturing failures. The human mind may not be capable of inverting a $50 \times 50$ matrix efficiently and quickly, but a computer is capable of doing it. Hence, with CIM, constant rescheduling or updating can be done. This means that the manager has control of the design and manufacturing functions, knows what is happening in a specific as well as a general sense, and is in communication with all aspects of the business and can make decisions based on accurate, real-time information.

Advantages of CIM Scheduling

Access to common data base $\begin{cases} \text{1. Handle large permutations easily} \\ \text{2. Restructuring of workstation loading and high level master} \\ \phantom{2.} \text{production scheduling accomplished quickly} \\ \text{3. "What if" scenarios can be examined} \\ \text{4. Real-time information easily available} \end{cases}$

CIM can be thought of as a synonym for linking all activities through a common data base. How is this accomplished? In theory, quite simply. All activities involved in manufacturing, engineering, finance, marketing, and employee relations are monitored or controlled by a computer or a series of computers, which in turn are linked together in a communications network. The system is designed so that the data related to each of these activities reside in a common memory, the data base. The data base may be stored in one computer or distributed among many. That is a system design question left to the computer systems engineers to obtain maximum efficiency.

With CIM, instead of multiple stand-alone activities, we have everything tied together. With computerized systems operating independently we have to go outside their domains to transfer information between the various steps of business activities. This results in vulnerability to late and missing information, which leads to errors. The common data base solution, the CIM solution, provides correct up-to-date information that leads to significantly more efficient operations.

For CIM to be used as discussed above, what must be contained within the common data base? Let us examine the index of information contained in the common data base and see which types of information affect scheduling.

1. *Design characteristics*: This is the design engineering data on the geometry of the parts and the materials they are made of. This data base is usually not consulted when scheduling changes are made.

2. *CNC tool paths*: These generic data, based on the design characteristics data base, are fed to the various machine tools to machine the parts. While this data base is not consulted in revising schedules, the basic tool path cycle time is included in the planning data base.

3. *Planning*: This is the input from the MP&WM time standard and planning activities showing how much time and on what workstation the parts, as defined in the design characteristics data base, will be made. Most planning data bases give alternative workstations with planned times for parts manufacturing. This is of interest for the master production schedule and rescheduling activities.

4. *Purchasing*: This data base contains information on vendors, the material requirements for parts taken from the design data base, contracts with vendors, promised dates from vendors, and quality assurance requirements for materials certification. The vendors' promised dates for receipt of materials will be of interest in rescheduling.

5. *Quality requirements*: This data base contains the quality plan for the specific parts and becomes a resource input for the planning and purchasing data bases. It also includes all quality failures and causes, and therefore becomes a feedback to the design characteristics and CNC tool path data bases. There is a direct link to rescheduling activities related to quality failures.

6. *Master production schedule\**: This is the key data base for planning the strategy of the company. It receives inputs from the planning data base merged with the strategic plan for sales and income, and keys the purchasing data base sequence as well as the CNC tool path, design characteristics, and planning data bases. It is the controlling data base as well as being controlled by the feedbacks from others. It usually includes the dispatching activity. This data base is intimately involved with the rescheduling activities.

7. *Data collection*: This is the recording data base that informs the master production schedule data base of the status of the progress of parts in accordance with the schedule. The reporting back of items accomplished is vital for rescheduling. This reporting back is done by all functions involved in design, procurement, planning, and production. Usually this data base is the weak link in a com-

---

*Many firms expand the master production schedule to be the Manufacturing Resources Planning (MRP II) data base. We define and and describe MRP II in the chapter devoted to that subject, chapter 14.

puterized production system. Much work has been done toward automating the reporting back of status and making it a real-time system. Chapter 12 is devoted to this subject.

8. *Cost analysis*: All of the costs from the various data bases are collected here, usually in terms of hours for in-house activities and dollars for external activities. This data base is essential for determining whether products are being made within budget. It offers opportunities for analysis to show where difficulties have occurred.

The basic data bases listed above make up the common data base. More data bases may exist to meet specific company needs, such as a maintenance data base to insert required machine tool downtimes into the master production schedule. Note that the common data base is nothing more than the system flow manufacturing production control in Chapter 3, Fig. 3.4, with the inputs from the various functions and subfunctions rearranged to create a computer-optimized set of programs.

There are many commercial programs called CIM common data base systems. They all work after a fashion. However, they are general in nature in order to fit a broad range of products and manufacturing techniques, so that in virtually every case they must be modified to fit specific factory needs, or the specific factory must change its way of doing things to fit the system. Most CIM common data base systems tend to be either modifications of commercially available systems or completely locally developed systems. Special management techniques are necessary to implement these types of systems, and will be discussed in a later section of this chapter.

Control of a factory through CIM comprises automated design retrieval, automated planning, control of manufacturing starts and stops by the computerized master production schedule, release of purchase orders for material as it is needed in accordance with a real-time schedule, insertion of quality monitoring and rework requirements into a real-time schedule, and control of machine tools by design-derived tool paths directly fed into the control logics. Such control leads to an order of magnitude improvement in productivity and profitability. By far the most important aspect of this system is the ability to control schedules, which enables the manager to minutely manipulate or "fine tune" the entire sequence. This means that contingencies in scheduling and corresponding cycle times can be significantly reduced, resulting in significant cost reductions. Labor costs are reduced because the work force is not kept waiting as much to be available when "the job breaks." As the cycle time is reduced, material carrying costs are reduced, since the queue time for material is shortened. Design and manufacturing engineering costs are lower because delay times due to unavailability to perform tests and verifications are minimized. The better the schedule is controlled, the less expensive it becomes to produce the company's product.

It is well known that 90 to 95% of all job shop cycle time is nonmachining time. A significant portion of the nonmachining time can be eliminated by achieving real-time scheduling through the use of CIM, as summarized in Fig. 11.7.

| CIM Means: | Most important and overlooked benefit of CIM: |
|---|---|
| 1. Automated design and retrieval | Control of Schedule |
| 2. Automated planning | Yields: Fine tuning of entire sequence |
| 3. Automated master schedule | 1. Contingency time reduced |
|   a. Starts and stops | 2. Cycle times reduced |
|   b. Material order | 3. Less labor cost |
|   c. Inventory control | 4. Reduced "work in process inventory" costs |
|   d. Real-time production monitoring | 5. Lower engineering costs |
| 4. Automated machine tool CNC tool path generation | |

**Figure 11.7** Reductions in cost factors achievable with CIM.

# A SYSTEMATIC APPROACH TO IMPLEMENTING CIM IN JOB SHOPS

Managers are often fascinated with new technological devices such as CNC machines and machining centers, robots, and flexible automated process lines. However, if these devices are purchased and implemented without a systematic integration plan they will not achieve their full potential. Much more than the purchase of these new technologies is needed to achieve the productivity improvements possible with CIM.

The "factory of the future" is not so much an automated factory as one where excellent, complete, instantaneous communications are standard, and decisions affecting schedules and production rates are made on a real-time basis. Whether the factory is fully automated, a mixture of computer-controlled and manual tools, or all manual tools is of secondary importance compared to excellence in communications and real-time decision making. The type of equipment used in a factory depends on the type of products produced. The use of CNC machines in a hydro generator factory, while important, is not nearly as important as in a reciprocating pump factory. The difference is due to the physical size of the products and the typical lot sizes and cycle times. But the communications/decision-making aspects are equally important in these two types of job shops.

During the past two decades we have seen CNC machine tools and processes pervade job shops, which have benefited significantly from their introduction, but not to the fullest extent possible. The reason for this is that we have concentrated too much on only one part of the CAD/CAM triad. Recently, more attention has been focused on the other two parts of the triad.

Figure 11.8 illustrates the CAD/CAM triad: (1) Machine/process control, (2) design and planning control, and (3) production and measurement control. All three of these parts are equal in importance, as indicated by the triangular arrangement in Fig. 11.8. All three must be present in order to truly control the factory and optimize the benefits obtainable through CIM. Note that the abbreviation CNC does not appear in Fig. 11.8. This is because all machines, not only CNC ones, can be effectively controlled, or directed, by CIM. For a manual machine this is done by giving the operator real-time instructions, perhaps from a video display terminal. All three parts of the triad derive the necessary information

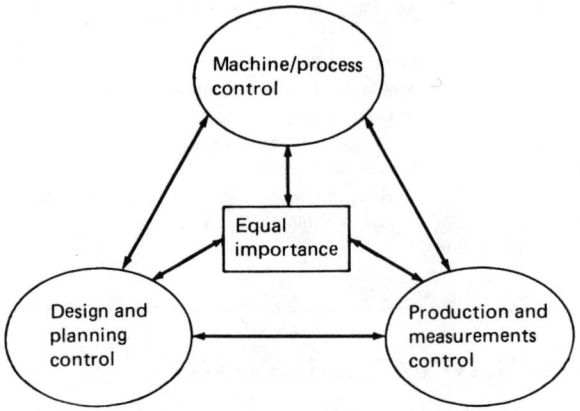

**Figure 11.8** The CAD/CAM triad.

from the common data base, and the efficiency of achieving and disseminating this information depends on the communications/decision-making system.

All too often companies make the mistake of thinking of CIM as only the machine/process control part of the triad. This thinking is typical of the company that rushes out and buys CNC equipment and fails to benefit fully from it.

> **Example:** A CNC lathe versus a manual lathe. The productivity realized with the CNC lathe should be three times that of the manual lathe, but the company only realizes a 1.5-fold improvement in productivity. This company failed to tie the new device into a controllable system. They forgot about the links to design, planning, production scheduling, and measurements. They have not developed a CIM system.

Therefore, designing a CIM system is really the first thing to do in order to implement CIM. This should be done before the first new equipment is purchased. Implementing CIM requires an understanding of the way all product or service producing companies work. To gain this understanding, we will introduce the concept of the manufacturing system and its seven integrated steps.

The manufacturing system is a methodology that has evolved since the beginning of the industrial revolution. It is simply a logical approach to obtaining information and using that information to manufacture the product in an optimal fashion. Realistically, all companies have to follow this system logic or else chaos occurs. All companies perform the seven steps of the manufacturing system in carrying out their specific process of making products or providing a service. Curiously, the seven steps occur whether or not the company is conscious of them. The seven steps, in chronological order, are:

1. Obtain product specification.
2. Design a method for producing the product, including design and purchase of equipment/processes to produce, if required.
3. Schedule to produce.

4. Purchase raw materials in accordance with the schedule.
5. Produce in the factory.
6. Monitor results for technical compliance and cost control.
7. Ship the completed product to the customer.

The task of applying CIM theory is to integrate the information flow between the seven steps and to automate the operations to minimize random errors in the processes. Information is needed for every step, and much of the information is redundant. Thus we can see the need for a common data base. Therefore, the process of implementing CIM becomes one of initiating a common data base to link the various operations together. In practice, this means, for example, having CNC machines (the machine/process control triad) receive their tool path instructions from the design data base (the design and planning control triad) and their sequencing of work order instructions from the production schedule data base (the production and measurement control triad). Both of these specific data bases would be components of the integrated common data base (like two books in a library).

Since the entire company is affected by the implementation of CIM, developing a CIM system must be a total company approach, with representatives of all functions being part of the planning process. A cross-functional organization must be established that has the responsibility for developing the criteria for the system and making sure the specifications are defined for use within the company. For lack of a better title, let us call this cross-functional team the systems council. The people on the council should be competent in computer usage or potential usage within their home organizations. They should be knowledgeable about the parts processed in their company. They should fully understand how an order from a customer is translated into a design, a producible design, manufacturing instructions, and finally a product shipped to the customer. If they do not have these prerequisites, then the systems council's first order of business is to educate themselves in these areas. Sometimes consultants are hired to help with this education, particularly in computer utilization, but often it is incumbent upon the systems council to train itself. However it is done, this training must be accomplished to avoid costly delays and possibly failure. The systems council may also find it necessary to train their superiors, the senior members of the company's management. Here consultants may be very worthwhile.

After the preliminary training, the systems council can get on with the task of developing the firm's CIM plan. The first thing to be worked on is the communications/decision-making strategy. The council must decide what kind of manufacturing scheduling system is needed for their product. How is the design engineering data put into the system? How is information on propositions and design transmitted from marketing to design? How is the design information translated to manufacturing information? How should the status of design, proposition activity, and manufacturing progress be reported? When should purchase orders for materials be released? These basic questions must be answered in a manner that is specific to the particular factory and that shows the overall relationships of the communications/decision-making patterns of the company. Once this information is obtained, the systems council must decide whether they like what they have, or whether change is desirable or even necessary to improve efficiency. This decision must be made within the systems council. It cannot be made by vendors of computer equipment or packaged CIM systems or by hired consultants. These sources can help by pointing out similarities to other companies and showing how others have arrived at decisions, but the systems council alone must make the decisions. Once the decisions have been made, the system analysts, programmers, and engineers can convert them into the familiar flowcharts. At this point we are ready to purchase implementing hardware and software. It then becomes the task of the specialists reporting to the systems council, who are usually technical people and often from manufacturing engineering, to specify the various types of equipment and software necessary to implement this tailored communications/decision-making system.

Notice that nothing has been said yet about engineering workstations or CNC or other new technologies. This is not an oversight. While these are very valuable for a manufacturing concern, their full potential cannot be realized unless the communications/decision-making system is in place. After it is operational, the engineering workstations, CNCs, and so forth may be added to further improve productivity. For manufacturing firms that already have engineering workstations and CNCs but do not have a communications/decision-making system, a hold should be put on further expansion of the new equipment until an adequate communications/decision-making system has been implemented.

| First Rule of Thumb | Second Rule of Thumb |
|---|---|
| Develop and install a communications/ decision-making system first for major productivity improvements. Follow up with "computer-automated equipment" for additional productivity improvements. | For companies already possessing "computer-automated equipment." Put a hold on further development of such equipment until a communication/ decision-making system is developed and installed. |

Components of a communications/decision-making system would be a shop floor data collection system, MRP II, or a design retrieval system. These projects would have high priority, while a CNC vertical boring mill would be a medium-priority project that should be approved for purchase only if the cash flow payback period is good, usually less than two years. The key here is setting the proper priorities. There must be a framework to build on, otherwise the company will end up with islands of automation that never live up to their potential. This leads to frustration and possibly even abandonment of CIM; in the latter case the company may be unknowingly deciding to leave the business.

In summary, then, the sequence to be followed in implementing CIM is:

1. Recognize the CAD/CAM triad relationships for the company's specific products.
2. Establish a systems council to determine overall strategy and specify the communications/decision-making system.
3. Implement the communications/decision-making system.
4. Add new technology to further improve productivity.

When implementing a CIM system, it must always be kept in mind that CIM is needed to improve productivity. If an aspect of the program developed by the systems council is vague on this, it must be looked at critically. No projects should be carried out unless the end result is an improvement in productivity, hence profitability.

## ARTIFICIAL INTELLIGENCE IN A CIM WORLD

An often overlooked but an increasingly important application within the CIM environment is the use of Artificial Intelligence (AI). One would logically ask what difference does it make if AI is used via common data bases or separate ones. After all, AI is a theoretical approach to help in decision making and not directly in line with the routine operations of a company. This may be so, but there is a subtle but very important difference between AI working with a common data base and a more narrowly defined data base. I will explain, but first we need to define and demonstrate the potential benefits of AI to make the differences relevant.

We use CIM to smooth out the flow of information in conducting daily activities related to the seven steps of the manufacturing system. Specifically, we use CIM modules to do the clerical and editorial activities, thus freeing ourselves to do the analytical and strategy activities. Even with these higher activities we use the CIM modules to access data and to manipulate data. But the creative thought or idea generating is entirely the domain of men and women, not machines. Now we have AI available to help us even with the creative process. AI can help by giving us logical supposition based on data but not related directly to the specifics of the data. This is the next evolutionary phase of computer use as the industrial revolution merges with the information age.

What is AI? It is not computers with cognizant awareness. Computers do not think in human reference terms. The computer, as we know it, reacts very rigidly to its programmed logic statements. For non-AI, the logic statements are the familiar If/Then routines. This means that if a certain condition exists, then the response (or action) will always be the same. Specific actions always elicit specific reactions. The major innovation AI provides in industry is accomplished by the AI subgroup called expert systems. Expert systems, I must emphasize, cannot make computers thinking machines any more than conventional software does. Expert systems are logical extensions to conventional programs that create an illusion of independent reasoning, more so than we usually associate with computers. The innovation we find useful is that the program appears to learn. This is accomplished by evolving from If/Then logic to If/Then Probable Then. In AI (we will refer in this way to expert systems) we use a clever set of programs that allow the data base to search for information and then use that information to add to the original logic rules. Hence we say that the computer "learned." This is very useful for many manufacturing activities. Examples of such activities are diagnosis of failures in maintenance activities, process planning when many potential solutions exist, and aspects of statistical process control in which specific results are required, hence "intelligent" monitoring routines are needed. Here, If/Then Probable Then logic expands our ability to find out if a course of action would result in desired results. We shall see how this process works when we review the techniques of expert systems.

Let's delve a little deeper into AI theory so we can build a more complete understanding of why AI capabilities are brought out more fully under a CIM environment. As would be expected, AI programs have terminology specific to them. It will help to explain and demonstrate some of these. There are three terms to be defined:

Working memory
Knowledge base
Inference engine

The *working memory* is not specifically part of the AI program, but without it the program has no way to learn. It would have no experience base. The working memory consists of the memoirs or knowledge of the people considered experts in the technology we are trying to emulate. Experts are queried in a logical manner to understand how they solve problems that are not necessarily solved using precise causative relationships. This means we are exploring intuition as a method of reaching conclusions about problems. The person doing the debriefing, the knowledge engineer, uses the following sequence to break down instinctive knowledge of the experts.

Problem definition
Symptoms
Root cause
Elimination of root cause (solution)

This is analogous to the traditional scientific method, which may require many iterations to come up with solutions. Thus we can see that knowledge engineers have to be very patient and very logical.

The *knowledge base* is the data base of the AI system. One of the advantages of CIM, of course, is that the data base can be very large. This holds true for AI data bases operating under a CIM system, the only proviso being that the data base be structured in a manner that is compatible with the knowledge base needs. This turns out to be an advantage for AI users in a CIM environment because the proviso is easy to comply with. Thus we have a gigantic universe of learning to draw from.

The knowledge engineer structures the information he or she gains from the experts into two categories: facts and rules. Fact are statements of conditions. An example of a fact would be:

"nail length = 1.011 inch"

Rules are a set of actions to be taken with sets of facts as directed by the AI program. An example of a rule related to a fact would be:

"if length = 0.75 inch to 1.25 inch, then finishing nail"

The *inference engine* is the AI program, similar to the blank format screens of familiar spreadsheet programs. These programs all have logic built into them for spreadsheet uses and we tailor it to our particular needs. So, too, with the inference engine. The inference engine is a generic set of If/Then Probable Then logic statements that can be purchased commercially and is often referred to as the generic shell. The generic shell loaded with the facts and rules of the knowledge base becomes the specific AI program. Shells (the inference engine) have the capability to make an inferred solution based on rules and facts.

What's the difference between AI programs and traditional programs? Traditional programs are numerically related and have only one allowable conclusion based on the facts. Scheduling algorithms are good examples of the traditional approach. AI programs differ in that they deal in and/or logic. And/or logic is not causative mathematical logic but deals more in the realm of probabilities. Hence, an answer is not always correct but close enough to 100% certainty to be valid for most problem solving. This type of reasoning is more like human thought patterns. So we say we have artificial intelligence.

The important factor about AI is its apparent ability to learn. This, of course, is not true in a human sense but it is definitely true in an applications sense. This occurs because the program arrives at its solution through forward or backward chaining. Essentially, with each solution we are adding another link to the chain—hence, the appearance of learning taking place. When the AI program reaches a logic dead end and no answer is apparent through chaining, the program is set up to ask the user (human interface) related questions.

In Fig. 11.9 we see an example of the inference engine content and the schematic of how new rules are added through interaction with the software. Note the list of facts and list of rules contained in the inference engine. From this set of facts and rules, by simple inductive reason, we can come up with new results or inferred solutions. We can see that the statement "stainless steel nails are silver and used for finishing" comes directly from the fact "finish nails, stainless steel" and the rule "nail is silver, then material is stainless steel." There is no other possible inferred solution. So far, no learning is taking place that is different from non-AI software. Let's look at the interactive action shown in the bottom half of the figure.

We see that the user (interactive input) enters the statement "nail is black." The computer, searching the inference engine by looking for relationships involving black nails, finds by fact that black nails are used for cabinets. That's all that exists at this point in time in the data base. So the computer states that there are no rules about length or material specifically for black nails used for cabinet making. Therefore the inference engine software has been preprogrammed to ask questions related to other rules it already knows. In this case it asks, "what is the

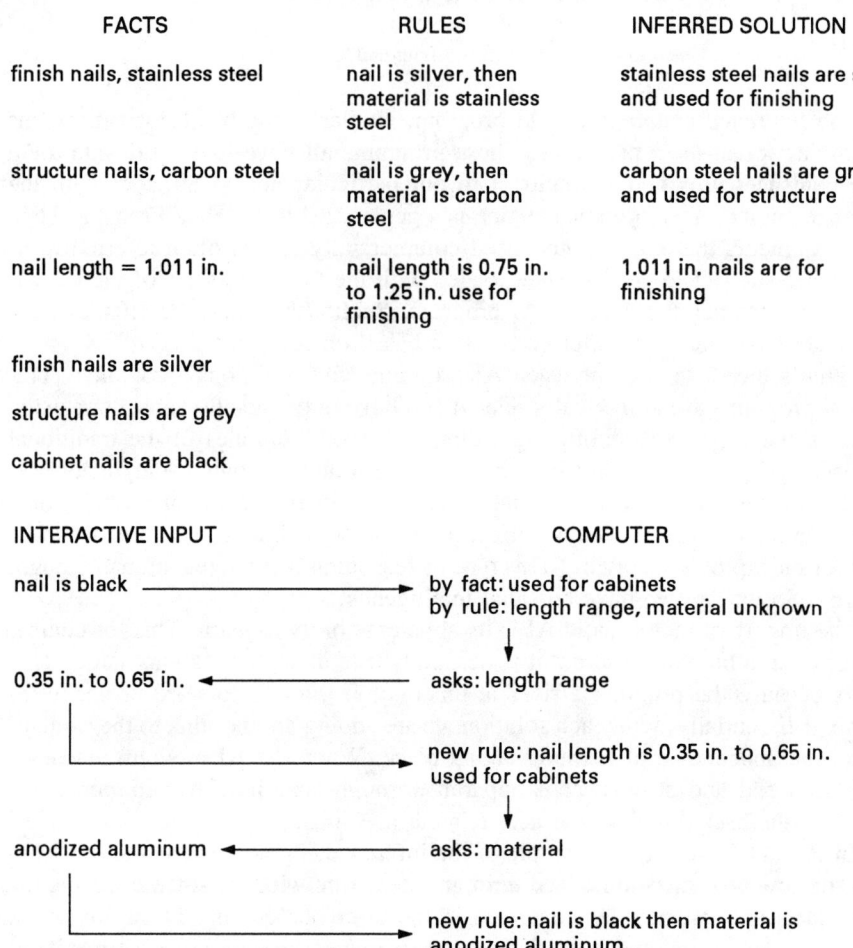

**Figure 11.9** Learning by chaining (based on inference engine logic).

length range?" The human user responds "0.35 in. to 0.65 in." From this information via chaining the computer software creates a new rule: "nail length is 0.35 in. to 0.65 in. used for cabinets." We say the program has learned, and it can be used in future situations.

   The importance of the learning ability is that the conclusions the AI program presents always take into account the most current validated information. This is not simply a different numerical value to solve the same preprogrammed equation. It is more akin to having the different numerical value and a new equation too. In the real world, not only may data inputs change but also the circumstances affecting those data may change. Humans can accommodate these changes quite readily. Rigid, non-AI programs cannot accommodate changes at all. It is this limited ability to accommodate change that is the breakthrough expert systems gives us for using computers in a more integrative manner.

AI offers the potential to solve hosts of manufacturing problems. Currently it is most useful for Statistical Quality Control (SQC) programs. Here, the learning capability is beneficial in evaluating whether or not a process is in control. We are also beginning to see AI used in planning programs. Determining how a product could be produced for optimum performance depends on complex logic of CNC availability; types of materials, which have different stress profiles; overall process cycle times; and conceivably the entire manufacturing process knowledge data base from forming to painting. Here we can readily see how powerful such a tool could be and how much more effective it would be if we had integrated data bases, that is, CIM.

The advent of If/Then Probable Then logic associated with AI opens up a host of opportunities for business. As discussed above, we can actually use the computer as an electronic partner of sorts to solve complex problems with nonrigid algorithms and less than specific related data. This puts a premium on the Probable aspect. Having the wherewithall to find related information for the "learning" to take place is at a premium. This is where CIM, the integration of information in a singular or linked set of data bases, really provides an advantage.

CIM has the ability to search for information from a much larger universe than stand-alone data bases. For example, if we were to employ AI for a planning program we would need manufacturing process information, scheduling information, and design geometry information. In a CIM system this information would be contained in several linked data bases and be readily accessible. In stand-alone systems we would have a design data base, a master schedule data base, an MRP system data base, and a manufacturing engineering data base. All would be independent and mute to each other. How would the AI program chain to know the next step in production based on machine availability and customer need data? It would be impossible; hence, AI would not be employed and we would not be able to have a generative planning system. With CIM with data search capabilities, AI can be used, and we would have a planning system with schedule and availability realism built in. Thus CIM expands the potential for AI use, from tightly bounded diagnostic programs using stand-alone data bases to much more generic scenario and control programs enjoying the resources of CIM linked data bases.

## SUMMARY

In this chapter we have defined and developed the fundamentals of CIM. We have seen that CIM is essentially a productivity improvement concept, one so powerful and important that all discrete parts job shops must eventually implement it. Companies that choose not to implement CIM will eventually cease to exist in their business.

The fact that significant productivity improvement is a direct result of CIM implementation has been stressed, and the mechanisms for this improvement have been demonstrated by spelling out in detail why we must use CIM and how it is implemented. We have discussed the common data base and learned why this integrating computer information system is necessary to achieve excellence in communications between the functions. We have also learned that CIM means much more than implementing CNC machines in the factory—that, in fact, it means communications/decision-making excellence and CNC machine tools are only a part of it. This led us to the important concept of the CAD/CAM triad: machine/process control, design and planning control, and production and measurements control. Thus, CIM is total control of the business, using real-time feedback systems, and communications excellence is the key factor. The philosophy of implementing CIM via a systems council has been explained. Finally, we looked at the future via applications of AI with CIM.

## REVIEW QUESTIONS

1. CIM drives productivity improvement, which in turn drives further improvement in CIM. Explain why this iterative process occurs.

2. Define "common data base." Explain why this integrating entity is necessary for productivity improvement.

3. Prior to common data bases, why was it impractical for design engineers to create engineering drawings that were suitable for CNC input?

4. Explain how finance, marketing, design engineering, and manufacturing would make use of the following information from a common data base:

Nuclear valve body,
   weight—8200 lb
   outside major diameter—4.25 ft
   valve seat diameter—28.00 in.
   overall length—8.40 ft
   wall thickness at valve seat—4.85 in.
   material: chrome, vanadium, alloy steel
   operating conditions: cyclic; open to close in 0.8 sec; 1000 PSI steam, 1000°F

5. Discuss how access to a common data base tends to minimize design and process development iterations. Use a hypothetical example to illustrate your answer.

6. Describe the advantages of group technology classification and coding for design and manufacturing engineering activities.

7. One of the early disadvantages of the use of CNC machines was that they caused increased queue times and bottlenecks after the CNC workstation. Explain how CIM mitigated this disadvantage.

8. Explain why excellence in communications is the single most important factor in improving productivity via CIM.

9. Discuss the differences between a static and a dynamic manufacturing plan. Explain the reason for establishing static plans instead of dynamic plans.

10. Explain how CIM can minimize the "accordion effect" caused by unforeseen or unscheduled events within a manufacturing schedule.

11. Why does the modern job shop traditionally generate 90 to 95% nonproductive time for the entire part manufacturing cycle? Why does CIM offer the potential to decrease this nonproductive time?

12. Explain why the purchase of CNC machines is not the highest priority in establishing an effective CIM system.

13. An in-house "systems council" is the preferred mechanism for defining the communications/decision-making system. For each of the following organizations, outline the point of view and expertise each will contribute to the systems council.

(a) Design engineering
(b) Manufacturing engineering
(c) Materials
(d) Shop operations
(e) Finance

14. Discuss the pros and cons of purchasing a communications/decision-making system from a vendor.

15. Explain why all organizations making products or providing services comply with the seven steps of the manufacturing system, even though some organizations may not even be conscious of the existence of the manufacturing system.

16. With reference to AI logic statements, what is the difference between If/Then and If/Then Probable Then? Which represents AI and which non-AI computer programs?

17. An expert system knowledge base contains the following statements. Which are rules and which are facts?

(a) ball bearings fail at 400°F minimum
(b) roller bearings fail at 300°F minimum
(c) roller bearings fail at lower temperatures than ball bearings
(d) roller bearings are used for loads over 250 lb
(e) if load is over 250 lb then roller bearings
(f) if load is under 250 lb then ball bearings
(g) failed bearing has load of 175 lb and is ball bearing

18. An expert system learning by chaining sequence is outlined below. What is the new rule inferred by the sequence shown? Hint: Current rules and facts are compatible with those shown in question 17.

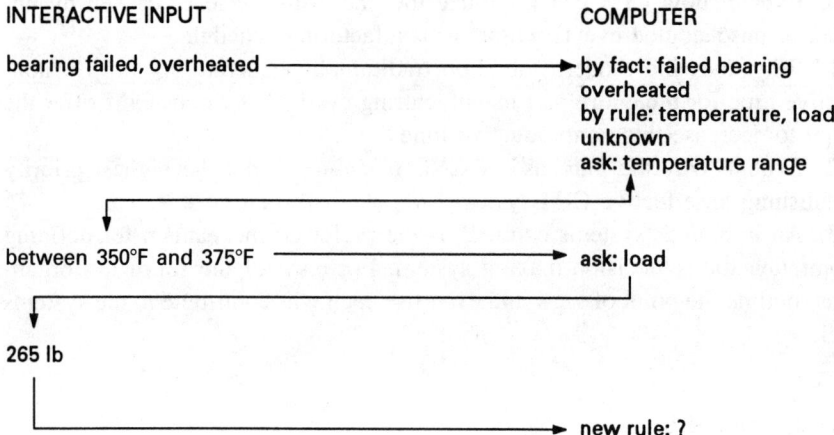

INTERACTIVE INPUT                                    COMPUTER

bearing failed, overheated ──────────────────▶ by fact: failed bearing
                                                 overheated
                                                 by rule: temperature, load
                                                 unknown
                                                 ask: temperature range

between 350°F and 375°F ──────────────────▶ ask: load

265 lb

                                                 new rule: ?

# COMPUTER-AIDED PROCESS PLANNING
# AND DATA COLLECTION

The subject of this chapter is the CIM activity sometimes known as Computer-Aided Manufacturing (CAM). Chapter 11 dealt with the excellence in communications necessary for achievement of the total CIM potential. In this chapter we will examine what is meant by excellence in communications for the optimal operation of a factory. We will develop the management philosophy behind the integrated factory communications system. The chapters concerning the responsibilities of the various manufacturing engineering units have amply demonstrated the complexity of designing and operating a manufacturing system. We are now ready to explore the use of CAM to significantly improve this communications capability.

## HISTORICAL DEVELOPMENT
## OF COMPUTER-AIDED MANUFACTURING

There have been three very distinct applications of computers in product-producing companies. They are different because they were championed by different organizations in the industrial world.

The first of these applications was scientific computing. Since the computer was developed to solve complex mathematical problems for research scientists, it is natural that the first industrial applications of computers were within the design engineering activity—the closest approximation to the work of research scientists within product-producing companies—and employed modifications of scientific programs to solve design problems. Hence, design engineering had computers before anyone else in industry.

The second application was in the financial function. This again was a natural application. Computers are excellent "number crunchers"; that is, they are good at addition and subtraction and, through programming manipulation, multiplication and division and, of course, other more complex functions. Since finance is basically an operation that evaluates numbers, the attraction of having a superfast method of computing great volumes of figures is understandable. Therefore, finance became a champion of computers for keeping financial accounts accurate and up to date. By the mid-1960s virtually every medium-size and large company and many of the smaller companies had extensive computerized finance sections.

The third application of computers in industry came in the manufacturing area. This was the use of process computers to control machine tools and other manufacturing processing equipment. It was evident that computers could capture the entire method for making a part on a machine, which is an exercise in three-dimensional geometry. If the path of a cutting tool could be mathematically calculated, it would be a relatively simple task to program a computer to direct a controller. The controller activates the drive systems of the machine, and we make a part under computer control. Manufacturing engineering introduced the computer to the factory floor with the advent of numerical control. In Chapter 10 we reviewed the impact of CNC on modern manufacturing.

Until the 1970s computers were used in industry in these three separate, clearly defined ways. Then, as described in Chapter 11, the boundaries became less rigid and we began to consider how the uses of computers could and should be transposed from one function to the others. From finance, manufacturing found that computers could be applied to production control, if production control data could be broken down to numerical values. As we know, these data can be broken down, and we have seen the birth of the computer data base that can handle vast quantities of production scheduling and materials purchasing data. This became Manufacturing Resources Planning (MRP II). Manufacturing also borrowed from design engineering, using the interactive graphics tool to define the CNC tool path with the same geometry used to analyze stress and strain. All this borrowing, of course, led to the CIM integration philosophy discussed in Chapter 11.

The one thing that was missing until recently was a way to improve the availability of information characterized by process planning and know what is happening on the factory floor on a real-time basis. These two areas of the manufacturing function's responsibility have no real parallel in the other two major functions that use computers; therefore, they have had to modify techniques borrowed from finance or engineering, or develop new techniques within the manufacturing function. Both of these approaches have been used.

The status of parts completion is of interest to finance so that it can project expenditures of funds and make plans to borrow capital. A time lag in obtaining this information is not a serious concern up to perhaps 1 month. But the basic financial programs already existed within the cost analysis functions. As long as there have been manufacturing firms, the production area has reported completion to the bill-paying area. Therefore, the financial function's computerization of the bill-paying area proved to be of some use to manufacturing. The end result has been a status reporting system developed by finance for finance into which manufacturing inputs data and from which it obtains some limited benefits. The primary benefit is information on how much time it took to complete a part compared to the original planned time, which is used to determine efficiency. For monitoring factory progress, the financial system did not help much.

The only communications programs that were developed entirely by manufacturing were the planning programs. These are an evolutionary development of the scientific time standards based on data related to motion analysis. Motion analysis, or time standards, can be refined to mathematical equations. Therefore, such relationships would be useful in determining how long it should take to make parts and the sequence in which they should be made. If such things can be determined mathematically, computer programs can be written to have the computer do the computations. As we will see, this has been done in computer-aided process planning.

Thus there is a definite historical prerequisite in attempting to automate the communications activities of manufacturing. The desire to have automated planning and the desire to know the results of that planning have been evident for many years. With the availability of CIM common data bases and excellence in communications this concept is now becoming a reality. The computer-aided process planning and real-time data collection systems have been the last major areas within manufacturing to be computerized. Let us see how these cause and effect activities (initiation and reporting of results) so basic to manufacturing success have become useful data bases and what this means to the success of an industrial organization.

## EARLY COMPUTER-AIDED PROCESS PLANNING SYSTEMS

Planning is the development of strategy to achieve an aim. It consists of a direct-line approach and contingency plans in case an accident occurs along the way. In manufacturing, this means having a primary set of instructions for shop operations and one or more alternatives if the primary method is not available. Good planners have an excellent knowledge of the parts to be made and the capability of the factory to make those parts.

As one would imagine, contingency planning on a large scale requires either an exceptional memory based on experience or a superb filing system with first-rate indexing. This is especially true in a job shop, where part lots being manufactured are seldom identical to previous ones. Therefore, the first idea for Computer-Aided Process Planning (CAPP) was to use the computer as an efficient file cabinet to hold information on the previous experiences for easy recall. No attempt was made to automate planning; the data were simply stored so that the methods engineer could review them and base new planning on similar previous experi-

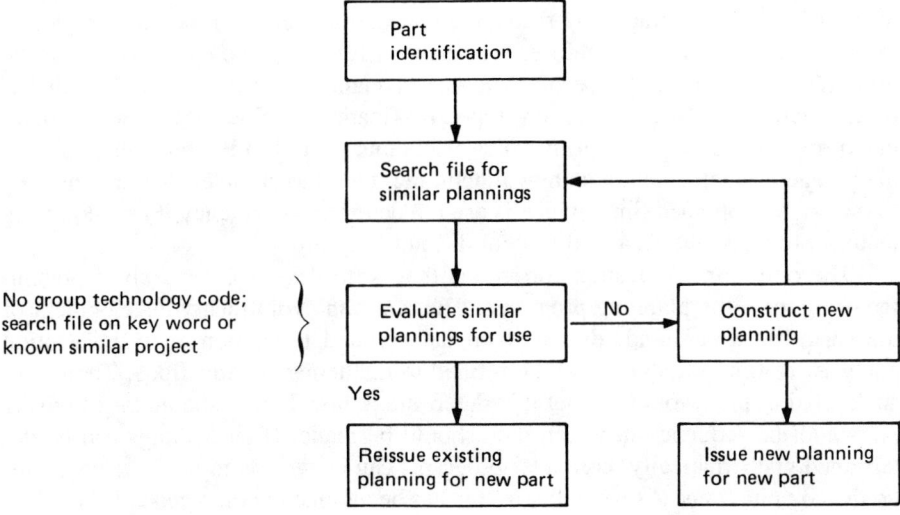

**Figure 12.1** Early CAPP systems flow.

ences. Figure 12.1 is a flow diagram of the early CAPP systems. It represents a true stand-alone system. There is no input from design engineering or any other function, or from any other manufacturing subfunction. The MP&WM unit is entirely on its own. It receives drawings from design engineering and reviews them for information needed to make the plans. The plans are then sent to the master scheduler to fit into the schedule.

Figure 12.1 represents a manual system with computer sorting. Parts were identified by drawing number and/or a key word or phrase. This meant that in looking for an exact match, the system was limited to finding the exact drawing number issued for an earlier contract, which is not a likely occurrence, or using the key word or phrase to call up the history of drawings with the same nomenclature.

Thus, the methods engineers and planners had a set of parts with similar names that they could review for planning sequences. They could call up each plan and compare it with the new part to be produced. If they had made the part before, it was a simple matter to reissue the plan and the job was done. If they had not made it before, they looked for plans that could be applied to the current part, perhaps with some modification. Finally, if no existing plans fit, the planner had to create new planning to get the job done. In all cases when the job was completed, the hard-copy planning was submitted to the master scheduler or production dispatcher to use when requested.

## THE CAM CAPP SYSTEM

The system described above is a computer-assisted system but not an integrated system. It uses a computer sort program to find similar plans for further evaluation but the sort is very coarse because there is no way to specify further what is being looked for. The next version of a CAPP system did a much better job of utilizing the computer to do the sorting and evaluating task, but it was still a stand-alone system. It was still only a CAM system and not CIM. Figure 12.2 represents a typical CAM CAPP system and shows the additional features related to communications and efficiency that it provides.

The CAM CAPP system has three major advances over the early CAPP system:

1. It uses Group Technology (GT) classification coding for finer selection of similar planning.
2. It is integrated with the master production schedule.
3. It provides alternative planning for use in scheduling.

These are significant improvements that we will elaborate on in the following subsections.

## Use of Group Technology Classification Coding System

Recall that the early CAPP system used a drawing number to find an exact match and, failing that, used key words or phrases to find similar parts to consider. With a coding system based on the geometry and/or processing sequence, it would be possible to have a prototype planning that could be easily accessed for all similar parts and have the computer select building blocks for the individual characteristics of the new part. Let us review the salient points of a GT classification and coding system. Figure 12.3 shows a very simplified example.

The classification and code number A343631 adequately describes the part called a valve stem. Each digit explains or defines a salient feature of the part. The letter A identifies the family or broadest classification. The first digit specifies the material the part is made of. The second, third, and fourth digits bound its physical size, in this case longest length and largest diameter. The fifth digit identifies the seating diameter. Since A identifies the part as a valve stem of the A family, the next to last digit would specify the seating diameter. If the letter had been a D, the family could have been motor rotor shafts, and the next to last digit could have represented a seal ring groove. Codes can be hierarchical, hence many thousands of parts can be represented via this seven-digit code. (This is a relatively simple code; in practice, some codes have as many as 32 digits.) The final digit in Fig. 12.3 indicates a special feature of the part. In this case the 1 indicates a groove. If the last digit was 0, it might mean that there were no unusual features. Or if it was 6, the part would have a threaded hole off the perpendicular to its longitudinal axis.

Clearly, a code can be a very powerful condensation of significant information about the part. It is certainly more revealing than the key word or phrase search technique used for the early CAPP system. The CAM CAPP system applies the group technology classification code as follows:

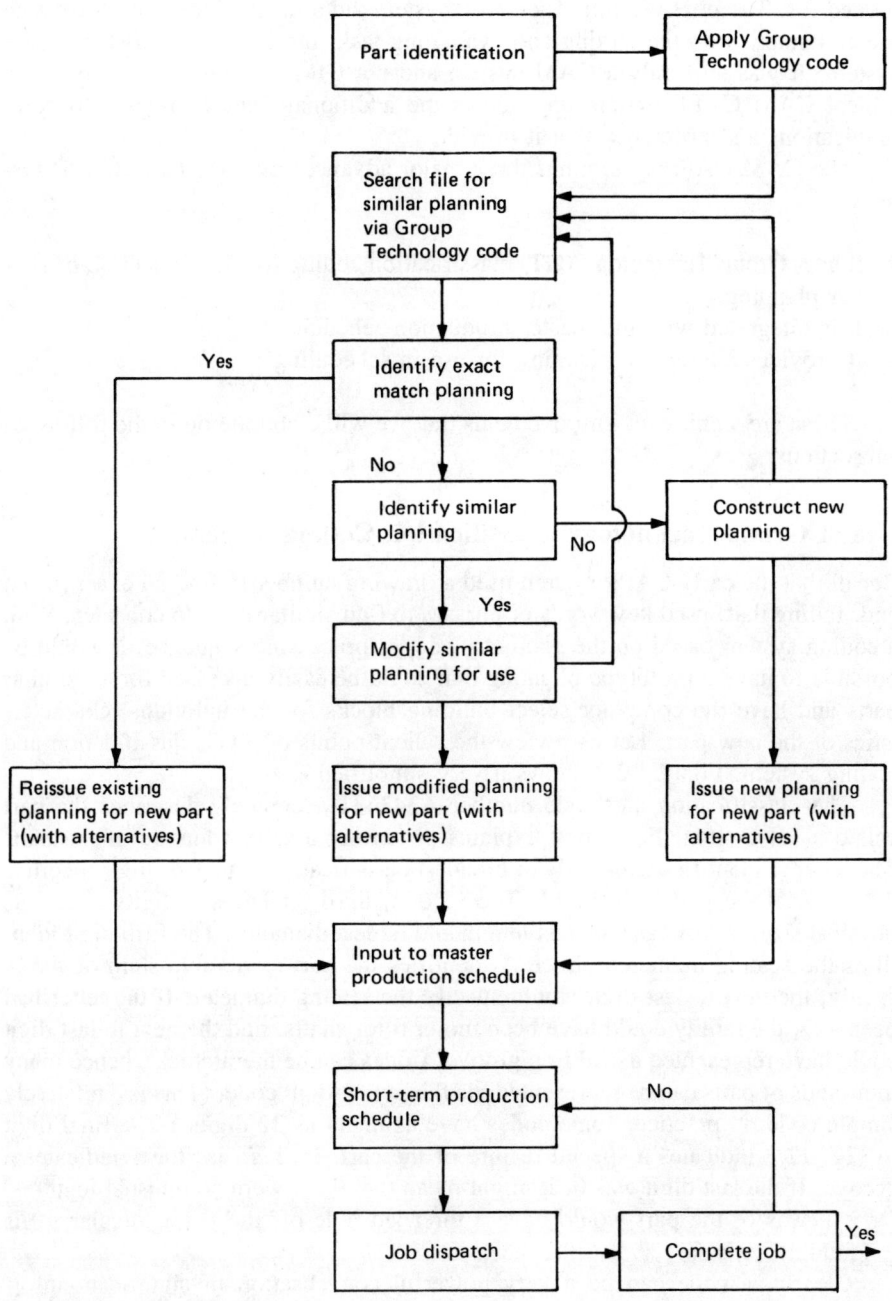

**Figure 12.2** CAM CAPP system.

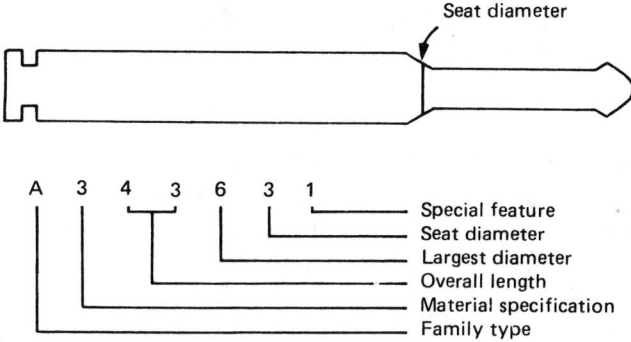

**Figure 12.3** Valve stem code.

1. A part drawing is received from design engineering. The planner, using a code matrix, codes the part.
2. Using the newly created code number, the planner accesses the data base and has it search for a match with other codes. This is started at the most general matching feature, in our example the A for the part family (valve stem), and continued down to the most specific feature, the rightmost digit 1, the special groove. The computer will find the closest match.
3. Once the closest match has been found, the planner again accesses the data base to find the planning for the part or series of parts that are the closest match.
4. Using the closest match as a guide, the planner can make minor adjustments to the previous planning and create new planning. The new planning is then input into the master production schedule. It is also fed back into the data base as a planning associated with this newly developed group technology code number.

This system has significant advantages over the early CAPP system. It allows a much finer search for similar parts, and it keeps looking until the closest match or matches have been found. This shortens the planner's review of parts considerably, and since the match is as close to an exact match as possible, planning the system presents as close to what is needed as possible. Hence only fine-tuning modifications will be required. In our example, perhaps the match is good except that the new part is to be 12 in. longer than the closest match from the data base. Therefore, by using the planning presented by the system and entering into the machining time formula for the planning, the planner can modify the length of time necessary to do the lathe roughing and finishing cuts.

The early CAPP system required a manual review of all plans presented regardless of how close a match they were. The CAM CAPP system requires a review only when the closest match has been found. The reduction in the time the planner must spend is a significant productivity improvement.

## Integration with the Master Production Schedule of the Manufacturing Resources Planning System

The output of the early CAPP system was a hard copy that was mailed or hand delivered to the master scheduler or production control specialist. Usually, a manual system was used to match it with the flow of work orders to the factory floor. When the production control specialist called for that part to be made, he or she would have the dispatcher take the planning from the file and match it with the drawings, then send it to the proper workstation. In the CAM CAPP system, the output from the planning activity is dispatched electronically to the master production schedule, a component of the MRP II system. Essentially, the master production schedule is told that the planning is complete and available. When the schedule says that the part is to be released to the factory floor—that is, material and machine space are available—the computer prints out the planning and associated reporting completion vouchers for shop use. The dispatcher then merges it with the drawings and dispatches them to the starting workstation.

## Provision of Alternative Planning for Use in Scheduling

Notice that on the flowchart in Fig. 12.2, the three boxes showing the issuance of planning state in parentheses that alternatives are also issued. This is contingency planning. If the preferred machine is down for maintenance, or a sales schedule need has created a bottleneck on the primary machine, a decision to wait is usually not a viable management choice. Alternative means must be planned to produce the part. An alternative can be planned when it is needed, which tends to be very inefficient, or the plan can exist before any need arises.

With the early CAPP system the capability to do alternative planning was limited. The planners had to spend considerably more time searching files for good matches of existing plans to new parts than with the CAM CAPP system for reasons explained previously. Obviously, before the early CAPP system, the time to do alternative planning was even more severely restricted.

## THE INTEGRATED CIM CAPP SYSTEM

A system in which the functions are fully integrated is an integrated CIM system. This system provides the excellence in communications needed to obtain all of the benefits available from CIM implementation. Figure 12.4 defines the flow of the integrated CIM CAPP system.

The integrated CIM CAPP system is a further improvement over the CAM CAPP system. It realizes the full potential of CAPP for automated information input and retrieval, which is the backbone of true CIM. Three significant features available in the integrated CIM CAPP system, which did not exist before but are vital for true CIM, are:

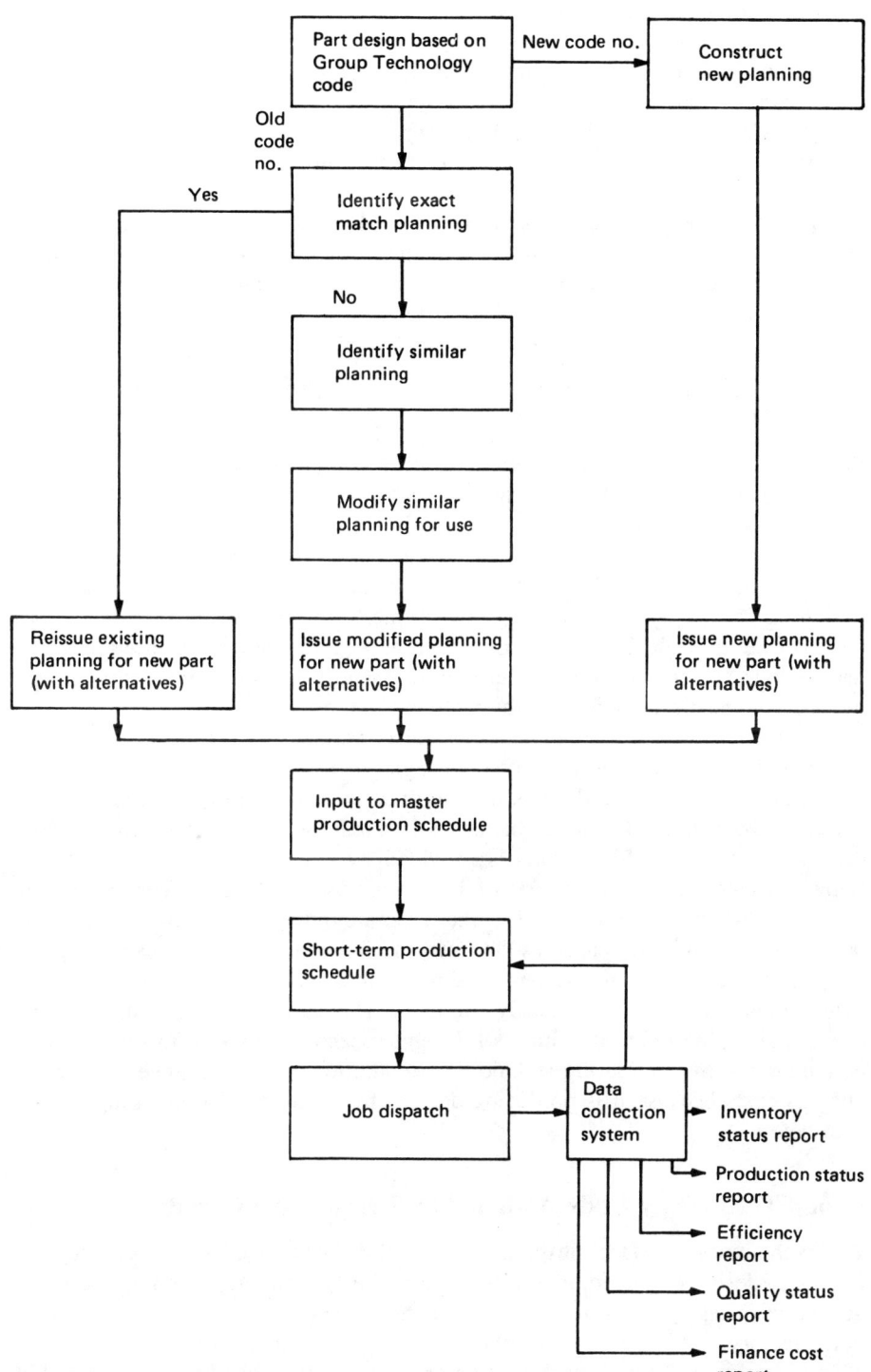

**Figure 12.4** Integrated CIM CAPP system.

303

1. Design engineering utilization of group technology to minimize proliferation of parts design.
2. GT code assigned by design engineering.
3. Data collection system used to control factory floor activities.

With these developments the activities of planning how parts are to be made are spread beyond the traditional manufacturing base, resulting in a more integrated manufacturing concern. Let us review in some detail these three new developments.

## Design Engineering Utilization of Group Technology to Minimize Proliferation of Parts Design

In our discussions of GT so far we have excluded all functions other than manufacturing engineering. We have shown how the MP&WM personnel apply a code matrix to parts drawings to find existing plans that can be used, or modified to be used, for the new part. Why must manufacturing engineering do the coding? Would there be a greater benefit if the classification and coding were done by design engineering? The answers to these questions were evident when the potential benefits to design engineering became known. Since design engineering had to code to achieve the benefits, there was no need for manufacturing engineering to repeat the coding for the planning purpose.

Like MP&WM, design engineering has considerable files of information on previous designs, and the number of designs that have to be produced can be reduced by using existing designs in total or as prototypes to support new needs. What has been needed is a mechanism for searching the file to find the proper design to reuse or modify. This is provided by GT classification and coding. Benefits similar to those experienced by manufacturing have been gained by design engineering. Through GT, the absolute number of new designs has been reduced, in some cases by as much as 30%. This allows design engineering to concentrate on existing designs and bring them to a mature state much sooner than was previously possible. This affects all of manufacturing as well, since the reduced volume of designs means that more effort can go into perfecting the manufacturing techniques. In addition, design engineering can reduce its costs because it is producing designs for products at a much higher rate than before.

## Group Technology Code Assigned by Design Engineering

This is the unifying aspect of the integrated CIM CAPP system. By assigning the GT code, design engineering continues its traditional role of designing the product, but also takes on the new role of recommending how the part will be made in the factory. However, we cannot expect the design engineer to become a manufacturing expert. Therefore, this second responsibility is fulfilled as described in the following sequence:

1. A layout of the new project is made by drafting, based on the project specifications detailed by the design engineer.
2. The design engineer evaluates the layout for its adequacy in meeting the customer's requirements and the liability requirement (i.e., checks that the product will perform as intended and will not fail) by performing the proper stress and strain analysis.
3. The layout is reviewed by the design engineers to determine what component parts drawings are needed.
4. The component parts requiring drawings are coded by using the prescribed GT classification code matrix.
5. The GT codes are entered into the data base to look for similarities and exact matches for drawings and plans.
6. If exact matches are found, drawing and planning are dispatched electronically to the master production schedule, according to Fig. 12.4.
7. If only similar matches are found, the design engineer evaluates whether they can be used if modified. Plans are referred to manufacturing engineering with the information that modified drawings are being produced that will probably require modified plans.
8. If there is no match, design engineering makes a new design and manufacturing engineering makes a new plan. The results of both functions will be added to the ever-expanding data base.

Thus, the basic manufacturing responsibility stays with manufacturing engineering, except that plans that are already good matches need not be referred to MP&WM for further evaluation. One important thing accomplished by such a sharing of data bases is that each function gains an appreciation of the other function's work and learns a great deal about that work.

Let us look at the GT codes once more. The code in Fig. 12.3 is very basic. Design engineering or manufacturing engineering can input the code, and the information contained is equally important to both functions. However, there are code portions that are only of interest to one or the other function. Therefore, each function usually retains a capability to input into the code as it sees fit by reserving certain digits for its own use. For example, manufacturing may reserve code space for individual material lots for the purpose of material utilization efficiency, or may use it for special quality control statistical sampling. Similarly, design engineering may wish to use digits of the code to classify the part for different uses of the part, or for allocation to laboratory evaluations. The GT code can be a powerful information gathering and processing tool that far transcends its original use.

## Data Collection System Used to Control Factory Floor Activities

With the CAM CAPP system, the electronic portion of the CAPP data information ended with the production scheduling step. Thereafter the system consisted of manual paperwork. The planning was written down step by step, although on a high-speed printer. Along with the planning came a stack of work vouchers that showed the time for each step to be completed and usually left space for the workstation operator to enter the actual time taken.

The computer-aided portion ended with the production of the paperwork and reverted back to precomputer systems. This was the system put in place by finance to obtain cost data. Data were collected manually, then converted by a keypunch operation into computer input, and finally run in batches once a week to obtain primarily cost data and secondarily manufacturing efficiency data. The problem from the shop manager's viewpoint was that the data were historical in nature and not of much use in correcting current problems.

With the integrated CIM CAPP system, manufacturing does not depend on the finance system to supply manufacturing data. We can now take CAPP to its logical conclusion and use it to measure the effectiveness of the plan. It is necessary to plan to accomplish anything consistently and necessary to measure to determine whether the plan is working. Prior to the integrated CIM CAPP system, the information flow back to the master production schedule to measure status based on the plan was severely limited. Therefore, a powerful real-time data collection system was required to make the integrated CIM CAPP system the communications system needed for the factory of the future. This system had to dispatch information to workstations electronically and receive information feedback in the same way. Such systems have been developed and will be discussed in a later section of this chapter.

## MANAGEMENT CONSIDERATIONS, CAPP SYSTEMS

In operating CAPP systems, the manager of MP&WM must first consider CAPP to be different from manual planning, as CNC machine support is different from manual machine support. In the CAPP mode, the common approach would be to modify good planning on file to suit a specific part or assembly. The key word here is "good." If the planning on file is not optimum, we can hardly expect the modified usage of it to be much better. In discussing CNC we made the point that the program fed into the machine's MCU was equivalent to the best, most experienced operator working with a manual machine; therefore, we obtained optimum performance every time the CNC machine was used. We must have the same goal with CAPP. All plans on file must be the best; otherwise, the fact that we have electronically available planning means little.

This best planning is achieved by doing two things: utilizing the most experienced planners in creating the original CAPP data base, and establishing standardized ways of performing repeatable operations. By standardized ways we mean a catalog of methods.

**Example:** If we wanted to paint a garage door, we could immediately start painting with a brush; or we could place drop cloths under the door and in the vicinity, then commence painting with a spray machine; or we could wash the door first, then put down the drop cloths, then tape the door handles and fixtures, then paint with a brush, and so forth. There are many ways of accomplishing the same thing, but usually there are better, less risky ways and inferior ways.

The task in standardization is to identify in the catalog the overall best ways. In CAPP this is extremely important, because the original data base input can be easily recalled and hence will be used over and over. Therefore, it should be the overall best way of doing the task if the company is to realize optimum productivity. In a manual system a mediocre plan would only affect the specific job it was developed for, but in a CAPP system the same mediocre plan can cause reverberations for a considerable time. For this reason standardization manuals and reference sources are important when developing a CAPP data base.

A second problem facing MP&WM management is that of computer paralysis. We find that the senior, more experienced planners in general tend to avoid the computer and therefore do not aggressively pursue CAPP opportunities. This leads management to assign the CAPP data base generation work to the younger, less experienced planners. The result is an even more critical dependence on the standardization manuals and usually some less than optimal planning entering the data base. The extent of the problem, of course, depends on the level of training in the MP&WM unit. Computer paralysis can be overcome by judicious use of computer familiarization training and by appealing to the pride of the senior planners. This is psychology, of course, but it is the basis of many facets of good management. It is not necessary to have all senior planners involved in CAPP data base development, but there should be enough of them to critically evaluate core plans for practicality and effectiveness. The pragmatic approach is more likely to be taken by the senior planners than the junior planners, and this approach most often results in optimum planning.

Optimizing techniques such as regression analysis have also been used to ensure that planning is optimal. These techniques are usually cumbersome and the improvements derived are generally not worth the effort. Utilizing the many years of experience of the senior planners is usually the best approach in developing an effective CAPP data base, one that is analogous to having captured the techniques of the best manual operator for the CNC machine.

## DATA COLLECTION SYSTEMS

With computer-aided process planning we have seen that a unified planning system can be made available to be matched with the master production schedule. The next logical step would be to develop an automated system to report back on how the company is doing in producing products in accordance with the planning. This is what a data collection system does. Let us review the history of such systems, then examine their operations, and finally discuss the management philosophy involved.

As stated earlier, the earliest data collection activities were imposed by finance for cost analysis purposes. Manufacturing told finance when a component part was finished, how many hours it took to complete the task, and what quantities of materials were used. Finance then totaled the cost and compared it to the selling price of the product or component to judge whether a profit had been made. If a loss occurred, manufacturing was asked why. This led to manufacturing asking to see the data collected by finance so that they could analyze the data for cause.

Manufacturing analyzed the data by identifying the high-cost portions of the process and determining the reasons for their cost, after the fact. Bright manufacturing managers then decided that since questions were bound to be asked, they should make use of the data before finance did, or at least concurrently. Hence the planning data were compared with the actual results to determine workplace efficiency and materials usage. As the use of computers grew, so did the availability of data from the finance system. It was inevitable that manufacturing would want real-time data, not batch data at least a week after the fact, in order to optimize operations. This led to the concept of factory data collection.

The goal of the factory data collection system is to develop an ability to determine the current status of all parts going through production. This means that managers should be able to ask a status question at any time and get an up-to-date reply. To satisfy such requests, the system for collecting data must be up to date at all times. The original finance cost accounting system is inefficient for such use. A real-time communications system is required.

In the finance-based data collection system used by manufacturing, paper vouchers for operator recording of status were distributed to the workstation along with the engineering drawings and planning for a particular job. Since a time delay was not considered significantly detrimental, these vouchers were collected at the end of each shift or perhaps weekly, sent to a keypunch operation, and finally fed into the computer 1 week to 10 days after the occurrence. Then the efficiency data were finally made available to the various manufacturing components. Managers knew what the efficiency trends were, and inventory control knew the status of materials usage as of the date of the report. The basic problem that still existed for manufacturing was that this report did not show the progress of each part compared to the schedule. This was partially solved by having many "production chasers" available to follow and report on critical components going through the factory. In fact, this is still the method of control employed in most nonautomated factory operations.

Nothing is wrong with the production chaser method if the factory is making simple components with few parts. Then it is possible to keep track of each part and report status when needed. However, most factories try to fully load their facilities to optimize profitability. Therefore, they have a continuing need to minimize inventory, minimize quality defect problems, and optimize efficiency. They also have too many parts going through at the same time to do all these things manually and do them well. Excellence in communication, which in this case means excellence in handling data, is needed, hence the need for a computerized data collection system as part of the CIM common data base.

A modern factory data collection system must overcome the deficiencies of the finance-based data collection format. A way must be developed to record completions and starts of work as they occur and to constantly compare status with the plan. Also, it must be possible to access this constantly updated status data base. Therefore, the development of factory data collection systems has been paced by the development of communications terminals suitable for the factory floor.

Terminals have been developed solely as communication devices and they have also been developed as part of CNC MCUs. Regardless of the type, the terminals must be able to easily instruct the operator and to let the operator ask clarifying questions and report status. Along with communication devices, dedicated processing computers had to be developed that were small enough to be efficiently dedicated to the data collection task, but large enough to contain sufficient data. User-friendly software also had to be developed.

In the case of factory data collection, the majority of data inputs via computer terminals are made by people who are not trained in computers. Those making the inputs have virtually no programming skills, nor could the company ever afford to provide such skills. This means that data collection systems must have built into their prime programming all possible contingencies. They must be extremely user friendly. The same characteristic is required for those who access the data base to learn current status. The purpose of the data collection system would be thwarted if a computer specialist were required to access the data base for the manager of manufacturing.

Now let us look at the flowchart of a generalized factory data collection system so that we can understand the various functions carried out within such systems, and where they interrelate with other communications/decision-making systems.

Figure 12.5 represents a generic factory data collection system. Note the predominance of workstation operator interaction with the computer system. The benefit of this interaction is that the operator becomes more knowledgeable about the master production schedule and the successes and failures of manufacturing. This familiarity leads to acceptance of greater responsibility, hence improved productivity.

Note that the system allows for errors to be made and corrected by the operator, recognizing that factory floor operators are not computer operators. The error correction routine prevents errors from getting into the data base. Typical of errors searched out by this program are:

Operator identification not on file.
Start day shown later than finish day.
Time interval between start and finish unrealistic.
Start time missing.
Stop time missing.
Quality defect data missing, if quality-related input.

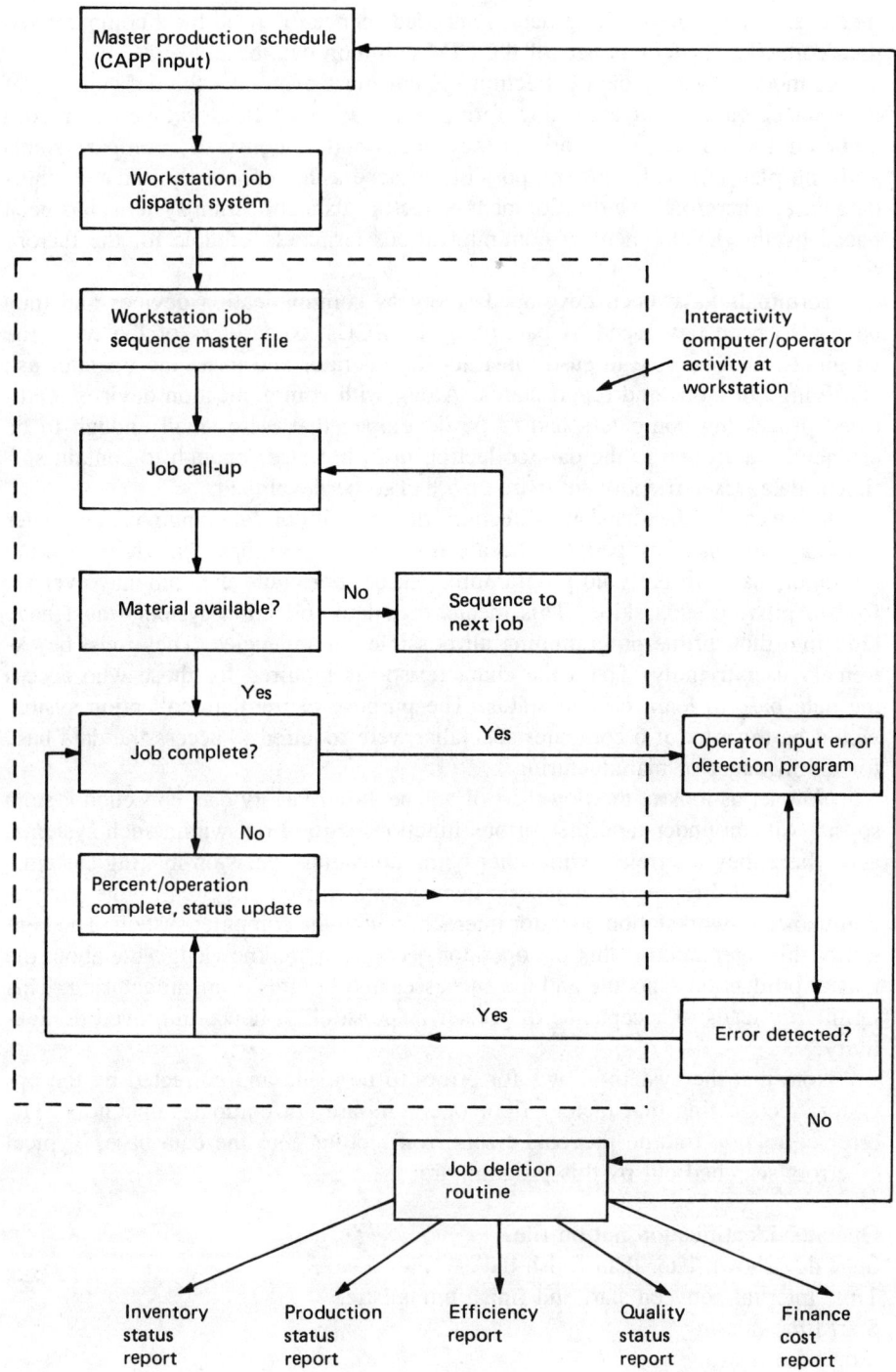

**Figure 12.5** Generic factory data collection system.

Drawing number missing.
Part number missing.
Operation sequence missing.
Percent complete missing.
Workstation number missing or incorrect.
Operation exceeds 100% claimed.
Labor category missing.

This, of course, is not an all-inclusive listing. A complete listing would depend on the specific factory and types of products being produced. The important fact is that the system is user friendly. If the operator makes an input error, the program states what it is and the operator can easily correct it.

Finally, note the job deletion routine. This is a series of programs that update the status data base continuously as input is received. The output is sent to the master production schedule to update the main production control data base. The output also goes to a host of information programs, all on a real-time basis. Therefore, all personnel who need the information can access these programs to determine the status of interest to them. The data collection system is a real-time communications system that serves a wide variety of users.

Data collection systems can be very useful for management control of operations. However, there are certain operating considerations that must be complied with. Let us discuss the major ones from a management viewpoint.

A data collection system is only as good as the system that supports it. Therefore, care must be taken that both the hardware and software are up to the rigors of a factory floor environment. If the equipment is not hardened, and continually breaks down because of such things as factory dust, the system will not be in use. If it is not in use a considerable part of the time, then to make its output believable, considerable time will have to be expended in manually feeding in data to keep the system current. A data collection system is worthless if it is not collecting data. Therefore, management must keep the servicing factor in mind when the decision is being made to implement such a system. This should not deter implementation of such systems, but should be recognized as an expense factor in running the system.

Another important factor for management to consider is the increased need for factory discipline to make data collection systems work. By this is meant the way and order in which work is accomplished. Before data collection systems, it was common for foremen to tell the operators to do work out of sequence if it made sense to do so—that is, to do operation 6 before operations 4 and 5 if the earlier operations could not be accomplished because of machine breakdown, unavailability of special tools, and so forth. Of course, these instructions would be given only if a subsequent operation would not unalterably change the ability to do the previous operations. With a data collection system this type of variation becomes more difficult to tolerate. If the system does not know that a step has been skipped it is likely to assume that the step has been completed, thus forever missing the operation and resulting in a quality defect later on. The need for communication discipline is essential. Supervisors understand that the lack of freedom to deviate from planning is more than compensated for by the overall greater efficiency made possible by the data collection system.

Another aspect of the processing discipline involves the planners feeding the CAPP system. For data collection to be meaningful, the planning sequences must be realistic. The plan cannot be a document that is ignored by the shop floor personnel because it is in serious variance with how things must be accomplished. One of the benefits of CAPP and data collections systems is that they tend to weed out irrelevant and erroneous plans, which might never have been corrected when manual control was the operating mode.

## SUMMARY

With CAPP and data collection systems implemented on the factory floor, we have a communication/decision-making system that permits the use of CIM.

In this chapter we have discussed the history of CAPP and data collection systems and seen the extent of the development that has taken place. Each evolutionary step has brought us closer to the goal of real-time control of the manufacturing process. We have developed an understanding of what is meant by excellence in communication and why it is a central theme in progressing toward the "factory of the future." Each step of this process has been linked to improved productivity—the only reason for implementing CAPP and data collection systems.

We have also discussed the concept of group technology coding and seen how coding makes possible easy retrieval of historical data. These data can be used to establish new plans. Another important development due to GT code use is that of integrating the design engineering tasks with the responsibilities of the planner. The idea that a function other than manufacturing could do planning work was introduced, and we saw our first practical example of an integrated function. This integrated approach can come about only through the use of CIM, portending significant productivity improvement opportunities for the factory of the future.

## REVIEW QUESTIONS

1. Explain why a time lag in gathering part status data during manufacture is acceptable to the finance function but unacceptable to the manufacturing function.

2. What are the two communications/decision-making computer systems initiated by the manufacturing function? How are their origins alike and how are they different?

3. Referring to Fig. 12.1, discuss the adequacy of the "key word" search function compared to a group technology classification and coding system.

4. What is the difference between a computer-assisted system and a computer-integrated system in relation to process planning?

5. In Fig. 12.3, the overall length is 43 in., the largest diameter is 6 in., and the part has a groove. Suppose we have a similar valve stem with an overall length of 25 in. and the largest diameter is 3 in. This similar valve stem does not have a groove or any other special feature. Using the code described by Fig. 12.3, create the group technology code number for the similar valve stem.

6. Why is it preferable to have alternative planning available when developing a master production schedule?

7. Describe how a computer-automated data collection system improves productivity.

8. List eight uses of GT classification coding: three for design engineering, three for manufacturing, and two for finance.

9. Discuss the differences in management concepts between manual planning and computer-aided process planning.

10. What do "real time" and "batch processing" mean with respect to data collection systems? Which is most beneficial to manufacturing? Explain why.

11. Discuss the interface between the operator and the data collection system. What major items must be dealt with to achieve a successful system?

12. What are the constraints that a data collection system imposes on management? Explain your answers.

# THIRTEEN

## THE GROUP TECHNOLOGY BASIS
## FOR PLANT LAYOUT

In Chapter 12 we discussed the use of the classification and coding schemes in design and planning. However, there is much more to Group Technology (GT) than coding for data base retrieval. Group technology concepts can be a powerful tool in plant layout, especially for manufacturers of miscellaneous parts. In this chapter we will describe in pragmatic terms how to justify and proceed with implementing a group technology layout. We will discuss the management concerns that must be successfully and affirmatively answered in order to proceed with such a layout, and we will demonstrate the differences between a traditional layout for job shops and a layout based on the principle of sameness utilized by group technology.

## THE PRINCIPLE OF SAMENESS

Mass production or flow-type manufacturing offers the greatest economies of scale. If we are involved in a business that mass produces its products, and its products are virtually identical, we have the opportunity to employ special-purpose machines optimized to produce the product. However, mass production accounts for only 30% of all manufacturing. The remaining 70% of manufacturing is accomplished in job shops; where it is done on relatively general-purpose machine tools, the cost of optimizing is usually uneconomic, and many unproductive compromises must be made. Therefore, much thought has been given to finding ways to make job shop techniques approach those of flow manufacturing. An approach based on the principle of sameness, the underlying theme of group technology, does just that. All coding, classification, production routing, batching, and layout techniques based on group technology employ the principle of sameness as their theoretical starting point. The principle of sameness refers to the grouping of parts to be manufactured according to similar characteristics, either geometric or processing, so that they can be produced on the same equipment in a batch mode.

Consider the universe of shapes used for an internal combustion engine. There are flat diametric shapes such as piston rings and gaskets; long dowel-like shapes such as piston rods and hold-down studs; cylindrical shapes such as pistons, cylinder liners, and carburetor bowles; flat milled shapes such as head gaskets and flapper carburetor valves; and so forth. Thus, there are many dissimilar parts that share a common geometry. We can also find numerous examples of dissimilar parts that do not have similar shapes, but are consistently manufactured by similar techniques. For example, the crankshaft of an automobile engine requires lathe work followed by grinding; so does a steam valve stem (Fig. 12.3). These parts have dissimilar geometries, but their processing steps are similar and they are probably both constructed of similar steel alloy material.

We have tabulated like characteristics of dissimilar parts because we would like to approach the economies of scale enjoyed in flow manufacturing. For example, if we can find enough parts that require a tapped hole 5 in. in diameter, we can prepare a drill press to do nothing more than make such holes in a flow production manner. Of course, we rarely have products that require only one operation, so our ability to find sameness or commonality will always be constrained. Nevertheless, every time we can make use of a setup more than once, we are saving money and becoming more productive. The principle of sameness encourages us to look for more than one use per setup, and it starts us thinking about other areas where grouping can be beneficial. We will see some of these benefits emerging as we investigate the use of group technology for layout purposes.

## GROUP TECHNOLOGY PLANT LAYOUT CONCEPT

According to group technology theory, the plant layout should take advantage of the sameness of parts being produced. This is in conflict with the common practice in job shops of basing the plant layout on machine tools and processes. Figure 13.1 is a block diagram of a typical job shop layout. With this type of layout we dispatch jobs to each area where work must be performed in the sequence dictated by process planning. Therefore, the number of possible combinations or different sequences going through the factory is 10 factional. In practice, this means that we have a traffic maze and that each part travels a great distance in its journey from raw material to finished product. It means that we need a great effort to load and control each station effectively, and that we will have a larger inventory of work in process. Finally, it means that a new setup is likely for each piece that

| Drills | 1 | Planner mills | 2 | Inspection station | 3 | Saws | 4 | Shipping and receiving | 5 |
|---|---|---|---|---|---|---|---|---|---|
| Lathes | 6 | Horizontal boring mills | 7 | Vertical boring mills | 8 | Welding | 9 | Heat treat | 10 |

**Figure 13.1** Block diagram, job shop layout.

crosses each workstation. Quite bluntly, the classical job shop workplace layout seems to have been designed for grouped maintenance rather than efficient production. Figure 13.2 shows what this flow is like with only 10 jobs. One can imagine what it would be like with closer to 10 factorial jobs.

A great deal can be done to improve the flows illustrated in Fig. 13.2. We can look for sameness in geometry or production equipment usage and then construct cells of dedicated equipment to produce the similar parts. These are not necessarily parts that are similar in end usage, but ones that are similar in geometry or in how they are manufactured. We are looking for ways to minimize the learning curve. Clearly, if it is necessary to recalibrate, rethink, and retool for each job, it is difficult to achieve expertise in making any of the parts that cross the workstation. On the other hand, if there is a high degree of similarity in operations to make a variety of different parts, we will become expert in making those parts.

If we can find sameness in a sufficient number of parts, we can afford to segregate particular facilities to make those parts. Replace "sameness" by "family of," and we have the common expression for group technology, the family of parts assigned to a common manufacturing cell. The manufacturing cell may consist of one complex CNC machining center, or many manual machines, or combinations of both. The common denominator is that each cell is dedicated to making one family of parts or perhaps a few adjacent families of parts. Figure 13.3 depicts a layout based on the GT cellular manufacturing principle.

Figure 13.3 shows that different facilities are mixed together to form the individual cells. Since the machine tools in each cell are dedicated to making the family of parts assigned to the cell, each tool can be optimized to reduce the setup time—in fact, to approach having a universal setup for each machine to match the family of parts it is dedicated to. This, of course, allows significantly reduced throughput times and therefore productivity gains.

The objection is sometimes raised that job shops cannot be set up along group technology manufacturing lines because general-purpose job shops do not know what they will be called upon to produce. However, this is not correct because job shops tend to specialize in certain areas, based on the type of equipment they own. Furthermore, virtually all parts can be broken down by size and shape characteristics, and therefore can be broadly categorized into families of parts even before they exist for job shop production.

Therefore, the question is not whether a group technology cellular layout is beneficial, but how one achieves such a layout. We will review the methodology

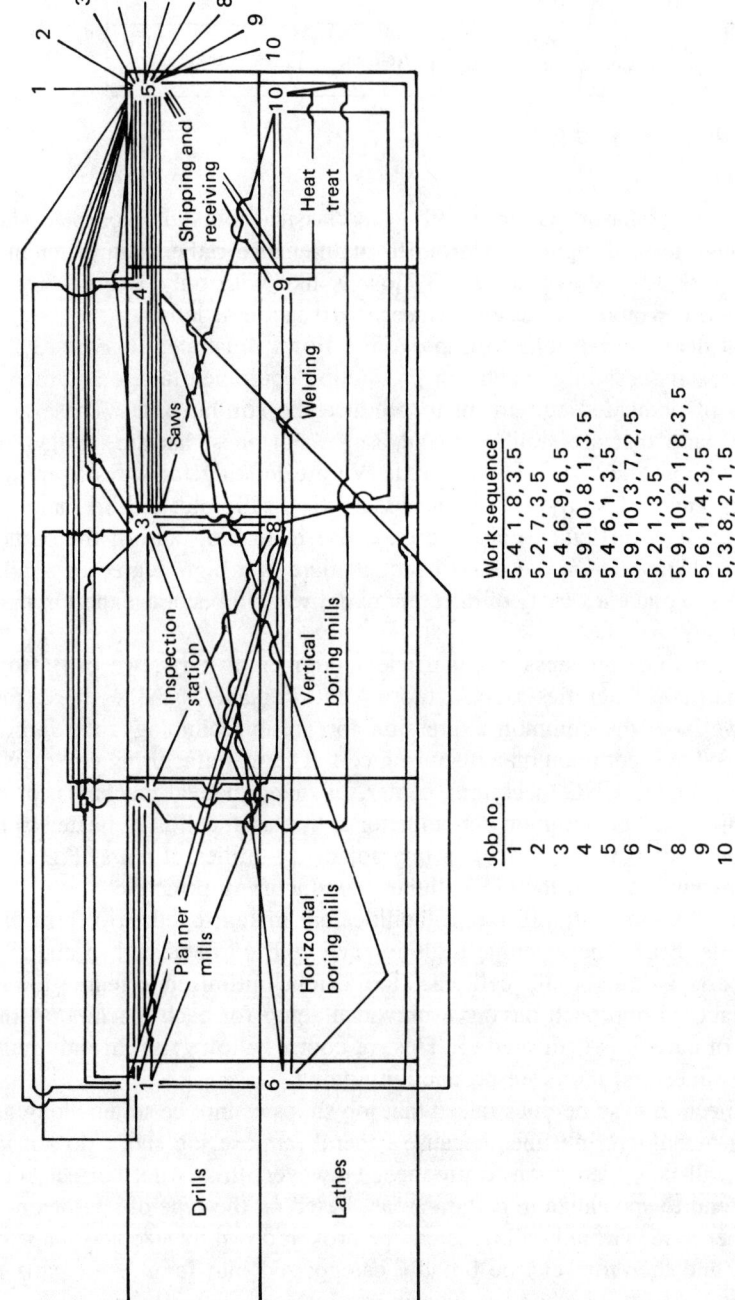

**Figure 13.2** Job shop classical layout, showing 10-job flow.

Drills

Lathes

Planner mills

Horizontal boring mills

Inspection station

Vertical boring mills

Saws

Welding

Shipping and receiving

Heat treat

| Job no. | Work sequence |
|---------|---------------|
| 1 | 5, 4, 1, 8, 3, 5 |
| 2 | 5, 2, 7, 3, 5 |
| 3 | 5, 4, 6, 9, 6, 5 |
| 4 | 5, 9, 10, 8, 1, 3, 5 |
| 5 | 5, 4, 6, 1, 3, 5 |
| 6 | 5, 9, 10, 3, 7, 2, 5 |
| 7 | 5, 2, 1, 3, 5 |
| 8 | 5, 9, 10, 2, 1, 8, 3, 5 |
| 9 | 5, 6, 1, 4, 3, 5 |
| 10 | 5, 3, 8, 2, 1, 5 |

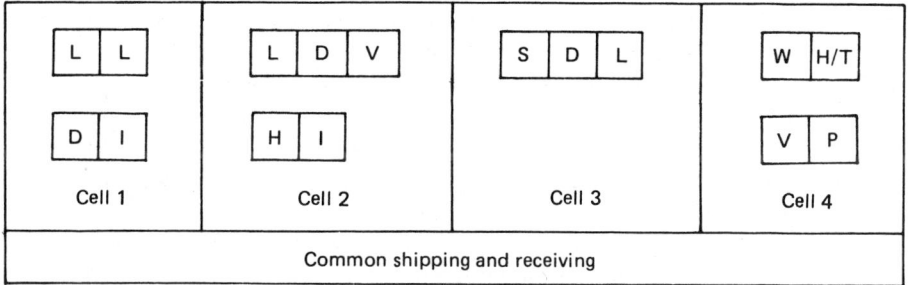

**Figure 13.3** Group technology cellular manufacturing layout.

of doing that, but first we must consider the economics of group technology cellular layouts. This is necessary in order to understand what is theoretically possible, which modified by realities will result in what is practically possible. As manufacturing engineers, we are interested in the practical, not the theoretical.

## THE ECONOMICS OF GROUP TECHNOLOGY CELLULAR LAYOUT

The benefits of group technology are many and varied. Most practitioners of group technology would list the following as reasons for adoption of the group technology cellular layout philosophy:

1. Reduces setup time
2. Improves direct labor productivity
3. Reduces indirect jobs such as production expediting
4. Reduces manufacturing losses
5. Reduces throughput time

The range of improvements varies, but my experience with such layouts indicates that a 15 to 30% overall improvement is possible. To know exactly what the improvements are, a company needs good historical records against which to measure the accomplishments. Most companies do not maintain sufficient records to obtain precise measurements of improvement. In fact, operations that are in trouble necessitating a change usually have the worst records. This means that changes are often based on experiences of other organizations, which involves some risk. Different factories with different management emphasis may or may not achieve improvements in the range indicated above. In view of this variability, it is mandatory that the feasibility of savings be evaluated early in the project and as pragmatically as possible.

Achieving a group technology cellular layout costs considerable sums of money. Therefore, before venturing into the project, the costs must be carefully evaluated. A cellular layout is not the only means of achieving the benefits of group technology. As a completely cellular layout is approached the potential for gains approaches an optimum. Therefore, every step on the path of implementing group technology will bring incrementally greater benefits. For example, in Chapter 12 we discussed the benefits of coding and classifying but did not mention cellular layouts. These are a good example of the incremental benefits achievable. If we continue and tie coding and classifying to the cellular concept, we can achieve even greater benefits.

In considering a group technology cellular layout, a company must evaluate the feasibility of successfully forming families of parts. Although cellular layouts can be made to work for virtually all job shops, some companies will be able to do this more easily than others. To be able to justify the costs involved to their superiors, those interested in forming a cellular manufacturing layout should be able to make the following statements:

1. The factory is making a wide variety of parts.
2. The parts are repeated over and over again with usually minor and occasionally major variations.

If these two criteria are met, the job shop is well suited to a GT cellular layout. There will be a large enough sample size to look for geometric or processing sameness, and there will be enough repeatability to justify family-of-parts universal fixtures. Whether the parts are presently made in small lot sizes is unimportant in deciding for or against a GT cellular layout.

In addition to the above considerations, there must be a strong economic incentive. If the traditional layout is considered satisfactory in terms of costs and profits by the senior management, they will not be willing to make a change. Since group technology benefits are described in terms of ranges, they are difficult to justify to senior management, especially since larger amounts of money must be spent to convert to the new layout. The fact that savings ranges and averages are difficult to enter into a meaningful cost evaluation means that GT cellular layout projects must often be portrayed as a secondary reason for making a layout change.

Let us examine the group technology benefits and the need for a strong economic reason in terms of a hypothetical miscellaneous parts manufacturing area. We assume that there are 50 major machine tools, lathes, vertical boring mills, horizontal boring mills, shapers, saws, welders, and so forth. Each of these is to be relocated in accordance with a proven group technology principle in order to form manufacturing cells. Suppose that it costs $12,000.00 on average to remove and reinstall one major machine tool. This includes rerouting all services (e.g., air, water, AC and DC electricity, and telephone) and a prorating of rearrangement of supporting areas such as inspection, material storage areas, and project engineering costs. This means that for the hypothetical 50-workstation miscellaneous parts manufacturing area, a cost of $600,000.00 would be incurred. Even assuming a conservative 10% reduction in the need for equipment with the new layout, the facilities move would still cost $540,000.00.

Now let us look at benefits to be achieved by group technology alone. Assume that each workstation can produce four parts per 8-hr shift and that each completed part averages five operations. Four completed parts per set of five workstations, or ten sets of workstations at four completed parts per shift, equals 40 parts per shift for the area. This makes 600 completed parts per week if there are three shifts per day, or 28,000 parts for a nominal 48-week work year. At a nominal $10.00 per hour labor cost, each part costs

$$\frac{(8 \text{ hr/shift})}{(4 \text{ parts/shift})} (\$10/\text{hr})(5 \text{ operations}) = \$100/\text{part}$$

The cost is thus $2,880,000.00 per year, excluding overhead. The next thing we must do is analyze the possible savings, using the five savings categories attributed to group technology.

1. Reduces setup time: We will use a 15 to 30% range of possible savings, as discussed earlier in this section. Average time on each machine for our example is 2.0 hr. Setup time is normally 15% of the total machine cycle time; hence it is 0.30 hr per machine or 1.50 hr per part. Assuming a full 30% reduction, we get 0.45 hr saved, or $4.50 per part. Realistically, one would plan conservatively, so the savings would be at the 15% reduction level, or $2.25 per part. For the remainder of the example I will use the low end of the range to allow for the numerous unknowns and intangibles in a project such as this.

2. Improves direct labor productivity: This, of course, must include setup. In order not to double-count, we will subtract the setup time from the total cycle time. Therefore, the theoretical improvement associated with this job enrichment factor is:

$$(5 \text{ operations/part})(2.0 \text{ hr} - 0.3 \text{ h})(0.15/\text{operations}) = 1.275 \text{ hr/part}$$

At a direct labor cost of $10 per hour this equals $12.75 per part. In my experience, this should be discounted by at least 80%. This productivity improvement comes about because the operator is making similar parts or performing similar operations on dissimilar parts, but parts in the prescribed group technology family. Therefore, the operator is familiar with them and does not have to progress through a learning curve. My observed 15 to 30% savings are for the total group technology process, not for each individual phase. I believe that the group technology benefits for this phase are exaggerated. In the traditional job shop layout, there is considerably more variety of work going through a workstation than in a GT cellular layout. Therefore, the job enrichment often attributed to an operator in a GT cellular layout situation seeing completed parts being made in the operator's cell must be discounted. The learning curve benefit must also be discounted. In practice, most job shops route similar parts to the same workstation, because the foreman usually knows the skill levels of the operators and assigns jobs accordingly. Also to be considered in discounting this particular saving attributed to group technology is the novelty factor. Many theoreticians believe that there is a long-term benefit in manufacturing completed parts in a specific cell. However, my observations indicate that such benefits diminish after the novelty disappears, which happens in approximately 3 months. Therefore, the calculated saving of $12.75 per part will really be in the neighborhood of $2.55 per part for improved direct labor productivity.

3. Reduces indirect jobs such as production expediting: Good managers will do this as a matter of course. Successful operations must be as efficient as possible; therefore, there is a constant effort to avoid using excess people. It is difficult to establish a layout as a proven reason for reductions in support personnel. Even though group technology cellular layouts are compatible with personnel reductions, it is difficult to show that this is not a saving that can be achieved through good management. Hence, in practice this saving cannot be proved to exist.

4. Reduces manufacturing losses: Making a completed part in a cell where raw stock is converted to a finished part will reduce losses caused by indifference. Some advocates of group technology claim that a 70% reduction of losses will occur. However, I believe that the novelty factor once more comes into play here, and based on a 15 to 30% savings attributable to group technology overall, I would choose a 20% reduction for this part of the calculation. Assuming a traditional 5% manufacturing loss budget, or $5.00 of cost per part associated with losses, we can save 20% of that, or $1.00 per part.

5. Reduces throughput time: This appears to be the only aspect of a group technology cellular layout for which a case can be made for tangible savings, the type of savings that manufacturing engineers would be able to place on a plant appropriation request. Therefore, we will assume a 20% reduction. This saving is due to reducing the distance traveled from workstation to workstation by having the succeeding workstations adjacent to each other. This is the core of the cellular concept.

In order not to double-count, we must subtract setup time, machine operating time, and queue time at the workstation from the total cycle and deduct 20% of the remaining time, the transit time, as the saving. Our example analysis shows the setup time to be 0.30 hr per workstation. The remaining 1.70 hr per part at each workstation is machining and waiting time. Since we lack firm data on the travel time between workstations, which is the case for most factories, we must make an assumption about what part of the 1.70 hr is machining and waiting time and attribute the remainder to travel time. Based on experience, I have chosen 20%.

This assumption of a percentage travel time keeps this category in the intangible area vis-à-vis plant appropriation requests. Some manufacturing engineers have dealt with this problem by commissioning studies of times for moving material from station to station. The studies usually employ timers and reminders to the operators to supply accurate data. Since there is no way to check the accuracy of such data, long periods of data collection are usually substituted for verifiability. There is always some uncertainty about the accuracy of the results.

In our example, assuming that we have a 20% material transport factor and

80% machining and waiting time, the reduction in throughput time per workstation is:

$$[1.70 \text{ hr/part} - (1.70 \text{ hr/part})(0.80)](0.20) = 0.068 \text{ hr/part}$$

Since there are five workstation operations per part, we have a throughput saving of 0.340 hr per part or $3.40 per part.

This is the most significant factor in the cost reduction calculation, as one would expect. A review of Figs. 13.2 and 13.3 indicates that a positive gain should be made by simplifying the flow path of the parts. The group technology cellular layout allows simple flows, while the traditional job shop layout does not.

Summing the savings attributed to the group technology concept on per-part basis, we have

| | |
|---|---|
| 1. Reduces setup time: | $2.25 |
| 2. Improves direct labor productivity: | $2.55 |
| 3. Reduces indirect jobs: | $0.00 |
| 4. Reduces manufacturing losses: | $1.00 |
| 5. Reduces throughput time: | $3.40 |
| Total | $9.20 |

At 28,800 parts per year, this corresponds to a $264,960.00 saving in operating expenses less overhead. This is a significant amount of money, equal to 9.2% of the original labor cost for the 28,800 parts.

Would senior management fund a project to rearrange a 50-machine miscellaneous parts area into a group technology cellular layout? Probably not. The reason senior management would give for rejecting such a project is that these are "soft" savings—that is, they are subjective. They are based on average results from other factories, and it is always difficult to determine how close a factory is to the average. Would manufacturing engineering offer such a project for approval? Let us consider the facts.

First, we would be asking for an investment to spend $540,000.00 to save $0.00 (firm) and $264,960.00 (soft or variable).

Second, we could state that savings are on the conservative side and that statistical knowledge tells us that the variable portion would be higher, but only if we do indeed have an average factory.

Third, we must also consider disruption. While the factory is being rearranged it is still necessary to produce products. This will have to be done at a cost greater than the current cost; if it were not, we would always produce by the alternative method. Let us look at these disruption costs in greater detail.

In our example we are contemplating the rearrangement of 50 machines or workstations into a cellular layout. We cannot afford to take all 50 out of service at the same time. Nor can we afford to do only one workstation at a time; using a 2-week rearrangement cycle time per machine, taking out only one machine at a time would result in a nominal 2 years to complete the project. Usually 3 to 4 months is the accepted cycle time for this type of project, and to extend it any longer would create an unacceptable cash flow. We can calculate disruption costs in two ways:

1. Calculate the cycle time per workstation, then add 2 weeks for each machine to be out of production. This is a tedious procedure and most companies cannot afford to support such detailed planning.
2. Use an experience-based factor to calculate disruption costs. My experience shows that for every 3 months of rearrangement project time, there will be 1 month of nonproduction. This means that the affected factory area will not produce during 1 month for every 3 months the area is undergoing rearrangement.

Since companies normally cannot tolerate a 1-month hiatus in production, the parts must be made somewhere else. This usually means subcontracting to another factory, which in this case means additional costs of about 25% over normal operating costs. Therefore, the disruption costs of our example are:

$$(2400 \text{ parts/month})(\$100/\text{part})(0.25) = \$60,000.00$$

We can now show the savings and costs of the proposed group technology cellular layout rearrangement project. The costs are:

| | |
|---|---|
| $540,000.00 | facility rearrangement |
| $ 60,000.00 | disruption cost |
| $600,000.00 | total costs |

and the savings:

| | |
|---|---|
| $        0.00 | firm |
| $264,960.00 | subjective (variable) |

Therefore, manufacturing engineering would not present this project in its current form for approval. It would be asking senior management to invest $600,000.00 to save a subjective $264,960.00, and since the savings are subjective, it would be easy to fault the logic for each of the four segments accounting for the savings.

Those who are proponents of group technology, and I count myself as one of those, have wrestled with this approval of project problem for a long time. The classification and coding portion of group technology fits in very well with integrated CIM and therefore can be financially justified with little difficulty. The classification and coding techniques can be used to form part families and lead to the conclusion that cellular layout is the way to proceed. However, cellular layout is very difficult to sell, as shown above. Therefore, group technology cellular layout projects must be piggybacked onto other projects showing firm or hard savings.

What are firm savings associated with rearrangement? Simply stated, firm savings are guaranteed savings. An example is energy savings. If equipment is moved into a tighter layout, less lighting equipment is required; therefore less money is spent on fixture maintenance and the power bill is reduced. We have energy savings. Another example of firm savings involves span of control. By grouping workstations closer together, less supervision (fewer foremen) is required to cover the area. The span of control is over a smaller area. Span-of-control savings always occur when multifloor layouts are converted to one-floor layouts. A third example is cost of factory space. By consolidating operations into fewer buildings, the costs of maintaining, renting, heating, lighting, and so forth are reduced. In each of these three examples of guaranteed savings, the saving is calculable and can be readily verified. The cost of energy and the number of lights eliminated are known. The span of control is known, so it is possible to calculate the reduction in supervision and the saving in salaries. The cost of occupancy per square foot is known; therefore, space vacated and no longer paid for becomes a firm saving. In all these examples, it is evident that if the company takes the required actions it will gain the stated savings with a high degree of confidence. With subjective savings there are no direct cause-and-effect relations, hence the savings may or may not come about. That is why they are called intangibles and why they are difficult to sell to senior management.

We can see that tying a subjective savings project to a firm savings project is the only way to obtain approval for the subjective project. Of course, the firm savings project must be a good one and its savings must somewhat cover the additional costs of the subjective project. Usually, if a firm project has a payback of 2 years or less, senior management will be willing to extend the payback period up to 25% to accomplish good potential subjective projects. Group technology cellular layout projects should fall into this category. This is especially true if the firm project is a facilities layout and not a new equipment project. Area consolidations, revamping product scopes, and setting up satellite manufacturing locations are examples of firm projects that group technology cellular layout projects can be tied to successfully.

Group technology cellular layout projects will normally not pass a critical financial review. Therefore, to achieve a cellular layout it is necessary to combine it with a firm savings project. Once the cellular layout project has been implemented and is successful, future GT cellular layout projects become easier to sell.

## TECHNIQUES FOR SELECTING FAMILIES OF PARTS

To form a group technology cellular layout, it is necessary to form families of parts and then determine what processes in what sequence are used to produce the parts in the family. This leads naturally to placement of machines in the cell, location of the inspection facilities, placement of entry and exit to the cell for materials and finished parts, and placement of the support facilities. The actual layout procedure is the same as that used in making a traditional layout: location of the principals and auxiliaries, flow parameters, access parameters, support facilities parameters, and so forth. We have discussed the techniques of making layouts in Chapter 5, and numerous supplemental readings can be found in good engineering libraries. The only new task we must examine is that of forming part families. Part families can be determined by three methods: (1) empirical decision, (2) classification code, or (3) part flow analysis. The method selected will depend on the needs of the individual business. The complexity of the business problem must be analyzed, and this analysis will determine the depth in which the company wants to become involved with group technology.

As stated before, the amount of group technology implemented is directly related to the amount of benefits to be received. Unfortunately, there is another ratio to consider: the more group technology implemented, the higher the implementing costs. Therefore, it is necessary for companies to know their own businesses and the state of the competition. Group technology, like other good cost-saving opportunities, must share in the total dollar pool available for technological improvements. Group technology projects, as demonstrated in the preceding section, are not always sure prospects for investment funds. Therefore, the business analysis must take into account the current performance of the business, the market share, the competition, and, most important, the profit level and potential, before any projects are approved. For example, a company may not have the funds to institute a full-scale group technology project, but may wish to have some hands-on experience so that it can expand into fuller implementation when funds become available. This company would probably choose a limited-area cellular layout and use part flow analysis in place of code to establish the part families. This would certainly be the case if their competition were at the same group technology implementation level. It would be a limited venture into group technology, costing limited funds, but only resulting in limited gains. Each company must evaluate its own needs and make its own decisions concerning group technology implementation. The good thing about this situation is that many levels of implementation can be commissioned, and virtually every implementation has an excellent chance of improving on the current status of operations.

Now let us explore the three methods of forming part families.

## Empirical Decision

This is the simplest of the three methods. Experienced planners decide which parts are similar either by shape or by operation sequence and thus create part families. This technique works for factories with minimal frequencies of part types and well-entrenched expertise—for example, long memories for parts made before but not seen in the production schedule for a long time. This technique cannot work for radically new parts, and it depends on the experts to place parts in the proper family.

## Classification Code

The classification code is the most complex and most beneficial of the methods. The context of a classification code was described in Chapter 12 for use with a computer-aided process planning and data collection system, where the code is a vital part of a successful data base communications system. Besides forming families of parts for cellular manufacturing, a classification code allows access to many other benefits of group technology such as Manufacturing Resource Planning (MRP II) and those mentioned in Chapter 12: CAPP, design analysis, and data collection. In the case of cellular manufacturing, the classification code yields either geometric similarities or processing similarities, which are easily used to create accurate part families for existing parts, and to assign new parts to existing families or new families.

The classification code method is the most expensive and time-consuming method in terms of personnel and cycle time to implement it. To use the classification code method, a company must first either invest in purchasing an existing code or develop its own code. It must then code all of its parts, or at least the active ones. This requires training analysts and then analyzing each part drawing. The classification code must also be customized to meet the specific needs of the particular factory. For a purchased code this can mean additional expense, depending on how well the commercially available code matches the factory's needs.

A group technology classification code will determine part families to suit the needs of the specific factory very well, thus allowing achievement of the cellular layout. The problem with classification codes is that in many cases they are unnecessarily sophisticated, unless management is totally committed to a group technology philosophy and is implementing CAPP and data collection as part of the company's communications system.

An advantage of cellular layouts is that they allow management to derive many benefits of group technology without a major commitment. Of course, the cellular layout must be such that a firm savings can be shown. Requiring that a classification code be used to define part families could easily add another $200,000.00 to the cost in the previous example. This would certainly ensure a negative decision by senior management.

## Part Flow Analysis

It is possible to create the part families for a cellular layout without a classification code and with a higher degree of objectivity than in the empirical decision method. Part flow analysis is the middle-of-the-road approach. It develops logical part families based on a planning sequence matrix, and thus has more credibility than a panel of experts making educated guesses. It also helps cellular layouts to work by routing families of parts defined by part flow analysis to the proper manufacturing cells. However, it does not afford the company any further gains in overall productivity because it is not a classification code and therefore does not readily tie into other company systems.

The parts flow analysis technique is simple in concept. We review the planning sequence sheet for a part and list in chronological order the operations necessary to produce the part. After doing enough of these reviews, we can develop a matrix of parts versus operations, from which we derive part families based on similar operation sequences. If this first iteration develops too many families, we then consider as a part family those parts having the same type of operations but not necessarily in the same order. This combining of families is necessary because too many families can result in many underutilized manufacturing cells. Experience shows that a second iteration is usually needed to develop a manageable number of part families.

Considering our example, 28,800 parts in 1 year would require analyzing approximately 120 plans a day just to finish the job in 1 year. Even if the company had the manpower to do this job with its permanent staff, it could not afford to do so. To get the job done, creative management must take over. We must look for a temporary staff that can read a planning sheet and make entries onto a matrix form; the actual analysis can be done by the permanent MP&WM staff. The solution, then, is to hire temporary clerical staff and quickly train them to review plans and record data on the appropriate matrix forms. However, even with a staff of temporaries to do the work, the task as stated in our example is enormous; 120 plans per day would require about 10 people doing this for an entire year. Something must be done to shorten the cycle, and that is statistical sampling. Most job shops would recognize a repetitious pattern in the work they do. Therefore, they should be able to define a cycle time after which they start to repeat similar jobs. For most miscellaneous parts job shops, where a cellular layout would be beneficial, the cycle turns out to be from 2 to 4 months, with 2 months being the most frequent result. For our example, then, this leads to 4800 parts to be reviewed. It should be possible to do this in 1 to 2 months.

Once a part flow analysis matrix is completed, we have a manufacturing parts family but not a true parts family according to a classification code. We have identified how these parts are made at present but we have not analyzed whether it is the best way to make the parts. Neither have we developed a capability to load significantly different parts on the same facility based on similar machining characteristics. All we have done is to achieve loading for the makeup of manufacturing cells without determining whether the plans are optimized.

We have seen that part families can be formed in several ways. The three primary ways are compared in Fig. 13.4. The choice of method depends primarily on the goals of senior management. If only a cellular layout is desired, then either empirical decision or part family analysis would be sufficient. If the number of the parts to be considered is relatively small, empirical decision would be favored. However, if the management team is going to continue to develop a fully integrated CIM business, then empirical decision and part family analysis are temporary techniques, and will eventually have to be replaced with a classification code. How long it will take to achieve CIM will dictate whether or not a classification code will be employed to achieve a group technology cellular layout. Management must keep in mind that once a method is decided upon, it will be difficult to change at a later date. Management must also decide whether a compromise approach is better than no action at all when contemplating initiation of any group technology-based project.

## SUMMARY

The argument for group technology cellular layouts has been presented in this chapter. The compelling reason for such layouts—a 15 to 30% average produc-

| | Simplicity | Handle large no. of parts | Shop knowledge required | Data applicable to other Group Technology concepts | Applicable to new and different parts | Cost | Time to complete |
|---|---|---|---|---|---|---|---|
| Empirical decision | Easiest | Poor | Yes | No | No | Least | Fastest |
| Classification code | Most difficult | Excellent | No | Yes | Yes | Most | Slowest |
| Part flow analysis | Medium | Good | No | Limited (process planning partially only) | Yes (limited) | Medium | Medium |

**Figure 13.4** Methods of forming GT cellular layout parts families.

tivity improvement—has been presented along with detailed examinations of the validity of the claims. We have also examined the strategy of economic analysis of savings and the approach for justifying group technology layout projects.

We saw that the principle of sameness underlies this layout concept and that a method for determining what constitutes a family of parts is essential for achieving any degree of success. Three methods for determining part families were introduced, and their selection was shown to be economically driven. Empirical decision is the simplest method and yields the least benefits, while classification code is the superior method and also the most expensive. The part flow analysis technique has been explained to show how this middle-of-the-road approach can be used. Finally, we saw that for use with an integrated CIM CAPP and data collection system, only the classification code is acceptable.

This chapter has defined the place for group technology layouts in the modern job shop. It is evident that a company must employ the group technology layout philosophy if its managers intend to keep it competitive.

## REVIEW QUESTIONS

1. Explain how the principle of sameness allows job shops to approach the economies of scale of flow manufacturing.

2. Compare the factory floor layouts shown in Figs. 13.1 and 13.2 with the layout shown in Fig. 13.3. Determine how the flows of jobs 1 through 10 of Fig. 13.2 would fit in the layout of Fig. 13.3. Describe the improvements possible.

3. What is a family of parts and how does it improve the probability of achieving economic production?

4. If a job shop, by definition, makes virtually any type of machined part, what logic is used to justify a group technology approach for its factory layout?

5. There are five reasons for adopting a group technology philosophy. With respect to layouts, state the order of applicability of these reasons. Give reasons for your ratings.

6. It was stated that a present practice of making parts in small lot sizes is not a relevant factor in determining whether a group technology cellular layout is desirable. Explain why.

7. Explain why group technology cellular layout projects are often difficult to have approved by senior management.

8. Describe the differences between subjective savings and firm savings with respect to appropriation requests. Give examples of each.

9. Discuss the reasons a senior manager may have for approving a subjective savings project.

10. Formation of families of parts is called the precursor event in developing a group technology cellular layout. Explain why.

11. In what business situation would the empirical decision method of forming families of parts be chosen?

12. Why is part flow analysis a good compromise method for determining families of parts?

# FOURTEEN

## MANUFACTURING ENGINEERING ASPECTS OF MANUFACTURING RESOURCES PLANNING

The topic of this chapter is manufacturing resources planning, commonly known by the acronym MRP II. It is the scheduling automation technique of choice in modern factories. We will examine this technology as a component of CIM, where it achieves its optimum success. MRP II will be described along with its development. Then we will discuss the manufacturing engineering techniques involved in implementing MRP II. We will also outline the procedures used for implementing and operating MRP II and the ongoing manufacturing engineering responsibilities. This chapter will further enhance our understanding of CIM and how information sharing via the common data base is critical for business success.

### DEFINITION AND DESCRIPTION OF MATERIALS REQUIREMENTS PLANNING, AND MANUFACTURING RESOURCES PLANNING

Materials Requirements Planning (MRP), sometimes called "little MRP" in deference to its expanded version, Manufacturing Resources Planning (MRP II), is the computer-driven purchasing and inventory management program. It evolved to satisfy the requirements of supplying materials to shop operations. The computer is used to sequence materials inputs in accordance with chronological need. This system optimizes the purchase and distribution of materials to the shop floor.

Unfortunately, by focusing on this, it is an island of automation approach and does not consider the entire business process. For this reason it was enhanced and expanded to include the other business processes requiring scheduling in creating a product. This includes shop scheduling of labor and availability of facilities, as well as design engineering due dates.

The evolution of MRP II from MRP is the logical outgrowth of the maturing of the use of computers in manufacturing. We have gone from stand-alone single-purpose computer applications to integrated and supportive uses. MRP was developed to solve the materials function's needs to cope with ever-increasing complexities of purchasing and inventory control. It was initially developed in the era before integration of all business activities was considered feasible or even desirable. Once awareness of the need for CIM became apparent, the logical extension of MRP into MRP II virtually exploded onto the scene. Let's describe further the power of this development and look at its applications and contributions to business optimization.

## MANUFACTURING RESOURCES PLANNING IN A CIM ENVIRONMENT

MRP II is one of the cornerstones of CIM. As we have seen, it derives from the materials function's concept of planning purchasing and allocation of materials in accordance with a computer-driven schedule. It encompasses at least four of the seven steps of the manufacturing system (see Chapter 11) directly and the remainder indirectly. MRP II is a major user of the integrated common data base and, when used correctly, embodies the philosophy of CIM to its fullest.

Figure 14.1 illustrates how MRP II works. Here we see how it relates to Just In Time (JIT) philosophy and the other modules of CIM. MRP II represents the Production and Measurements Control triad of the CAD/CAM triad. However, it is obvious from Fig. 14.1 that more than production and measurements is involved. In this CIM diagram of MRP II, we can observe the relationships between MRP II and other components of CIM. We see the Design and Planning Control triad represented by the master schedule. The perform work representation can easily be seen to be the Machine/Process Control triad. MRP II is truly a representation of the CIM philosophy of communications excellence applied to real-world activities. The entire essence of CIM is brought to bear in this activity. Let's summarize how MRP II works and then the basic strategy involved in successful implementation of this valuable CIM activity.

MRP II is the major innovation for dynamically allocating resources made possible by the introduction of the computer into manufacturing. It starts with the Bill Of Materials (BOM) prepared by Design Engineering. This lists in assembly hierarchy the component parts of the product. Some parts are purchased. Some are made in the factory. The BOM defines both and also shows assembly sequences. The BOM is the basis of the plan for making the product. Along with the BOM, the MRP II system is also fed by the routing. The routing is the chro-

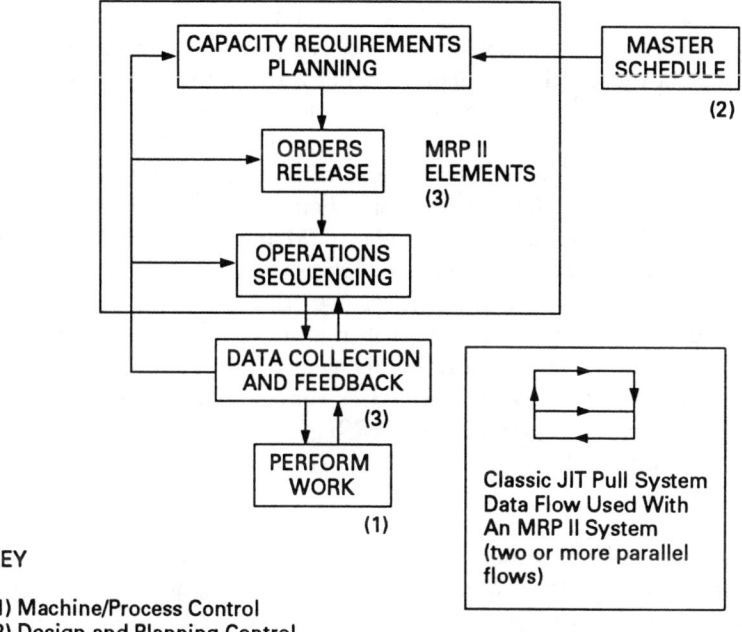

KEY

(1) Machine/Process Control
(2) Design and Planning Control
(3) Production and Measurements Control

**Figure 14.1** MRP II in a CIM world.

nological sequence of processing steps and the corresponding workstations where the work of making the specific part or assembly takes place.

MRP II, via computer algorithms, schedules multitudes of operations throughout the factory based on their chronological due dates. However, it still needs one more piece of information—how long it should take to perform each operation. This cycle time information is fed into the MRP II algorithm (and the routings) from the process planning activity. This too can be a CIM module, in which case it is normally called Computer-Aided Process Planning (CAPP). As we discussed in Chapter 12, CAPP systems such as the integrated CIM CAPP system (Fig. 12.4) are designed to work with other information data bases. So we see that MRP II is an enormously large reservation scheduler that performs its function via inputs from other data bases. Thus, the benefit of having an integrated common data base accessible to all cognizant parties becomes apparent. Also, it becomes obvious that MRP II is only as effective as the information it receives. The more dynamically correct that information is, the more accurate MRP II is. Since CIM is designed to operate in a dynamic situation, we see the power of CIM over stand-alone modules.

In stand-alone applications of computers, MRP II usage is entirely possible but hardly optimum. We can definitely establish a stand-alone data base for all production applications in the traditional "goes into pattern" logic. We can even have shop operations prepare schedules for the machine tools and workstations

334 MANUFACTURING ENGINEERING: PRINCIPLES FOR OPTIMIZATION

via MRP II based on a static schedule. And it would work, after a fashion. But what happens when the master schedule changes at the behest of the general manager? How long does it take the new data to be delivered to shop operations? How many no longer required parts are made in the interim? What happens if manufacturing engineering removes a machine tool for maintenance? How does MRP II know not to plan work for that workstation without an integrated data base? Obviously, it will eventually, but at what cost of bottleneck confusion? It becomes obvious that if all this information from other CIM modules were linked via a common data base, we would avoid confusion and optimize the utilization of these computer-based systems. A change made in one triad would automatically be reflected in other modules regardless of what triad they happen to be part of.

MRP II is so fundamental to business today that it is hard to imagine any company striving to be a successful world-class competitor surviving without it. Let's look at the steps for implementing MRP II.

The plan for implementation is quite straightforward. The basic chronological listing is shown below.

1. Understand how MRP II works and educate all employees.
2. Evaluate the existing Bill Of Materials (BOMs) and routings for accuracy as compared with actual practice.
3. Establish and implement a program to bring the BOM and routing documentation up to a minimum of 98% accuracy as compared to actual practice.
4. Identify and document the cycle time for each step of the routing. This information will be used to establish finite capacity for each workstation.
5. Establish materials ordering, master scheduling, and production planning policies that reflect how your company actually carries out these activities.
6. Investigate commercially available MRP II software packages for compatibility with your company's needs.
7. Implement and debug the selected software in a pilot location.
8. Once the pilot location is operational and debugged, spread one area at a time to the rest of the factory.

The implementation process is indeed straightforward, but there are dangers to be avoided. The primary trap to be aware of is the absolute need to have accurate information. If the BOM is wrong, the algorithms will miss sequence steps needed to build the product. Similarly, the routings must be correct. Otherwise the sequence of producing the product will not coincide with reality and the schedule will be inaccurate. Also, if the cycle times for each operation are wrong, the capacity to do work will be in error. The results will be either too much or not enough cycle time allocated at each workstation. This, again, means the schedule will be in error.

Finally, when MRP II is up and running, I must stress the importance of accurate recording of data pertaining to actual happenings at the various workstations. If this is not done, the system will not know where material resides and in what quantity. Remember, in actuality the MRP II algorithm is a large reservation system. It knows the entire chronological sequence of manufacturing and dispatches materials and labor and reserves facilities accordingly. It modifies the dispatching of resources in accordance with reports on actual accomplishments. If operators report erroneously, the MRP II system cannot order work in the proper manner. Every MRP II system has means for correcting errors. However, the more diligent our engineers and operators are in making correct data inputs, the less likely the MRP II system is to be in error and thus misleading.

One more tip about implementing MRP II. Many well-meaning people suggest that a consultant be engaged to guide the company through the implementing stage. In general, this is sound advice. However, keep in mind that the consultant does not know the product or the proper sequence of fabricating it. This means consultants will be of little help in the critical evaluation of the BOMs, routings, and cycle times. Even in the education process consultants cannot do it alone. They will undoubtedly understand the theory of MRP II and will do justice in that phase of the education of the implementing company's personnel. But consultants cannot be expected to go from the general to the particular of the company. This instruction will require a person familiar with MRP and very familiar with the company's way of doing things. The end result is that a "bootstrap" approach is required for specific instruction needs. The one area in which a consultant is effective is in the software selection arena. Here his or her experience can be very useful and can eliminate some of the "reinventing the wheel" syndrome.

Implementing an MRP II system is an "all hands" evolution. This cannot be overemphasized. The need to educate, educate, and educate some more is significant. People will find that although the benefits of dynamically accurate and timely information are great, there is a much higher need for self-discipline in working the system than in a non–MRP II environment. Understanding the intricacies of the system is very important. MRP II allows companies to cope effectively with the dynamic nature of manufacturing. But there is a price to pay. The laissez-faire approach does not work. Everyone must be constantly aware of the need to input accurate data and resist the temptation to freelance decisions and actions outside the system. Data accuracy is vital for the system to work. However, when it is working, an MRP II–driven factory is a wonder to behold. Promise dates are met. Inventory and production costs are down. Quality is enhanced. All in all, the company functions better and is an improved competitor. The effort to implement MRP II is intense, but the rewards are definitely on par with that effort. No modern manufacturing company can be without an MRP II system.

## MANUFACTURING ENGINEERING RESPONSIBILITIES FOR IMPLEMENTING MANUFACTURING RESOURCES PLANNING

Implementing MRP II has a technical content as well as the human factors portion. Manufacturing engineers are responsible for the technical content development, but with such a people-oriented activity the ergonomics must also be considered.

MRP II is a well-developed concept in a generic sense. However, since no two companies are identical in products or approaches in manufacturing, the job of the manufacturing engineer along with other user groups is to tailor generic MRP II into specific MRP II. Using Fig. 14.1, let's dissect MRP II to understand the process.

In Fig. 14.1 we see that MRP II consists of three modules—capacity requirements planning, orders release planning, and operations sequencing—plus complementing master scheduling and data collection. We discussed master scheduling and data collection in earlier chapters, and since they are not technically part of MRP II we will ignore them for purposes of this discussion. Also, JIT is discussed in a separate chapter and hence not covered here. "Perform work" in Fig. 14.1 implies value-adding activities at the various workstations in accordance with the production schedule.

## Capacity Requirements Planning

Let's start with Capacity Requirements Planning (CRP). This is the highest-order module of MRP II. It receives as its inputs the chronological sequence of work orders from the master production schedule. Typically this would be 6 months to a year's worth of orders, the planning period. This module then evaluates the work to be done versus the capability to do so in the time frame allowed. It shows areas of over- and under capacity at various workstations at various times throughout the planning period.

CRP is a refinement procedure for the chronological listing of orders from the master production schedule. The master production schedule has to be correct in a macro sense. That is, there can be no gross errors in capacity and capability made in the master production schedule, but there could conceivably be errors due to estimating during the quest for orders from customers. Since the master production schedule is used by the general manager to fill the company's order books, it has to reflect real capability, but only in an overall overview sense. It is the capacity requirements module of MRP II that refines the master production schedule to ensure that the chronological sequence of orders commitments can be met. CRP generally evaluates data from the master production schedule and then recommends what schedule adjustments should be made for capacity with respect to production equipment. This also leads to suggestions for labor loading and for identifying subcontract of work.

It is possible for CRP to refine the master production schedule simply because of the shorter time frame, which allows the vagaries of the company's facilities and labor situation to be better known. The master production schedule usually chronologically loads the factory for one or more years, up to a typical maximum of 3 years. The CRP rarely goes over six months, so the conditions of the factory are more readily known. Hence the company can take into account more precisely known factors affecting production capabilities. It stands to reason that specifics of labor trends and machine performances will be more easily predicted for 6 months than for 1 to 3 years. Obviously, the effectiveness of close-in improvement activities can be more accurately predicted than that of activities 1 to 3 years out. So CRP is a definite refining activity of the master production schedule.

CRP is also the module that initiates long-lead-time production materials. Any materials that require a long period of time must be ordered at this juncture. Since we do not want to have excessive materials on hand for just-in-case insurance, one of the primary functions of CRP is to make sure that this category of materials is ordered. To do so means that the engineering specifications generation required for ordering these materials is also scheduled via the CRP. This is one of the important aspects of MRP II. It coordinates the entire business function, not just the manufacturing portion. The output of the CRP module becomes the input of the orders release planning module.

## Orders Release Planning

This is the important intermediate level of planning that takes the chronological sequence of orders corrected for capacity and provides details for near-term production. It provides details for planning of subcontract work and in-house work and releases all but long-cycle-time purchase orders. Orders release planning is the logistics module of MRP II for what is called near-term planning. This can be anywhere from 1 week to a full production quarter, 13 weeks. It is the issuance of work orders for either in-house work or purchased services and materials. It is the stage at which strategic planning from the capacity requirements planning module makes a transition to tactical activities to achieve the production schedule.

Orders release planning is the vital function that modifies well-conceived strategies that may be floundering because they do not meet the current tactical situation. It does this by modifying strategies into achievable plans that are compatible with the current reality. It makes adjustments in order release dates if capacity cannot be scheduled economically or unexpected problems occur with production equipment. This module also takes into account unexpected labor situations, such as strikes or influxes of untrained labor into the factory. The module derives its basic sequence of information from the output of the CRP module; thus it depends on the accuracy of CRP to create the details necessary to go to this next, more precise, level.

## Operations Sequencing

This module performs the functions commonly known as production control. It controls the queuing sequence of work at every workstation. Thus it provides the instructions about what job is to be done and in what order at the specific workstation. In conjunction with the data collection module it performs the feedback loop so necessary to have a dynamic CIM system.

Production control tries to load workstations to their fullest capacity as measured by the overall optimization of the entire production facility. This module, supported by the CRP and order sequencing planning modules, allows the production controllers to understand fully the integrated needs of the entire factory to optimize output. They accomplish their task by using the computer data bases to meet due dates by utilizing available labor and facilities resources, including reserve equipment, overtime, and temporary out sourcing to qualified vendors as required. This correctly implies complex software networking with purchasing, inventory control, and manufacturing engineering.

This is the most dynamic of the MRP II modules. The planning horizon is the work shift out to about one week. The time increments are in minutes, not days or even hours. Here we see that the ability to operate in such an environment requires a constant stream of accurate and current information. This we achieve through application of data collection systems, such as described in previous chapters. We see that MRP II requires cohesive integration with the other module of CIM, and doing so optimizes the whole of the manufacturing enterprise. The relationship of the operations sequencing module to the rest of the MRP II modules, as seen in Fig. 14.2, demonstrates how the increments of MRP II, and for

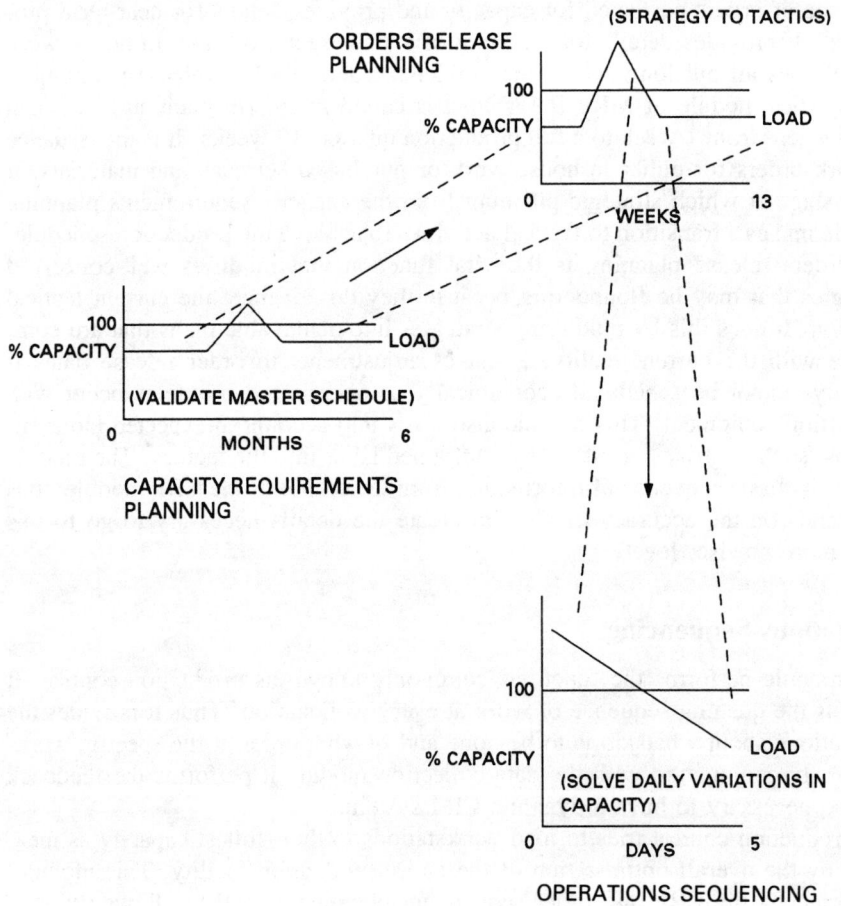

**Figure 14.2** MRP II, cascading of information from general to specific.

that matter all of the CIM modules, need each other and work off each other to provide an optimum output of products.

# TECHNIQUES FOR ENHANCING IMPLEMENTATION OF MRP II

While manufacturing engineers are involved in the entirety of implementing MRP II, there are specific activities in which they have the prime responsibility. Let us look at these prime activities and investigate the manufacturing engineering techniques involved.

Eight generic activities are involved in implementing MRP II, as listed earlier in this chapter. Manufacturing engineers would be expected to take the lead in developing three of these activities. They are:

2. Evaluate the existing BOM and routings for accuracy as compared with actual practice.
3. Establish and implement programs to bring BOM and routing accuracy up to 98%.
4. Identify and document cycle time for each step of the routing, and use this information to establish finite capacity for each workstation.

Two of the three deal with methods and planning activities, and the third deals with factory loading and capacity activities.

## Cycle Time, Factory Loading, and Capacity Issues

Factory loading and capacity techniques are discussed in Chapter 3. The requirements for MRP II are the same as for normal considerations. The engineer has to determine the load level that each workstation can handle over specific periods of time. For MRP II purposes this is usually by work shift, a nominal 8 hours. All the interference and nonproductive times are factored in, just as described in Chapter 3. The only additional task the manufacturing engineer has is to input the workstation capacities into the CRP module. This is done at the initiation of the MRP II project and every time there is a significant change in workstation capacity that is considered permanent. By permanent, I meant that the change represents a planned difference that will stay in place for the foreseeable future. For temporary changes in capacity, for example, as a result of equipment failure or inexperienced workers, the engineer will not enter the data into the CRP module. Instead the data will be transmitted to production control for it to be input into the operation sequencing module.

The overall responsibility of manufacturing engineering with respect to MRP II is to implement data to and maintain the CRP module. This means they manage the module. They make sure that data from master scheduling are input in accordance with the policy established by management. These data are then merged with the capacity data and the results are made available to materials management for the logistic planning and scheduling of the orders release planning module. Although this appears to be an ongoing work load of some significance, in reality it is not. Most MRP II systems handle these transactions as a matter of course all within the computer software with little human interaction. The only human interaction is the initial loading of the data base and change inputs. Thus, the engineering work is heavy during implementation and start up and then quickly becomes routine.

## Bill of Materials and Routings

The two methods and planning activities evolve about the BOM and routings. Let's discuss the BOM first. This document has come into prime use with MRP II, because it is a very organized way to transmit materials and labor needs in chronological order, in a manner easy for systems software to emulate.

The methods and planning activity defines how a part will be fabricated and assembled and how long it should take to do so. The work is based on the engineering drawing and a set of instructions that gives the sequence of assembly. This sequence is called the bill of materials. The bill, as it is called, is sufficient for sequencing operations. It shows how the design is to be made. But it is insufficient for determining how long it should take to do so. Determining the time to do an operation, hence an entire production sequence, is the work of manufacturing engineering. This process is described in Chapter 7. This cycle time information is merged with the BOM to become the "routing," sometimes called the planning. Let's now look at the details of the BOM and then discuss how engineers work with the document to ensure the accuracy of the MRP II module.

Figure 14.3 is an example of an assembly drawing of a piano leg.* In order to produce the leg, it is necessary to identify the components that go into the leg assembly and the sequence of fabrication and assembly steps.

The BOM lists these assemblies and parts in two basic formats: the indented bill of materials and the single level bill of materials for a part. These are shown in Figs. 14.4 and 14.5, respectively. The indented bill of materials, commonly known as the indented bill, is the "goes-into" structure, and the single level bill of materials is a part list for making a single part contained in the indented bill. The information in both versions of the BOM refers to information derived from the part drawing, Fig. 14.3. Let's review the information found in both of these documents. Although the details are specific to a piano component part, the headings found in the documents are generic and applicable to any user. For explanation purposes, Figs. 14.4 and 14.5 have their headings labeled Ⓐ through Ⓗ and Ⓐ through Ⓙ, respectively.

---

*All figures demonstrating drawings, bill of materials, and routings in this chapter are from Steinway & Sons, Inc. and are used with permission.

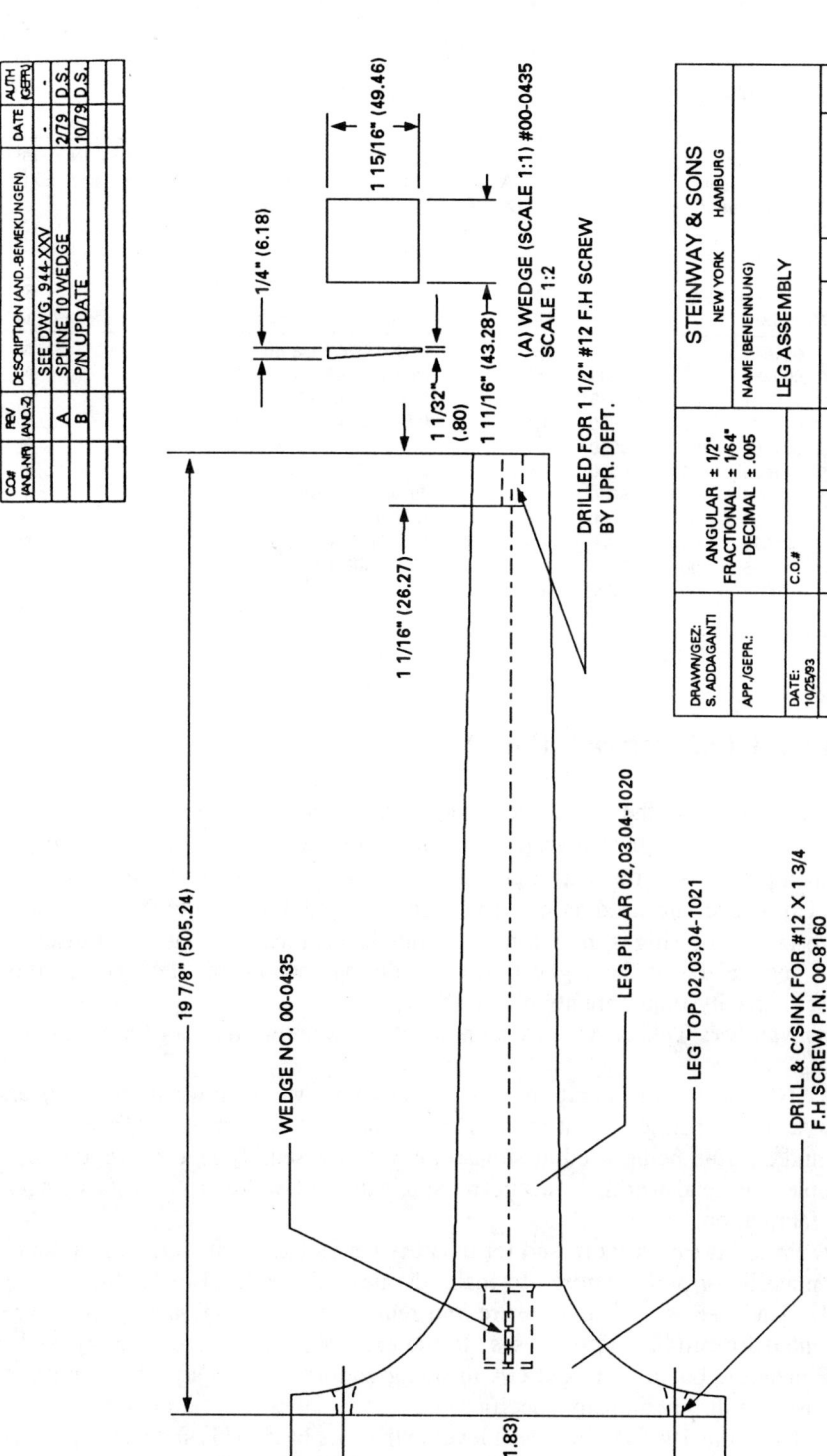

**Figure 14.3** Piano Leg Assembly for Demonstration of BOM Principles.

341

| | | | | | | | REV INDT BILL |
|---|---|---|---|---|---|---|---|
| Indented Bill of Materials | | | | | | | |
| Review | Review | Review | Add | Stop | Printed | Both | EXIT |
| WU | Route | Bill | Struct | | Report | | |

Part Number 021019

| Ⓐ. Level | Ⓑ. Part Number | Ⓒ. CC | Ⓓ. Qty Per | Ⓔ. UM | Ⓕ. LL | Ⓖ. Description | Ⓗ. Phan |
|---|---|---|---|---|---|---|---|
| 0 | 021019 | | | EA | 3 | LEG ASSY 1098 EB | N |
| 1 | 000435 | 00 | 1.0000 | EA | 4 | LEG WEDGE 1098 & K | N |
| 2 | 119021 | 00 | .0015 | BF | 8 | 5/4 BIRCH MLB | Y |
| 3 | 019021 | 00 | 1.3699 | BF | 9 | BIRCH 5/4 | N |
| 1 | 007758 | 00 | .0030 | GA | 8 | WOOD-LOK 40-0272/7045 (5 GAL) | N |
| 1 | 021020 | 00 | 1.0000 | EA | 4 | LEG PILLAR 1098 E | N |
| 2 | 119024 | 00 | .8680 | BF | 9 | 10/4 BIRCH MLB | Y |
| 3 | 019024 | 00 | 2.0000 | BF | 10 | BIRCH 10/4 | N |
| 1 | 021021 | 00 | 1.0000 | EA | 4 | LEG TOP 1098 & K EB | N |
| 2 | 119025 | 00 | .5859 | BF | 7 | 12/4 BIRCH MLB | Y |
| 3 | 019025 | 00 | 2.2222 | BF | 8 | BIRCH 12/4 | N |
| End of retrieval | | | | | | | 1016 |

**Figure 14.4** Indented BOM.

## Indented Bill of Materials (Fig. 14.4)

Ⓐ. *Level*: his shows the "goes-into" sequence of the parts. In this example we see that there are three levels. The construction starts with the lowest level. It then proceeds to the next higher level, where completed components of the previous level are used as material inputs for this level. Note that there is a zero level too. This is the master schedule level, meaning that it is a discrete part recognized at the highest part scheduling module of MRP II—in this case, capacity requirements planning.

Ⓑ. *Part Number*: This is the drawing number, or reference number for purchased parts.

Ⓒ. *CC*: CC means configuration code. If a number were shown it would mean that a part is being used at more than one level of construction. An example would be glue being used at subassembly and assembly levels. Another example, in metal working, would be application of welding at various stages of fabrication.

Ⓓ. *Qty Per*: Qty Per is shorthand for quantity per assembly. It shows the amount of material or parts required to make the next higher level of assembly.

Ⓔ. *UM*: UM means unit of measure and represents how materials or parts are counted for production purposes. In the example we see EA meaning each, BF meaning board feet, and GA meaning gallons. Obviously, these units of measure will be company specific.

Ⓕ. *LL*: LL is the designation lowest level in the data base. This shows the parent-component relationship in the entire MRP II BOM. In this example we see Birch 10/4 as 10 under LL but 3 for this particular indented bill of materials. Note that most of the numbers under LL are greater than 3, which indicates that this specific indented BOM is relatively low level within the entire MRP II BOM.

Ⓖ. *Description*: The description identifies the component parts with enough details so that every cognizant user of the indented BOM understands what is being called for.

Ⓗ. *Phan*: Phan is shorthand for phantom flag. This column tells whether or not a component is made or not made (purchased). N means nonphantom, which means it is a component that is scheduled to be made in the factory. Conversely, Y means phantom; hence the part will not be scheduled to be made but has to be identified as internal to the process—for example, so that an inspection can be conducted.

## Single Level Bill of Material for a Part (Fig. 14.5)

Ⓐ. *Component*: This is the drawing number identification.

Ⓑ. *Config Code*: This is shorthand for configuration code and is the same as the CC that appears in the indented BOM, item Ⓒ.

Ⓒ. *Description*: This is the same as item Ⓖ of the indented BOM.

Ⓓ. *UM*: This is the same as item Ⓔ of the indented BOM.

Ⓔ. *Qty Per*: This is the same as item Ⓓ of the indented BOM.

Ⓕ. *Phantom Flag*: This is the same as item Ⓗ of the indented BOM.

Ⓖ. *Pick Seq*: This is shorthand for pick sequence. This identifies the sequence material needs to be available for making the part. Since no numbers are shown in Fig. 14.5 in this column, we know that all materials are required at the same time.

Ⓗ. *Lead Adj*: Lead Adj means material lead time adjustment for different manufacturing cycle times. Sometimes it is necessary to start a particular operation before others for making the same part. For example, in working with

| Single Level Bill of Materials for a Part | | | | | | | REVIEW BILL |
|---|---|---|---|---|---|---|---|
| Review WU | Review Alt | Add Struct | Add Part | Stop | Printed Report | Both | EXIT |

| | Part Number: 021019 | | | |
|---|---|---|---|---|
| Description | Unit of Measure | Part Class | Part Status | } Ⓘ |
| LEG ASSY 1098 EB | EA | F | | |

| Ⓐ. Component | Ⓑ. Config Code | Ⓒ. Description | Ⓓ. UM | Ⓔ. Qty Per | Ⓕ. Phantom Flag | Ⓖ. Pick Seq | Ⓗ. Lead Adj | |
|---|---|---|---|---|---|---|---|---|
| 000435 | 00 | LEG WEDGE 1098 & K | EA | 1.0000 | N | | 0 | ⎫ |
| 007758 | 00 | WOOD-LOK 40-0272/7045 (5 GAL) | GA | .0030 | N | | 0 | ⎬ Ⓙ |
| 021020 | 00 | LEG PILLAR 1098 E | EA | 1.0000 | N | | 0 | |
| 021021 | 00 | LEG TOP 1098 & K EB | EA | 1.0000 | N | | 0 | ⎭ |
| End of retrieval | | | | | | | 1016 | |

**Figure 14.5** Single level BOM for a part.

wood components it may be necessary to kiln dry the wood before it is machined. If this column is blank, there are no lead time adjustments for this particular single level BOM for a part.

Ⓘ. *Upper Heading*: The single bill, unlike the indented bill, has a defined upper heading area. This area defines what is called the parent part. We see a description, unit of measure, part classification, and part status fields. Description and unit of measure are the same as described above. Part classification is either F for fabricated (as shown) or P for purchased. Part status is a field reserved for whatever the specific factory requires. It could indicate make for inventory stock. Or it could be used to reserve this particular part to a specific order. Whatever makes sense for the specific factory is an appropriate use for this field.

Ⓙ. *Components for the Parent Part*: These are all the items Ⓐ through Ⓗ of the single level BOM for a part. They form the details supporting item Ⓘ and give all the necessary information about the material needed to make the specific part.

The BOM is the component part data base of MRP II. It has to be very accurate, otherwise the scheduling algorithms at the three MRP II levels will not reflect the true situation. For this reason we strive for 100% accuracy in defining the "goes into" pattern. In fact, we say that the margin of error is very small, only 2%. Thus we need 98% accuracy as a minimum for MRP II to work. It is easy to visualize what kind of havoc would be caused if the BOM were wrong and the data base does not reflect the way the factory assembles the product. If this happens, the ordering and dispatching of materials would not occur correctly, and in the extreme case the factory could be idled because of no material at the workstation.

Engineers must be cognizant of these facts and diligent in checking the accuracy of the BOM before these data are entered into the data base. There is no easy way to do this. Usually it requires verification that the planning is being followed on the factory floor. Where it isn't, the reason for the diversion has to be understood and a decision made either to get the factory to follow the plan or to have the plan modified to match the current conditions. The word current is chosen carefully. The BOM always has to reflect the current way the product is fabricated and assembled. This means that every time there is a change in how work is performed, the BOM has to reflect that change, because above all MRP II accuracy has to be maintained.

The process of verifying the BOM has to be accomplished before the MRP II module of CIM is implemented. Typically this is done work center by work center. The manufacturing engineer must physically observe the work sequence and resolve all discrepancies before implementation of the system. As a practical matter, this means that we implement MRP II one work area at a time.

The companion data base to the BOM is the routing document. This is the end result of the planning activity, as described in Chapters 7 and 12. It shows the sequence of operations, the work center where each operation occurs, and the time it takes to do each operation.

The operations sequencing scheduling algorithm is the overseer data base to the routing. Whereas the routing is a set of instructions as to where a part will be made, the operations sequencing is a compendium of the information found in all the routings in the overall chronological sequence. These are compatible data bases. Operation sequencing tells the shop what work to do at each workstation and in what chronological order. The routing defines the method to use at the workstation for the particular part or assembly and where the next operation is to be carried out. So, too, does the operations sequencing data base. But it also contains the relationships of the particular part or assembly to all the other parts and assemblies. The routing is concerned only with the specific part or assembly.

Figure 14.6 is an example of a routing. Let's review the information contained in a routing. As with the explanations of Figs. 14.4 and 14.5, circled capital letters are used to define the fields for ease of explanation.

## Review All Routings for a Part (Fig. 14.6)

Ⓐ. *Route Seq No*: The route sequence number is from an arbitrary numbering system used to list the manufacturing steps in chronological order. The fact

| Review All Routings for a Part | | | | | | REV ROUTE |
| | | Stop | Printed Report | Both | | EXIT |
| | | Part Number: 021020 | | | | |

| Description | | | | Common Route ID | | Ⓗ. |
| LEG PILLAR 1098 E | | | | | | |
| Ⓐ. Route Seq No | Ⓑ. Wctr ID | Ⓒ. Oper Number | Ⓓ. Operation Description | Ⓔ. Cur Setup Hour | Ⓕ. Cur Unit Run Hour | Ⓖ. T F |
|---|---|---|---|---|---|---|
| 0600 | X-CUT | 1302100 | RIP TO WIDTH CUT TO 20″ × 2-1/2″ | .000 | .0000 | Y |
| 0900 | MILL | 1410400 | FACE ONE SIDE—GENERAL NUMBER FACE OPERATION | .067 | .0040 | N |
| 1200 | MILL | 1411200 | CUT PARTS TO SIZE—TENONER CUT TO LENGTH 19-3/4″ | .333 | .0060 | Y |
| 1500 | FRAISE | 1635400 | SHAPE LEG PILLAR 10-98—K EB TURN TO SHAPE | 2.000 | .0225 | Y |
| 1700 | FRAISE | 1636400 | TURN DOWEL LEG PILLAR 1098 TURN DOWEL 1-1/8″D × 1-5/8″L | 2.000 | .0116 | N |
| 1900 | FRAISE | 1623800 | RIP ON VARIETY SAW—GEN OPER CUT BOTTOM END | .250 | .0097 | N |
| 2100 | FRIASE | 1622100 | FRAISE TO SHAPE CUT DOWEL TO LENGTH & CUT WEDGE SLOT | .340 | .0019 | Y |
| 2200 | FRAISE | 1603200 | M SAND LEG PILLAR 1098 K EB SAND 4 FACES MACH SAND 4 FACES ROUGH & FINE AND BREAL CORNERS.CHANGE#3066. | .000 | .0269 | Y |

**Figure 14.6** Example of a routing.

that the sequence starts at 0600 is meaningless. So, too, is the exact numerical distance between steps. The numerical spaces between steps simply allow additional steps to be added later, perhaps as a result of process improvements or design changes. The numbering sequence has only one purpose, to show the chronological order of the operations.

Ⓑ. *Wctr ID*: This is shorthand for work center identification. It defines where the operation is to be performed. In the example we see department names that are specific for the intended factory.

Ⓒ. *Oper Number*: The operation number catalogues the operation being performed. This can be a group technology–derived number.

Ⓓ. *Operation Description*: The operation description defines the work to be accomplished at the particular workstation. The description is normally divided into two levels of information: general and detailed. For example, operation number 1411200 has "cut parts to size—tenoner" on the first line. This is the general information. It identifies the generic procedure to be accomplished and in this case the machine type being used. The second line states the specific length the part is to be cut to. The format of the top line being general information and the bottom line specific information is typical.

Ⓔ. *Cur Setup Hour*: This column show the time allowed for setting up the part on the workstation value-added equipment. Following the example of operation 1411200, we see that 0.333 hr are required to set up the tenoner. It should be noted that this setup is for one or many identical parts.

Ⓕ. *Cur Unit Run Hour*: This is the amount of time it takes to perform the value-added work per part. It is the time standard for doing this particular operation. For operation 1411200 the run time per part (unit) is shown as 0.0060 hr.

Ⓖ. *TF*: TF is the abbreviation for track flag. This term tells which operations will be tracked by the MRP II system. The Y or N tells whether the operation is (Y) or is not (N) being monitored. As a practical matter, not all operations need be tracked by the MRP II system. For example, if several operations have very short setup and run times, the practicality of recording the completions may be nonexistent. It makes no sense to measure an operation if the time to do so is virtually the same as the time for the activity itself. We see that operation 1410400 has a combined setup and run time of 0.0674 hr or approximately 4 minutes. We can certainly afford to lose track of the part for that amount of time before it is tracked at the next operation. If we tracked all of these short-term operations we would simply be cluttering the system with data that would have little practical use. The proper choice is to select representative operations that give us good information about where the part is in the production cycle.

Ⓗ. *Common Route ID*: Common route identification tells us whether this planning is unique. If there were other parts that followed the same routing sequence, an identification number, probably a GT classification code number, would be listed here.

As with the BOM, engineers have to ensure that the routing is accurate. The BOM defines the materials that make up the problem and the relationship and the sequence of assembly of these sets of materials. The routing tells where the work is to take place and in what sequence. The routing also states how the work is to be performed. For these reasons both the BOM and the routing have to be at least 98% accurate. If not, the MRP II system would not be a useful tool in monitoring and offering suggestions on how to manage the dynamics of manufacturing.

Routing design and authentication are the responsibility of manufacturing engineering. Usually, engineers design the routing as part of the methods and planning activity, as defined and described in Chapters 7 and 12. The method is transcribed into the routing via the CAPP system, which is a companion module of MRP II in the CIM system. The role of the manufacturing engineer is to ensure that the method described is actually the method followed. As is the case with BOM accuracy, it is vital that the routing be accurate, otherwise the MRP II system would end up scheduling a fictitious situation and the factory could not function properly.

## SUMMARY

In this chapter we have reviewed the theory of MRP II, and the role of manufacturing engineering in developing and operating the system.

We see that Manufacturing Resources Planning is an expansion of Materials Requirements Planning, the so-called little MRP. MRP planned sequencing of materials purchases and dispatching to the factory. MRP II expands on this idea to handle not only materials but also labor planning and facilities allocations in accordance with a strategic manufacturing plan.

In the study of MRP II, it became evident that the software was necessarily complex so it could handle many iterations of large numbers of discretely scheduled parts. To do this MRP is divided into three distinct but linked scheduling algorithms: capacity requirements planning for long-range planning, orders release planning for intermediate range, and operations sequencing for daily scheduling. The working of these modules was explained.

The process of implementing MRP is presented in the chapter. Key recommendations for successful adoption of the system, in the context of the personnel of the company, were discussed. We see that training and explanation as well as team formation are important for successful implementation.

The chapter concludes with a detailed explanation of the physical manifestations of MRP II, the bill of materials and the routing. These are the outputs of the software, the products of the scheduling algorithms that are used to control manufacturing operations.

## REVIEW QUESTIONS

1. Explain why MRP became MRP II. Take into account "islands of automation" versus CIM in your answer.

2. Show which of the seven steps of the manufacturing system are primarily MRP II steps and explain why.

3. Demonstrate the relationship between Computer-Aided Process Planning (CAPP) and MRP II in a CIM environment versus a non-CIM environment.

4. Implementing MRP II is a complex affair involving eight interrelated steps. Step 4 requires the identification and documentation of the cycle time for each step of the routing. Explain how the manufacturing engineer accomplishes this task and how this information is included in the appropriate data bases. Also identify those data bases with reference to a CIM environment.

5. Explain why step 1, "understand how MRP II works and educate all employees," is vital to the success of MRP II implementation and operation.

6. Capacity requirements planning evaluates for long-term ability to perform to schedule. It evaluates the work center load levels for a sufficient period of time to identify trends. Explain how CRP relates to the BOM and routings to perform this function.

7. Orders release planning is the intermediate-level scheduling algorithm of MRP II. List and define the differences in approaches to scheduling priority items between it and the other two levels of scheduling.

8. Production control, sometimes known as operations sequencing, is the detailed scheduling activity of MRP II. Discuss what dynamic versus static scheduling means at this level. Consider how MRP II handles changes in schedule at this level in your answer.

9. The BOM is the document used to input materials and sequence information into the MRP II scheduling algorithm. Development of the BOM is a design engineering function but is considered to be insufficient when completed by the design function. Explain why this is so and what the necessary additions are and how they are accomplished.

10. Define the differences between the indented BOM and the single level BOM.

11. The routing gives the MRP II scheduling algorithm the definition of where and how long it should take to make or assemble the product. Discuss how the manufacturing engineer assembles the necessary data and inputs it into the scheduling data base.

12. Discuss the reason for mandating at least 98% accuracy between actual and planned operations sequencing of the routing.

13. Explain why it is necessary for the routing to have the setup and processing times as part of the information contained in the routing document.

14. Operations sequencing is the lowest level of scheduling algorithm. How does the information presented at this level compare to the information contained in the routing?

## JUST IN TIME: A PRAGMATIC APPLICATION OF MANUFACTURING ENGINEERING PHILOSOPHY

Just In Time (JIT) is fast becoming as familiar and identifiable in the lexicon of manufacturing terminology as mass production. Unfortunately, although the term is familiar to many, its definition remains clouded and not thoroughly understood. The purpose of this chapter is to remove the clouds and to give a pristine and clear explanation of what JIT is and how it is applied.

The most common misunderstanding of JIT is that it means delivery of materials to the factory just when they are needed. This implies reduced inventory levels maintained in the factory. Although there is a modicum of truth in this belief, it is not the true definition of JIT. JIT is one thing and one thing only: an awareness that true optimum manufacturing performance revolves about the dictate to eliminate waste in all of its many manifestations. In this chapter we will examine the various ways we can eliminate waste in accordance with the commonly accepted JIT goals. We will see that these goals are compatible with, in fact identical to, similar ideals we have already seen attributable to manufacturing engineering practice. We will also see that JIT theory, like manufacturing engineering theory, works better with and is very compatible with CIM systems.

### JUST IN TIME FROM THE VIEWPOINT OF MATERIALS MANAGEMENT

The first use of JIT appeared in the materials arena, not manufacturing engineering. This may be surprising to some, because JIT is really another name for good engineering application in a factory situation. Considering that JIT has been popularized by nonengineers, this is understandable. Because of this history we also see that the first claims of manufacturing improvement had a materials management flavor. Let's look at the improvements claimed and how they occur. Some of these claims are more psychological and philosophical than scientific. As we shall see, even though this methodology is based to a high degree on motivational techniques, they can be of use in achieving positive results. But we must be aware that they are not the primary method of achieving results.

First of all, the goal of eliminating waste is very much in play. Anything that does not add value to the product is considered waste. The more visible items that must be analyzed for excesses are the moves and waits that lead to queues, stock levels, setups that could be taking too much time, making more parts than required for sales, and losses due to poor quality. These are but samples of activities that should be evaluated for better methods and policies.

Note that each of the visible items listed for increased scrutiny for waste has to do directly or indirectly with inventory management. In fact, the exhortation is to reduce inventory levels so that we can expose insufficient methodology. The theory is that sloppy practice is being compensated for with large inventory reserves. This is the just in case versus just in time argument, which we will discuss later.

It is interesting to note that nonengineer proponents of JIT always refer to the water level analogy for inventory. Figure 15.1 shows this analogy. We can see that a high water level (analogous to high levels of inventory) floats the boat (the process of manufacturing) high above the rocky bottom (problems retarding manufacturing progress and raising costs). If we lower the water level the boat has problems navigating, and we must remove the rocks (solve the problems). Notice the obsession with inventory levels. The concept is that high inventory levels hide problems. High inventory levels ensure that customer orders can be filled but at the expense of excessive cost. It is much better, the philosophy states, to be more efficient in using materials and thus still meet the customer's needs while doing so at lower cost levels. This makes the company more profitable, hence better able to compete. The end goal is positive, and manufacturing engineering theory certainly supports it.

Managing inventory is a prime facet of materials management; therefore it is not unusual to see this slant given to JIT. Let's look a little closer. It is claimed that we can improve performance if we take the crutches away. This psychological approach parallels the famous Hawthorne experiments in which spotlighting particular activities brought about improvements simply by exposure. These famous experiments showed that placing importance on a technique and how people performed it brought out the best in people, and they did better. This too is the case with "lowering the water." Here we purposely make it more difficult by making it known that there is no or perhaps just a little reserve material; therefore we need to do it right the first time. The proponents say that this allows us to gain significant improvements, necessity being the mother of invention. This is hardly the case. The improvements come about because the need is assigned as a project

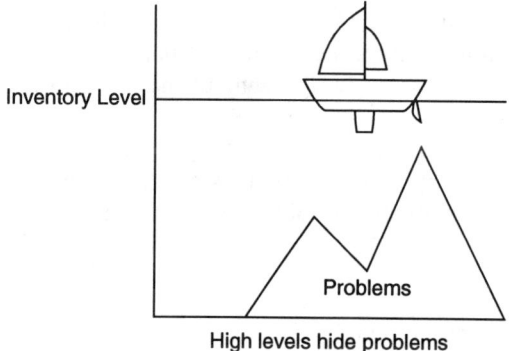

High levels hide problems

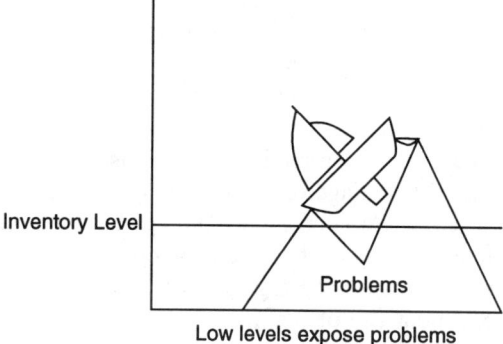

Low levels expose problems

**Figure 15.1** How lowering inventory levels exposes problems.

for improvement, usually requiring manufacturing engineering participation. What we really have is a way of establishing project priority. The goal of eliminating waste is valid. However, putting the company at risk through precipitous lowering of inventory to expose problems should not be done without serious consideration of possible consequences. Just in case is not always inferior to just in time. Let's look at an example of the lowering-the-water technique.

The theory says that we should reduce inventory to expose problems brought about by poor quality of materials and workmanship. Once the inventory is low enough, we cannot depend on high levels of inventory to ameliorate the unsatisfactory quality level. Thus we force the engineers, managers, and operators to take action to fix the problem or else the company will suffer.

One might say this is ridiculous—why create a crisis? Because we want quick action. We want the company to react to a problem that, when solved, will lead to performance improvements. This in turn lets us compete more effectively. The trick is to make sure the artificial crisis created is not too overwhelming. There is no doubt that if the company still demands the production, the pressure is put on the staff to solve the quality problem. The question is, is this the best way to make progress in eliminating waste? Maybe in some situations the answer is yes. However, I contend that rational approaches to problem resolution usually make more sense. Establishing priorities for problem resolution would just as easily get the same results. All we need to do is define the target need date for the project and assign measurable goals and responsibility as described in Chapter 2. I think it is reasonable to assume we can get the same results without the trauma of an artificial crisis.

Eliminating waste is a worthwhile goal and it can be achieved via good manufacturing engineering techniques. The hype and glamor associated with techniques such as lowering the water are really counterproductive and not what JIT is all about.

## JUST IN CASE VERSUS JUST IN TIME

The just in case approach versus the JIT approach has been mentioned, and occasionally the former turns out to be a better strategy, as we saw above. But just in case does not emphasize striving for perfection via elimination of all wasted efforts. For this reason JIT is predominantly the correct approach.

Just in case requires extra inventory at strategic locations to be used in certain situations. This could be to take advantage of sudden marketing situations in which the product cycle time is too long to take advantage of unforeseen opportunities. This is a proper application of just in case. Companies can plan to have buffer stocks of materials or partially complete products at strategic locations. They then use this stock when there is an opportunity for very fast delivery times to customers. This sprint capability is an excellent use of just in case theory.

When just in case is used as a crutch for inferior performance, it ought not be used. Here manufacturing engineering theory agrees with JIT philosophy. Just in case is not for covering up mistakes. It is only for creating opportunity. When we purposely make extra parts to cover deficiencies in people's performance, tooling, equipment capabilities, and vendor deliveries of materials, we are not getting to the root cause of the problem. This is simply perpetuating inefficiencies. We are using resources for purposes other than satisfying customer requirements. We are using resources to cover up problems. The proper procedure would be to establish projects to solve the problems, not cover them up.

The other problem associated with just in case is the propensity to make too much product at a workstation in order to take advantage of complicated setups. Many companies are enamored with economies of scale. They erroneously think that because the setup takes a long time, they must make many more parts with the same setup to lower the cost per piece. By simple arithmetic this is correct, but only for that workstation. If the company does not have any immediate use for the extra parts, the true cost is really higher than the simple arithmetic indicates. We have the cost of storing and caring for excess inventory and the lost opportunity costs of using the capacity to do something else. This is definitely creating waste. A better approach would be to find a way to shorten the time of the setup, thus making it easier to make only the parts that are required to meet current needs. This would eliminate all the extra costs of having excess parts in inventory.

JIT takes an opposite tack from just in case. Here the philosophy is to be miserly with what we produce in the factory. We will make only what is identifiably needed to satisfy requirements and nothing more. By producing only what is needed when it is needed, we preclude the opportunity to produce excess parts. This in turn precludes the opportunity for waste to be generated and passed on to succeeding workstations. Just in case, by putting more material in the production cycle, creates a more difficult to control production schedule because there are more parts and assemblies to account for. JIT does just the opposite. By promoting the miserly approach, we have less parts and assemblies to contend with, hence a more accurate production control system. This accuracy eliminates mistakes and leads to reduction of waste in the manufacturing process.

So far, the case for JIT has been built with pragmatic logic. We have seen that JIT is usually the right approach, but that there can be excessive claims by zealots in its pursuit. Some of these excesses can even lead to damage to the practicing companies if they comply blindly. An example would be lowering of inventory to expose problems. We saw that problems have to be identified and solved, and then the inventory can be lowered. This points out that like any good technique, JIT is not a totally perfect approach. We must understand why JIT becomes an enabling philosophy for accomplishing optimum manufacturing. The techniques for making it happen are tried and true engineering techniques, which have been discussed in depth throughout this book. Let's now examine the true goals of JIT and see how engineering techniques are used go about achieving them.

## THE TRUE OVERALL GOAL OF JUST IN TIME

The goal of JIT is entirely compatible with that of manufacturing engineering. In the practice of manufacturing engineering, the focus is to optimize the manufacturing process to the greatest extent possible. The goal is to create the highest level of profit with concurrent highest levels of product quality. JIT has the same goal but it is expressed in terms more readily understandable by those not familiar with manufacturing engineering techniques. The JIT goal, simply stated, is to eliminate all waste in the manufacturing process. There is no hidden meaning here. Root out all waste wherever found over the entire manufacturing process from material receipt, to value-added procedures, through quality control, and finally to shipment to satisfied customers.

Some non-engineer proponents of JIT support the goal of elimination of waste in a spirit that would have one believe that JIT is something new under the sun. Actually, it is not. It is good manufacturing practice that appears to have been rediscovered by those perhaps not aware of the manufacturing engineering discipline. This has caused many engineers to belittle JIT as an intruder into their domain. However, this is an improper response. JIT is, as engineers say, just a repackaging of manufacturing engineering, but it is a very good repackaging. It focuses on what is critical to serving customer needs and does so in a way that is understandable to people not normally involved in manufacturing. For this purpose alone it deserves to be explored, understood, and exploited. JIT is a correct, responsible approach, and when combined with CIM, it is a powerful tool for manufacturing optimization.

The goal of JIT is to eliminate all manufacturing waste. But for this to be useful, we must be more precise. Several authors have taken on the task of doing just this. One author who I believe has succeeded in expanding the concept of eliminating waste into terms that can be properly defined is Robert W. Hall in his book *Zero Inventory*, published by Dow Jones–Irwin. His six points defining elimination of waste are listed below (reproduced with permission of Dow Jones–Irwin).

- Produce the product the customer wants.
- Produce products only at the rate the customer wants them.
- Produce with perfect quality.
- Produce instantly—zero unnecessary lead times.
- Produce with no waste of labor, material, or equipment.
- Produce by methods which allow for development of people.

JIT is called by its promoters a fundamental way of thinking to transform overall manufacturing in the simplest way possible and to generate new and original techniques for doing so. I would dispute that there are any original techniques. However, there is no dispute that the points listed above are certainly attributes we can agree on. JIT theory states we should strive to achieve these goals, and many authors set out to show how this can be done, which in most cases boils down to good manufacturing engineering practice. However, if that's all JIT is, then it is not necessary. Our traditional engineering techniques work well without the hype. The true optimum achievement of the goals, such as those stated above, can be achieved only via the CIM approach.

## THE CIM APPROACH TOWARD REALIZATION OF JIT GOALS

The manufacturing engineering precepts espoused by JIT proponents can be approached in a traditional manual mode, but there is a much better way. That way is the CIM approach. We will look at the various productivity enhancers available via CIM and show that they are compatible with the six JIT points leading to accomplishment of the overall goal of eliminating waste in the manufacturing process.

### Flexible Automation

Flexible automation via the Flexible Manufacturing System (FMS) is the ultimate achievement of computer-controlled machines. These machines are multitooled turret types that can do many different types of operations with one setup. Operations such as drilling, milling, cutting, and shaping are commonly done in one setup on these types of machines. Flexible automation is a key goal of a CIM system in order to achieve efficiencies for job shops of the same magnitude as flow manufacturing.

This strategy is definitely compatible with the JIT precept of producing the product the customer wants. Flexible automation allows great customization of production, enhancing a company's opportunity to produce a specific product for a customer without significant increase of cost. These types of machines definitely allow factories to approach the long-standing goal of economic order quantity of unity, usually expressed as

$$EOQ = 1$$

This implies that a company is capable of serving customers with the ultimate in customer satisfaction: the ability to provide customized products at the cost of mass-produced products. This obviously is the JIT goal of producing the product the customer wants.

Flexible automation also means very short setup time because of the many operations that can be performed on the material while it is on the work platform. This is the achievement of the JIT goal of producing instantly—zero unnecessary lead time. In order to achieve very short setup times manufacturing engineers have devised Computer-Aided Process Planning (CAPP) along with group technology to find and plan families of parts. This makes achievement of short setup times much more likely, and we see the goal of producing instantly much more likely to be achieved. We can see that manufacturing engineering techniques are a very major factor in achieving JIT goals.

## Manufacturing Resources Planning

Many proponents of JIT mistakenly think MRP II is a noncompatible strategy. This is not the case. Let us examine how MRP II makes it possible to achieve the JIT goals of "producing only at the rate the customer desires" and "with no waste of labor, material, or equipment."

One of the most widely known terms associated with JIT is *kanban*. Kanban is a Japanese term loosely translated to mean a pull production system. Pioneered by Toyota Motors Corp., it is a system of production control that dispatches work based on the needs of the succeeding workstation. This is in contrast to the more traditional push system. The push system enters the work order at the beginning of the production cycle and depends on sequencing developed by the planning and methods documents to push the product through the factory to conclusion.

Proponents of the pull system claim it to be superior to the push system. They say it is analogous to pulling on a rope, whereas the push system is like pushing on that rope. Obviously, pulling the rope is more effective than pushing it. By pulling along material only when needed at the next workstation we can minimize making unnecessary parts, thus keep work in process inventory down. This would also keep the factory from expending any unnecessary labor. Here we see an example of the JIT goal of eliminating waste being achieved.

The kanban system works by the needing workstation dispatching requirements via an order ticket to the supplier workstation to fill. The supplier would literally be asked to make (or assemble) sufficient quantities for the immediate needs. This order ticket is also an authorization to proceed. It tells the workstation to make only what is required and no more. This is in contrast to the push system, which demands the workstation to make product as long as there is material available to do so, regardless of whether or not the succeeding workstation needs it. Quite often, as the dynamics of the situation change, a succeeding workstation could be glutted with work and not need any more entering the queue. This could be caused by changes in order rates or bottlenecks caused by machinery or quality problems. But the push system cannot easily shut off the flow, and material continues to be made and arrives oblivious to the current situational needs. The pull system has a much better chance of not allowing material to enter the next workstation if it is not needed.

The problem with the kanban system is that it depends on brute force to make it work. It depends on the receiving workstation accurately planning its needs sufficiently on time to keep pulling work at the proper rate. This is very difficult to do for all but the simplest production sequences. When there are more than one or two immediately preceding workstations to contend with, the dynamics of work in process can become very difficult to handle. Not only do we have to know the present needs, we also have to understand the succeeding demands being placed on the entire chain of workstations. We have to be able to predict all the needs on a real-time basis. Only a computerized system can keep track of all the variables with sufficient reaction time, and that system is MRP II. As we saw in Chapter 14, MRP II has the ability to dispatch and control the flow of work through multitudes of workstations and react to changes in situations very quickly.

Figure 14.1 shows how MRP II is designed to operate in a pull mode. Here we see that through data collection the situation of each workstation can be evaluated, and thus we can create the order tickets for each preceding workstation to meet the needs for every succeeding workstation. MRP II develops daily schedules that are, in effect, the pull requirements the kanban system demands.

The pull system cannot work effectively without MRP II. The push system, which is overall less effective than the pull concept, can. All it takes to make a push system work is to enter the orders at the beginning of the production cycle and let it flow. As each step of the process is completed, it will move to the next step, whether or not the next operation is prepared to receive it. Eventually the order will be finished, but there is no guarantee it will occur at the time originally intended. This is why push systems were used before the existence of MRP II. There is no need to make schedule corrections because there is no ability to do so. JIT-inspired pull systems are impractical without MRP II, and the axiom to produce with no waste is only a dream.

## Computer-Aided Process Planning

When we think about the JIT principles, one of the most vivid examples portrayed to the public is that of producing with no waste of material. This means striving to convert all the material to useful products. In order to do this effectively, we need a methodology for detailed planning for the material usage. This can be done manually, as we have seen in earlier chapters. But we also know from the chapter on CAPP that we can plan for manufacturing and coordinate sequencing faster and more accurately with computers.

In order to use materials correctly—in this case to use the optimum amount possible per unit of measure—we need to construct the goes-into patterns as precisely as possible and combine them with other parallel uses of the specific material. In this way we use the material to its fullest for all possible products (or parts). Obviously, this can be done manually by use of engineered time standards and methods-driven operation sequencing. But, as we have seen in the chapter devoted to methods, planning, and time standards, this is a tedious and meticulous process. It will be done only if there is sufficient time to do it and if the payback is considered significant enough to justify the effort. This means that most plans are approximations and material usage is seldom optimized. This is a poor strategy because the product cost ends up being higher than it should be.

With CAPP it is entirely possible to apply engineered time standards and methods-driven operations sequencing to the job. This will define the labor and materials needed to perform the job, and because of the link to the common data base the material can be designed to be portioned out to several jobs. It is analogous to saving nuts and bolts with the hope that we will some day find a use for all the excess hardware. As we know, that is hardly the case because we cannot find the right bolt or nut when we want it. But what if the computer was monitoring use and location of excess fasteners? Of course, we would use the excess fasteners instead of buying more. With a computerized system, e.g., CAPP, this means more of the material will be consumed in product manufacture and not wasted. The precept of JIT is achieved, whereas it is highly unlikely that it would be in a manual system.

## Computer-Aided Statistical Process Control

One of the most prominent of JIT precepts is "produce with perfect quality." A very desirable aid in achieving this precept is Statistical Process Control (SPC). We must be able to evaluate the condition of any process in order to ensure that we can control it and produce parts within desired tolerance ranges. If these evaluations can be made in a pragmatic manner, then we can achieve this precept.

The desire to produce with perfect quality leads to the need to measure process performance and understand the causes of process variance. SPC techniques as described in Chapter 8 are the methodology for measuring process performance. As we have seen, using control chart techniques is an exercise in minutiae and perseverance. It takes collection of data and calculation. Both of these tasks can be done better in a computer-integrated manufacturing environment than by the manual traditional approach.

We saw in the chapter devoted to CAPP that integrated collection of data to update production scheduling dynamically greatly enhances productivity. With computer-aided SPC that same data are entered to perform the control chart calculations, thus achieving virtually real-time SPC monitoring of the process. With this capability, real-time corrections to the process are possible and practical. This is the only way factories can approach zero defects. If we are forced to perform SPC in a manual mode, we never get the chance to make process corrections in real time. We simply cannot collect data and calculate fast enough to do anything but make after-the-fact corrections. The advantage of merging computer technology with SPC is what makes production with perfect quality a practical goal to pursue.

## Common Data Base

Perhaps the most unusual precept of JIT is "produce by methods which allow for development of people." This would imply enrichment programs such as team manufacturing and quality circles. In both of these cases we see empowerment of workers to make production decisions and thus develop their skill sets. Why would the proponents of JIT believe this is important?

The concepts of manufacturing have changed over the years, but all the concepts have had as their core goal to produce at optimum levels to maximize profits. In the early 1900s, it was thought that precisely spelling out details of how each person was to do his or her work was absolutely necessary. Most likely this came about because the engineers did not think the workers could comprehend the requirements the product had to attain. This was probably a logical conclusion at the time, based on the general technical education levels of the work force. However, times have changed, and for the better. The general level of technical education today is superior to that of almost 100 years ago. Unfortunately, the general regard for workers' intelligence for the most part has not changed in 100 years. This has led to stifling of creative initiatives, which means that many potential product and manufacturing methods improvements have not been realized.

The precept that companies should produce through improvement of people is aimed at getting us to realize that more productivity improvement is there for the taking. It is there if we would only make use of the talents of all the people. This means involving the entire work force, not just those who happen to work as supervisors and engineers. Obviously, this means investing resources in training workers in techniques not usually associated with their typical by-rote job responsibilities. Where that has been done the results have been very good. We saw in Chapter 7 that techniques for these participatory programs can be used to advantage. The question is, how do we optimize such programs?

Participatory programs generally require workers to take responsibility for suggesting changes in processes. In some more advanced programs workers do much more. They actually plan, procure materials, and schedule their facilities. It is claimed that these teams having total factory control are very productive and make high profits for their companies. There is certainly evidence that this is the case in specific circumstances. It may also be true for all circumstances to a certain degree. But it cannot be more than a side show unless these teams have ways to obtain information efficiently. This is where common data bases are useful and, in fact, make this precept a major factor for productivity improvement.

We know that information is necessary for the participatory teams to make effective decisions. We also know that getting correct and up-to-date information can be difficult. Use of common data bases in a CIM environment ameliorates this problem considerably. With the CIM axiom of communications excellence, we are continuously generating current information about the status of manufacturing. This means that the teams have access to the most up-to-date information. Any trials or tests they may be employing will benefit, in that results will be obtained faster and be more accurate. The results correlate with much more rapid achievement of gains.

Without common data bases, team projects usually take much longer and consequently the probability of loss of interest and distrust rises. This is the main reason why "produce by methods which allow for development of people" too often is a dormant precept. We need to dispel apathy and to keep the interest level up in participatory groups if we intend to achieve significant results. Keep in mind the people in these groups are predominantly production line employees. Their main responsibility is to perform the methods of transforming raw materials into products. Making improvements in how they do their job is unfortunately a secondary responsibility. Much as we would like it to be otherwise, the nature of production does not allow it. Thus, the best we can hope for is to stimulate the work force to develop improvement ideas and provide some time and resources to assist in this process. We cannot allow this task to become their primary function and take a large proportion of their time. This means that whatever ideas the team has, they need to be implemented and evaluated with as little queue time as possible. The common data base usage, which makes communications excellence practical, is the best way to manipulate data for monitoring progress and making evaluations.

We have seen in these few examples that JIT, like manufacturing engineering theory, is advanced toward practical reality through CIM. It became evident that JIT is great in theory but not totally practical without significant help from CIM. CIM is the catalyst necessary to make JIT work.

## SUMMARY

Just in time is a popularized version of good manufacturing engineering practice. It promotes the philosophy of doing all phases of manufacturing in a mutually supportive optimum manner. We see that elimination of all waste is the goal of just in time, which leads to practices of examining how we do things to look for improvement opportunities. This is not a new theory of engineering; it is simply a repackaging in a popular fashion, especially in a fashion that appeals to non-engineers.

We also examined how to bring JIT principles into practical reality. Here we saw that only through CIM techniques can JIT become anything more than a good attempt to improve manufacturing productivity. We saw the reasons why teaming with CIM philosophy and technology was the only way to achieve tangible results. This showed once more that good manufacturing theory requires computer-based procedures to be truly optimized.

JIT offers us the opportunity to popularize sound manufacturing engineering practices. It is not important for non-engineers to recognize the source of the theory. It is important, however, that the principles be practiced in manufacturing operations. To do so makes it more probable that the process can approach optimization.

## REVIEW QUESTIONS

1. Explain why JIT is more than delivery of materials to the required workstations when they are needed.

2. Identify the claims for improvement attributed to JIT that are considered psychological and explain why. As part of your answer, evaluate whether a psychological approach can be used to make permanent changes in operations.

3. Discuss how reduction in inventory levels leads to elimination of wastes in the manufacturing processes. Also discuss the dangers of reducing inventory levels in inappropriate ways. Give an example of inappropriate methods.

4. Define just in case and compare it to just in time. Give an example of a proper use of just in case.

5. Explain the push production control system and the pull production control system with respect to just in case and just in time.

6. Why is JIT considered an enabling philosophy for achieving manufacturing optimization?

7. Explain why the precept "produce by methods which allow for the development of people" is difficult to achieve in a manner that creates a permanent change in practice.

8. Show why flexible automation within a CIM system is compatible with JIT philosophy.

9. Explain why kanban will not work in a complex manufacturing operation without being tied to MRP II. Show how the pull system incorporated in MRP II software allows elimination of waste.

10. Discuss the relationship between CAPP and JIT. Show how CAPP is an enabler for JIT.

11. Explain what the precept "produce with perfect quality" means in a CIM environment.

12. What is the relationship of participatory programs to the goal of waste elimination? Show how common data bases enhance the ability of participatory programs to succeed.

# ENVIRONMENTAL CONTROL AND SAFETY

Awareness of safety and environmental hazards has led to a greater concern by management for impacts on employees and the company. Self-preservation is a human trait. However, the factory is an artificial environment where more than natural instincts are required to maintain a safe habitat. Hence a body of knowledge called environmental control and safety has been developed.

The safety aspects came first. By dealing directly with causes and effects, it was possible to codify how work should be done to minimize the always present danger. When materials are cut, bent, welded, sintered, melted, formed, ground, milled, and so forth, the very nature of the activity involves an almost violent action against the workpiece. If that action is misdirected, it can end up being applied to the operator or plant equipment, resulting in some degree of harm. Safety codes were designed to minimize and at best eliminate these causes and effects.

The environmental aspects came much later, when it was learned from circumstantial evidence that some of the processes being used at the workplace could result in harm to the operators and equipment after a long period of exposure. This has led to the continuing development of a body of knowledge concerning environmental hazards.

Accordingly, the topic of safety is divided here into two categories: one dealing with direct cause and effect, usually of an immediate or short-term nature, called "workplace safety," and the second dealing with indirect cause and effect, usually over a medium to long time period, called "workplace environmental control." A significant part of the understanding and corrective actions required in these two divisions involves applications of engineering disciplines. Therefore, as the technical arm of manufacturing, manufacturing engineering has been assigned the responsibility of providing for a safe workplace. This chapter will develop the managerial strategy involved in carrying out this responsibility.

## THE NEED FOR AN ENVIRONMENTAL CONTROL AND SAFETY CONCERN

It would be a mistake to think that safety and environmental programs are wasteful of capital and of little value. Properly conducted, these morale-building programs lead to greater profitability. Factories with good safety records also have good productivity records. Safe operation means that both the operator and the equipment are protected from harm. If this is not accomplished, the machines will operate less of the time, and profits will be correspondingly reduced.

Let us look at the safety aspect as it affects the operator. Operators are employed to control certain functions of the manufacturing procedure so that the purpose of the factory is accomplished. If they do their work effectively the company can make a profit. It is the manufacturing engineer's responsibility to plan how the operator will do his or her task. As we have seen in previous chapters, manufacturing engineering is continuously designing and redesigning the way work is to be performed with the objective of achieving the highest level of productivity. To achieve the productivity goal, the manufacturing engineer needs a good process, one that does not break down. An accident is a breakdown that causes a loss of productivity and, all too often, human suffering. It also causes property damage and an unanticipated additional cost. Machine breakdowns due to components wearing out and requiring replacement are expected and can be planned for to a certain degree; that is, they can be budgeted and staffed for. But accidents are breakdowns that should not happen, and when they do they are typically expensive in damage and in lost profitability. Therefore, it is to the advantage of management to institute programs to minimize all and eliminate most accidents.

Making the workplace safe for operators is the best way to eliminate the productivity deflator that we call accidents. In effect, we are protecting the operator from the machine and vice versa. Machines are designed to perform certain functions. The designer has taken into account the laws of physics and chemistry and has made the assumption that the function will be carried out and controlled as intended. If the function cannot be carried out and controlled as intended, the machine will not perform its function. The best that can happen is that the machine will not work; the worst is that an accident will occur. In that case, the operator or the machine or both will be injured. Therefore, operator safety is achieved by making sure that the operator can only operate the machine in a safe manner. If this is done successfully, then the operator is working in harmony with the goal of the factory, that is, producing a product in accordance with the plan, hence profitably.

The objective of the safety program is therefore to protect the operator from the machine and the machine from the operator. By doing this, we will substantially increase the factory's capability of producing at the planned productivity level. Accidents stop portions of the factory from working, and therefore they are contrary to our productivity goals.

The discussion above gives the economic foundation of the safety program. But the safety program also has a psychological basis. An operator who is afraid of being injured will work at a slower pace than desired, and we will have less than desired output levels. On the other hand, if the factory has a good accident prevention record, the operator will not fear injury and will probably work at the desired pace. There are many reasons for not working at the desired pace, and many programs have been conceived to eliminate these problems. Unfortunately, many otherwise astute managers forget the safety aspect. We must be aware of the human instinct for self-preservation and make it work to our advantage. A safe workstation is more likely to be a productive workstation.


## THE CASE FOR ESTABLISHING FORMAL ENVIRONMENTAL AND SAFETY PROGRAMS

Let us look at environmental and safety issues from the viewpoint of the public at large. Goodwill is an intangible factor that cannot be measured in dollars and cents like capital investment projects, but it is real nevertheless. Consider a company that has a good image and one that does not. If both companies are competing for a contract, the one with the good image will have an advantage in the evaluation process. Safety and environmental programs are vital parts of such image building. From senior managers down to the lowest paid workers, people like to be associated with organizations that show concern for their employees.

Thus, there are more than economic benefits to be derived from the environmental and safety program. We can expect to obtain morale benefits that enhance the company's ability to perform as a team. The goodwill obtained is beneficial in dealing with outside agents such as government regulators influencing rule-making activities. It facilitates public acceptance, which helps in selling the company's products. And it helps to attract competent workers by indicating that the company will be a good place to work. These are but a few of the intangible benefits of an effective environmental and safety program.

We have two powerful reasons for establishing environmental control and safety programs: economics and image building. Both lead to enhanced profits. Manufacturing engineering is responsible for the technical aspects of this program and frequently for the nontechnical portions as well. Some companies place total control of the program under the employee relations function. Some share control between employee relations and manufacturing engineering. Other firms place the entire program within manufacturing engineering with only advisory consultation from employee relations. Whatever the management concept employed, the program consists of technical and nontechnical aspects.

## KNOWLEDGE REQUIREMENTS FOR FACTORY ENVIRONMENTAL CONTROL AND SAFETY PROGRAMS

The typical environmental control and safety program is divided into awareness (publicity) activities and planned action activities. But before we delve into these particulars, let us review the knowledge necessary to effectively carry out such programs.

In most manufacturing engineering organizations, advanced manufacturing engineering will be assigned responsibility for keeping informed about regulatory policies related to workplace safety and environmental control. The specific items covered by such policies are described in the following subsections.

### Fire Codes

Provisions for building fire safety vary from jurisdiction to jurisdiction. So too do codes regulating electric wiring and controls for machines and equipment. Hence it is necessary to know the prevailing codes. Some companies also have restrictive codes that they impose on themselves in order to receive the lowest possible insurance rates. Manufacturing engineering must be cognizant of all these rules and regulations and make sure they are applied to all capital-related projects.

There are aids to achieving knowledge in these cases. Most insurance companies belong to underwriters associations which are willing to document correct safety procedures and assist customers in making site surveys. Manufacturing engineering is expected to take advantage of these sources and make the proper liaisons.

### Safety of Structures above the Work Floor

These structures include ladders and elevated work platforms, both temporary and permanent. Over the years the Occupational Safety and Health Administration (OSHA) has codified standards for such platforms and ladders. It is manufacturing engineering's responsibility to make sure that the factory is in compliance with these standards. Noncompliance is not only unsafe, it is also against the law and can result in criminal as well as civil penalties. Manufacturing engineering must protect the company against this liability.

Also falling in this category would be the placement of apparatus used in a workplace. For example, the placement of machinery, fire extinguishers, and electrical apparatus above the work floor requires guarding and visual identification procedures. Manufacturing engineering must ensure that the appropriate codes are complied with.

### Control of Hazardous and Toxic Substances

This is fast becoming one of the most complex and emotional issues facing manufacturing engineering. Toxic and hazardous substances present an immediate and a long-term health problem. Toxic substances are construed as poisons, acids, and other materials that can cause immediate harm. Hazardous substances are materials that, while not having an immediate effect on personnel, represent a statistical health hazard over the long term. Where toxic substances are present, it is necessary to have quick emergency procedures to be followed in case of an accident. Hazardous materials require administrative procedures to ensure that exposures are below the allowable levels mandated by the U.S. Environmental Protection Agency (EPA).

The problem facing manufacturing engineering is in the identification of all toxic and hazardous substances in the factory. Herein lies the complexity of controlling such substances. To ensure that the company does not inadvertently violate an EPA regulation, manufacturing engineering must determine the matrix of all chemicals present. This is a difficult task because most fluids and solids are purchased under trade names and the exact chemical composition is not known. Therefore manufacturing engineering must solicit information from vendors or conduct analytical tests, probably doing some of both. This is costly and time consuming.

The emotional issue arises when toxic and hazardous substances in the factory are identified, as they almost inevitably will be. Government regulations require notifying the employee of the potential hazard and of the preventive and emergency control procedures to be used to minimize the hazard. This, coupled with the somewhat sensational coverage of previous problems of this nature by the communication media, encourages emotional rather than objective reactions on the part of factory personnel. Therefore, manufacturing engineering must keep abreast of available information on these problems.

Manufacturing engineering is usually the only technical source readily available to interpret regulations and to properly inform management about the potential problems and their solution.

## Noise Abatement

A particularly vexing problem whose solution requires considerable knowledge and expertise is that of meeting the factory noise standard of 85 dBA per 8-hour exposure. It has been determined that a consistent noise level above 85 dBA can lead to loss of hearing over a long period of time. Therefore, current OSHA regulations require an engineering solution to reduce the noise level in areas where it is over 85 dBA. Manufacturing engineering must be the source of knowledge for the company's compliance with this regulation.

The fact that an engineering solution is mandated is important. This can be understood most readily by explaining what is *not* an engineering solution. The use of protective equipment for factory personnel is not an engineering solution. OSHA has allowed protective equipment only as an interim solution until an engineering solution can be devised and implemented.

An engineering solution to noise abatement requires a detailed understanding of the physics of sound and vibrations. In the past, this has not been a part of the primary expertise of the manufacturing engineer. But with a requirement to protect factory personnel from the danger of high noise energy, manufacturing engineers must develop expertise in acoustics and related fields.

I have found that searching for engineering solutions to noise problems leads to a greater understanding of the offending tools themselves, and this can have some interesting additional benefits. By understanding the tools better, we can design better applications and hence improve factory productivity. An example of this is an effort carried out jointly with a nearby university to reduce the noise level of pneumatic grinders for metal polishing from 100 dBA to 85 dBA or lower. In this project, we learned that a noisy tool was an inefficient tool. By using a proper muffler design and making some modifications of the air flow in the tool, we could develop a more efficient tool and one that met the noise standard as well. We were pleasantly surprised to find that the noise abatement project turned out to be a productivity improvement project.

I mention this example to point out the need to avoid thinking of noise abatement projects or any other safety and environmental projects only in terms of compliance. Managers must always encourage the attitude that knowledge applied to these projects should also lead to improvements in producing the company's products. There is no reason to be satisfied with merely complying with regulations. The approach should be one of complying and improving profitability at the same time. This is not always a simple task, but it is a necessary approach in highly competitive industries.

## Safety Apparel

Manufacturing engineering is often called upon to specify clothing and protective devices that operators must wear to protect themselves from work hazards. Examples are welding jackets to protect welders from flash and sparks, steel-toed shoes to protect against broken bones, safety eyeglasses with impact-resistant lenses and wire mesh side shields to protect against foreign particles becoming embedded in the eyes, and noise suppression earmuffs.

Knowledge of safety apparel is essential to effective planning of workstation layouts. Manufacturing engineers must not only be competent in designing effective workstations, they must also be knowledgeable about the safety considerations and the proper apparel required.

The designation of proper safety apparel definitely affects the cost of operations. For example, if safety glasses are required, a company must provide prescription as well as plain safety glasses to all employees. This includes new prescription safety glasses if an employee's prescription changes. The fact that safety apparel is a business expense means that manufacturing engineering must consider its choice of such apparel as it would any other device or commodity it specifies.

## Reporting Requirements

Understanding government regulations is essential. When government regulations are encountered, they usually require that compliance reports be submitted. This requires time to prepare the reports and an in-depth understanding of the regulations. Manufacturing engineering must be cognizant of the reporting requirements and must issue the reports on time. Many companies have suffered needless embarrassment and fines by failing to report correctly and on time. Management usually assigns this additional compliance responsibility to manufacturing engineering.

These are some of the areas of special knowledge related to safety and environmental control that manufacturing engineering must master. Mastery of these topics leads to the awareness activities and planned action activities that make up the environmental and safety program.

## THE ENVIRONMENTAL CONTROL AND SAFETY PROGRAM

This program is divided into awareness activities and planned action activities. The major concept in both parts of the program is a dedication to the principle of "work smart, work safe." We will review a typical two-phase program that exemplifies this commonsense approach and show how such a program improves the safety of the workplace.

Awareness activities and planned action activities are administered by manufacturing engineering or employee relations. If the majority of the activities will involve applications of technology to solve safety and environmental problems, then manufacturing engineering is the administrating agency. If the prime thrust is that of maintaining an established safety and environmental program, where the technical problems are minimal, then employee relations administers the program. Since we are primarily interested in manufacturing engineering management, we will explore a program administered by manufacturing engineering. This will be an environmental control and safety program skewed in favor of solving the associated technical problems.

The awareness activities in the program are designed to make all employees more aware of the need for safety and environmental care while at work. Awareness is an intangible thing, and those responsible will always have difficulty in determining whether their effort has been successful. Perhaps this can only be known with certainty when an accident occurs and personnel react quickly and correctly. This, however, is counter to the precepts of the program, which is successful only if accidents and mishaps are prevented. Measurement of a successful awareness program must therefore remain essentially intangible. We can measure effectiveness subjectively, however, by noting conditions in the workplace, looking for signs of good housekeeping, lack of safety hazards, properly cared for emergency equipment, and so forth. Awareness programs are, in fact, evaluated by a lack of negative factors being discovered during safety inspections. This gives an indication of the overall awareness level.

The planned action activities are those initiated to correct safety and environmental deficiencies that have been discovered and those required by law to maintain compliance standards. A significant proportion of planned action activities result from awareness activities, that is, subjective audits and inspections.

> **Example:** A safety and housekeeping audit may uncover a frayed fire hose, a chance occurrence. This would result in a planned action activity to replace the defective fire hose. This activity is measurable. We can measure the time to place an order for a new hose, monitor for its delivery, and determine whether the hose is installed by the targeted date.

Whereas awareness activities are usually measured in a subjective manner, all planned action activities are measurable in an objective fashion. Whether a safety article is written for the plant newspaper is a measurement of an awareness activity. The measurement is not concerned with whether the article had any impact on its readers in establishing a safer workplace. That is the purpose of the awareness activity, but it is beyond the scope of the measurement. Planned action activities can be measured for their effect. For example, making a sound contour map of the workplace to determine whether it is in compliance with the 85-dBA, 8-hour standard is a planned action activity. The effect of the work gives the answer. After the map is completed, we can easily tell whether the area is in compliance. If it is not, we progress to another measurable planned action activity, that of engineering a solution to lower the noise to an acceptable level.

Both awareness activities and planned action activities are merged to form an overall measurement. This could be a safety standing with reference to a goal set internally by the company or to a compliance level set by a controlling agent, such as an insurance company or government jurisdiction. Such a goal must be very broad based. It might involve the accident frequency per time period, the accident severity level, or combinations of both. In order to have an environmental control and safety program it is necessary to agree on a goal that can be understood and measured. Figure 16.1 is a flowchart of a typical technically oriented environmental control and safety program. We will review the various components of this program in the next two sections of this chapter. However, the first area to understand is the overall goal designated by the box titled "OSHA Incident Rate Goal." This is the culmination of the awareness activities and the planned action activities. If these activities are carried out effectively, we will see an improvement in the OSHA incident rate. An OSHA incident rate is a combined measure of accident severity and frequency, devised by the Occupational Health and Safety Administration, which measures overall safety improvement over a period of time. The important thing is that it is a measure of the effectiveness of the environmental control and safety program.

## AWARENESS ACTIVITIES PROGRAM

### Hourly Safety Meetings

Hourly safety meetings are designed to create awareness among the operators. The term hourly refers to personnel who are usually paid by the hour and are

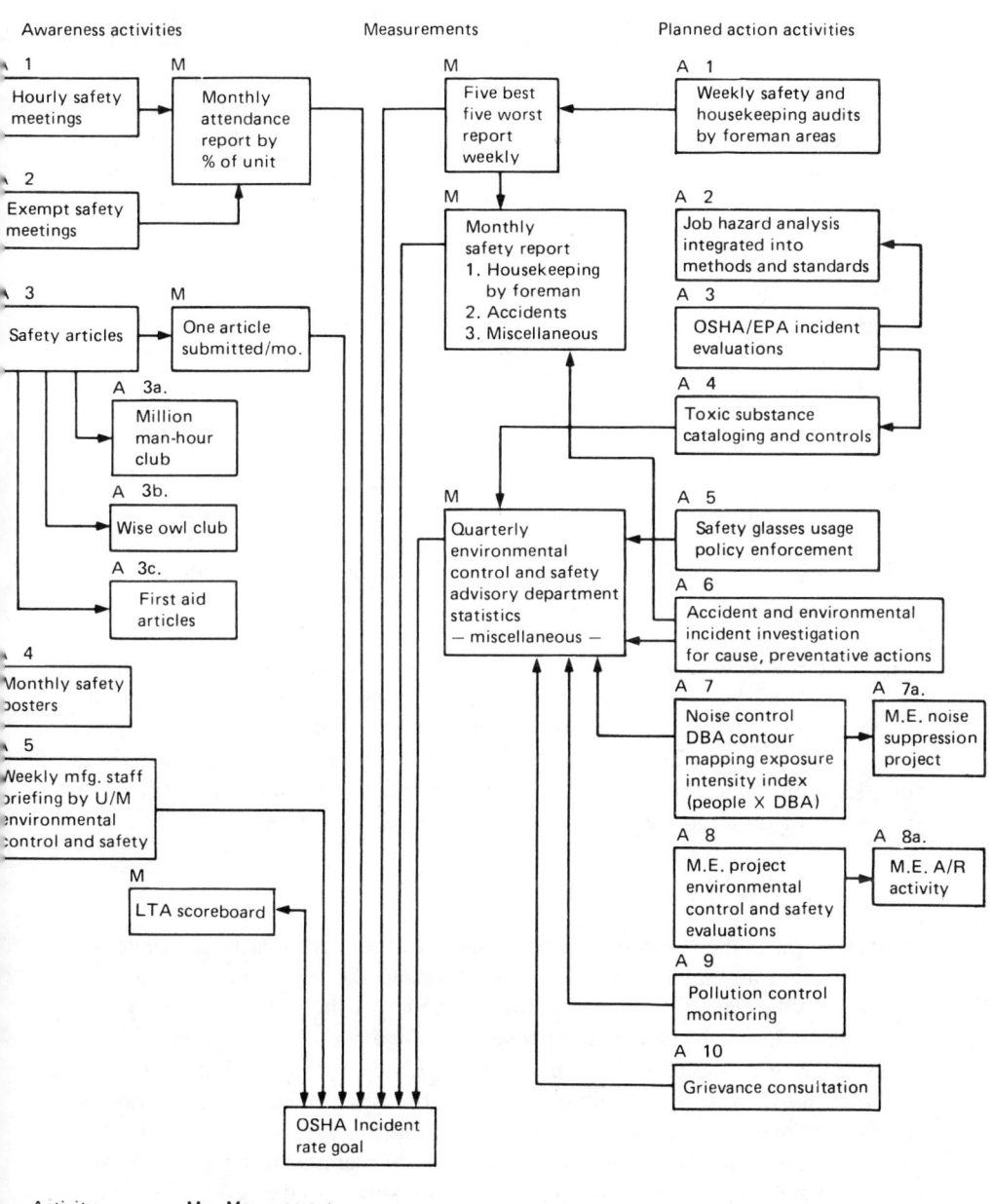

**Figure 16.1** Environmental controls and safety programs.

often members of a recognized labor union. These meetings are conducted by the foreman or other member of management associated with shop operations.

The topic for each meeting and the script to be followed is provided by manufacturing engineering. The AME or MP&WM unit prepares a program for the entire year consisting of the general and specific topics of major concern to their company. Items such as fire safety, first aid, toxic substance control, proper handling and storage of materials, electrical equipment hazards, overhead crane safety, and vehicular safety are covered in a lesson plan format. The format is very explicit in counseling instructors on how the lesson is to be presented, to the level of identifying the key message to be transmitted to the class.

Hourly safety meetings are held during the normal work week to emphasize that safety adherence is an important part of the company's business. Since these meetings interrupt the production of products, there is a tendency on the part of lower level management to be lax in adhering to the schedule. Occasionally meetings are skipped in order to complete an order or to catch up on production. Manufacturing engineering must be very critical of all excuses and must bring pressure to bear to minimize the skipping of meetings or the absence from meetings of segments of the hourly work population. Surveys have shown that areas that are lax in safety meetings tend to have a higher accident frequency rate. Over the long term, skipping meetings is counterproductive with respect to overall shop productivity levels. Besides the obvious humanitarian reason, maintaining a good productivity level is a motivation for insisting that these meetings be held.

Manufacturing engineering must also ensure that safety meetings do not become an obstacle to maintaining the rhythm of production. This means that meetings should not be scheduled too frequently or last too long. The frequency of meetings should be a fair compromise between the requirements to maintain production and to keep a high level of awareness of the need to work safely. Therefore, most environmental safety programs schedule hourly paid labor safety meetings to occur on a monthly basis and to last for approximately 30 minutes.

## Salaried Safety Meetings

In most respects these meetings are the same as those for hourly personnel. Salaried refers to all employees who do not fall into the hourly classification. The topics for salaried safety meetings are the same as those for the hourly safety meetings. It is good practice to expose office and professional personnel to the same program as the shop floor personnel. This enables the salaried employees to act as unofficial safety monitors during the times when they are in or traversing the factory operations area. Everyone in the entire business organization is expected to report unsafe conditions or practices to manufacturing engineering for corrective actions. The more people who are sensitized to the need for safety, the greater the number of potential dangers that will be reported and corrected.

In addition to the series of safety meeting topics, professional personnel must be made aware of the fact that the results of their work could cause a safety or environmental hazard on the factory floor so that they can guard against this.

**Example (a):** A weld fabrication design that requires a welder to assume a cramped or awkward position to perform the work would expose the welder to a higher accident risk.

**Example (b):** An engineer might decide to use a highly toxic solvent for cleaning parts before assembly. By doing some research the engineer could find a much safer solvent that accomplishes the same purpose.

**Example (c):** A maintenance engineer's desire to save power by turning out lights in areas of the factory might pose a hazard to personnel who have to traverse the area.

**Example (d):** A company personnel policy that limits crane personnel and substitutes operators in making certain lifts does not consider the increased safety hazard incurred by using inexperienced personnel.

The salaried safety program must have as one of its aims increasing the awareness of the safety consequences of the professionals' work. In this way it can reduce the incidence of accident by proxy. An hourly operator who is careless may injure himself. A professional who is careless may cause injury to others.

## Safety Articles

This is a companion program to the safety meetings. Again, the purpose is to create a high level of awareness among personnel so that they will carry out their duties in a safe manner.

Most manufacturing organizations have newspapers to impart information of interest to employees. These house organs have a goal of improving morale and thereby improving productivity. They do this by extolling good performance by an employee or groups of employees. They also carry human interest stories that elicit sympathy for the company or a desire on the part of the employees to do their jobs better. Part of this program should be the safety message. Since a good safety goal is a step toward achievement of the productivity goal, it is proper for the plant newspaper to carry safety articles.

Safety articles are usually positive. It would do little good to report an incident that resulted in an injury to an employee as a lesson to other employees to exercise greater care. Safety articles usually report actions through which injury was avoided, or instruct personnel in how to avoid accidents. A third type of safety article is one that commends employees for achieving a recognized goal in minimizing accident frequency and severity. The last type of article usually ends with a challenge to do even better, to reach the next plateau. Let us look at some examples of these types of articles, as outlined in Fig. 16.1.

**"Million Man-Hour Club."** Articles on the "Million Man-Hour Club" salute the achievement of an area of the shop, or the entire factory, working for a time equal to or better than 1 million man-hours without an accident that causes an injury. They usually compare the area's previous best performance with this milestone and quote workers on the importance of the achievement and how proud they are to work safely and to work with people who care about safety. This is usually followed by a statement of appreciation by the manager of manufacturing with a challenge to the group to continue the accident-free performance and to other groups to emulate this achievement. If the plant newspaper contains photographs it will print one showing the unit manager receiving the safety award plaque on behalf of the unit. Usually people within the unit will be encouraged to be in the picture.

"Wise Owl Club." The "Wise Owl Club" is a national institution that recognizes individuals for preventing the occurrence of serious eye injury. Articles on this subject commend individuals for avoiding an accident by conducting themselves in accordance with the safety rules of an area. Usually this means that the individual had the good sense to be wearing impact-resistant safety glasses that prevented a missile from entering his eye. The article shows a picture of the operator holding the safety glasses with the shattered but intact lens. The text explains the series of events that led to the safety glasses protecting the employee's eyesight. The purpose of this article is to show that adherence to safety rules does, in fact, protect people from harm. Such articles also state that membership in the Wise Owl Club is only given to those with the foresight to work safely.

First-aid articles. These articles are for positive instruction. Their purpose is to inform all personnel on how to react to the need to apply first aid in a particular set of circumstances. The articles are intended not to supplant first-aid manuals or first-aid classes, but to heighten awareness of the need to apply first aid quickly and correctly if the occasion arises. There are two general types of first-aid articles. The general announcement type states that a first-aid session of a particular kind is to be conducted, gives the time and place, and encourages readers to attend. The second type actually explains a first-aid technique and usually includes instructions on how the reader can obtain further information.

The purpose of all these types of safety articles is to make all personnel more aware of safety. It is easy not to write such articles. Therefore, most managers of manufacturing engineering will take steps to encourage the writing of such articles. They will give a quota of articles to be written to selected personnel and measure their performance in meeting the quotas.

## Monthly Safety Posters

Another good way to create safety awareness is through the use of safety posters. Good sources of posters include government agencies, private safety agencies, professional societies, unions, and producers of safety equipment. They produce posters that are witty and eye-catching and convey the specific safety message. It is up to management to use them correctly.

One effective way to use safety posters is to post them prominently and change them periodically, preferably monthly. They should be posted where personnel have to see them but are not distracted by them while working. Good locations are at the time clocks, on the notice bulletin boards, or beside the entrance to the plant cafeteria. They should be changed monthly so that they continue to attract attention. If a poster stays up for a long period of time, it becomes part of the background or landscape and is no longer read. Once the employees read the poster, its design should get the message across.

## Manufacturing Staff Meeting Environmental Control and Safety Briefing

This final component of the awareness activities phase may be the most important one because it is aimed at the senior levels of management. The briefing consists of a manufacturing engineer presenting a summary of the previous period's environmental control and safety activities and occurrences. The engineer also informs the staff of upcoming compliance requirements and what must be done by management to meet them.

The briefing draws attention to the safety and environmental issues facing the manufacturing function. It also reminds senior managers to carry out their environmental control and safety program duties. The manager of manufacturing uses these briefings as a mechanism for showing support for the entire program.

The briefing usually lasts 5 minutes and is followed by a question-and-answer session on the briefing or any related subject the managers may bring up. Frequently, discussions occur that require the manufacturing engineer to act as the voice of authority. All this aids the awareness program. If senior managers discuss safety and environmental concerns, so will their subordinates. The briefing and discussion may only last 15 minutes, but this brief time is valuable for raising the safety awareness factor.

# PLANNED ACTION ACTIVITIES PROGRAM

## Weekly Safety and Housekeeping Audits

This is often considered an awareness activity because the action of taking an audit creates a heightened awareness of proper safety and environmental procedures. However, it is discussed here under planned action activities because it creates action plans to correct discrepancies.

Audits are common activities in manufacturing. We audit for adherence to quality plans, for adherence to procedures, for the status of parts in the production cycle, for adequacy of efficiency measurements, and so forth. Thus it is natural to audit for compliance with safety and environmental control rules.

The audit for housekeeping and safety is a detailed examination of things in a specific area to uncover environmental and safety deficiencies. It is usually carried out by a team consisting of the manufacturing engineer, the area foreman, and a shop manager from another area, the latter being included to instill a sense of competition. Managers auditing each other's areas are more likely to uncover deficiencies than if they were auditing their own areas. They will also strive to find more deficiencies than their counterpart finds. To enhance the sense of competition, manufacturing engineering will also publish a best and worst report. Areas exhibiting the best housekeeping will be publicly proclaimed along with those exhibiting the worst performance.

The result of the safety and housekeeping audit is a planned action report showing the discrepancies found, who is responsible for the corrective action, and when the action will be completed. Figure 16.2 is a sample of such a report. The report is compiled by the manufacturing engineer assigned to the tour, who is also responsible for updating it as actions are completed or new items are added.

The results of the weekly safety and housekeeping planned action report and the weekly best and worst report are used to prepare the monthly safety report. This report is a detailed summary of the planned action activities of the month and a compilation of the safety statistics. The statistics are used to evaluate progress in achieving a safer working environment and to prepare the compliance reports required by government agencies and occasionally insurance underwriters companies. Virtually all planned action activities at some time or another create inputs to this report and its quarterly summary, which is known as the quarterly environmental control and safety advisory department statistics summary. In essence, these reports reflect the effectiveness of planned action activities.

The weekly safety and housekeeping audits are the only specifically scheduled items among the planned action activities. All other planned action activities are responses to a specific need, and after they are accomplished they will not be carried out again until another need arises. A need is usually revealed by the weekly safety and housekeeping audit, but is sometimes generated by an accident.

Item 29 in Fig. 16.2 is an example of the weekly safety and housekeeping audit creating a need to initiate another planned action activity. It notes that a job hazard analysis was not posted on vertical boring mill 2608. This implies that the

| Item no. | Item description | Responsible for completion | Date entered | Promised completion | Actual completion |
|---|---|---|---|---|---|
| 27 | Frayed fire hose, column A-12E | J. Jones | 1/26 | 1/27 | 1/27 |
| 28 | Pallets in aisle near lathe 2604 | F. Smith | 1/19 | 1/20 | 1/20 |
| 29 | Job hazard analysis not posted VBM 2608 | J. Jones | 1/26 | 1/29 | |
| 30 | D.C. circuit box door won't close, column A-17W | J. Jones | 1/19 | 1/20 | |

Total items added this report: 32
Total items remaining open since last report: 6
Accidents in area since last report: 0
Total accidents YTD 1

Figure 16.2 Weekly safety and housekeeping planned action report.

action is to post the analysis. But a competent engineer would review the current job hazard analysis before reposting it. If it no longer agrees with the methods and standards, it will have to be redone; hence the job hazard analysis will be integrated into a methods and standards planned action activity. Once this activity is completed it will be dormant as far as the environmental control and safety program is concerned, until it is reactivated by a need shown by the weekly safety and housekeeping audit.

## Job Hazard Analysis Integrated into Methods and Standards

The job hazard analysis is the safety check on the job design. It is a system for determining whether the job method conceived by MP&WM can be carried out safely and without any long-term environmental hazard to the operator. It is the final check before a job can be safely turned over to shop operations. The procedure for conducting the job hazard analysis is always the same. It starts with the methods sheet and the workplace layout, which the engineer will review for obvious safety hazards.

Let us look at the workplace layout (Chapter 7, Fig. 7.2). Note that it does not show any protective railing to separate the operator from the hazard, perhaps because it is not possible to install one and still have the machine perform correctly. This would be designated as a hazard. By identifying it as a hazard we are creating a heightened awareness of a possibly dangerous condition, but not necessarily an unsafe condition. (Similarly, a railroad station platform is hazardous because we could be hurt if we fall off the platform. Since we know the hazard exists, our awareness is heightened and an unsafe condition is mitigated.)

Now let us look at each methods (Chapter 7, Fig. 7.1) step and ask, "Is the operator being asked to perform an unsafe action?" Item 3 of the methods sheet reads: "Put in table T-slot nuts. Assemble studs to nuts and clamps." This is not a hazardous operation; only normal and reasonable care is required to do it. Item 2 calls for cleaning the table of the machine. This is probably not unsafe, but it introduces a new element to be checked by the engineer making the evaluation. Since we do not designate the solvent to be used in the cleaning operation, the engineer must investigate to see what is being used. If the solvent is known to be a hazardous chemical, the engineer will declare this step hazardous, and have added to the methods sheet the proper precautionary note on how the solvent is to be handled. This step of the method would also appear on the job hazard analysis sheet posted at the workstation.

Continuing our review of the methods sheet, we see that most steps are as simple as item 3 and appear to be nonhazardous. However, we must look a little further and ask ourselves what steps are related to activities that have caused accidents in the past, perhaps not at this workstation but at similar workstations. For example, accidents might occur at a point where an operator is addressing a tool to a rotating part. If the tool is not secured properly it could be twisted violently off the tool holder, resulting in damage to the workstation and possibly injury to the operator. Item 18 is a potential hazard by this definition. It instructs

| Item no. | Operation/location | Hazard | Preventative action |
|---|---|---|---|
| 1. | No railing between operator and workpiece. | Operator can come in contact with rotating part. | Observe caution while operating machine. |
| 2. | Operation 3, clean machine plate with solvent. | Rash or burns on unprotected hands can occur. | Wear gloves, long sleeve shirt, safety glasses. Observe instructions for solvent use. |
| 3. | Operation 18, engage borozon wheel to workpiece. | Fast engage can cause wheel damage, broken wheel, flying missiles. | Engage per instruction. Do not stand in front of wheel. Wear safety glasses. |

**Figure 16.3** Job hazard analysis.

the operator to engage the borozon wheel into the sealing ring. Note that it says to do it slowly, so that the methods engineer had in mind possible damage to the machine, workpiece, or even the operator if crash engaged. This step would appear on the job hazard analysis sheet.

Figure 16.3 shows part of a job hazard analysis sheet made up of the items discussed above. A job hazard analysis sheet should be posted at each workstation. Also, any change in a method calls for a review of the job hazard analysis sheet to see if it is still valid.

## OSHA/EPA Incident Evaluations

This planned action activity occurs only if there is an accident or an OSHA or EPA audit discrepancy. In either case the incident must be investigated for cause and for corrective action taken to remedy the immediate situation and to prevent future occurrences. These investigations are carried out by technically qualified personnel and are reviewed with the manager of manufacturing engineering. Before a report is submitted to the respective agency, the manager of manufacturing, general manager, and perhaps legal counsel will also review it to ensure that the company is not continuing an unsafe and illegal practice and that senior management is aware of the consequences.

## Toxic Substance Cataloging and Control

As mentioned earlier in this chapter, the emotional nature of the reaction to possible hazardous chemicals in the workplace must be controlled. Manufacturing engineering has the task of identifying all chemicals and substances used in operations and determining whether they are hazardous to health. This planned action activity is divided into two parts: identifying the chemicals and substances used, then determining whether they are harmful.

The identification step consists of reviewing all purchase invoices and determining what chemicals are contained in products purchased under trade names. This can be done either by laboratory analysis or by requiring vendors to specify the contents of their products. Asking the vendors for such information is preferable to laboratory analysis from a cost viewpoint. However, vendors may not want to disclose what is in their products, because they may be selling something that could be purchased more cheaply under the generic name, or because they feel they must protect proprietary information. This type of withholding of information is becoming rare due to hosts of public disclosure laws. In the case where disclosure is not forthcoming, the company must determine what chemicals are in the products by laboratory analysis.

Once the chemicals have been identified, manufacturing engineering must determine whether they are hazardous or toxic. Fortunately, this is not a subjective evaluation. The Environmental Protection Agency has published and continues to update a list of what are now legally designated hazardous and toxic substances. Therefore, manufacturing engineering consults the EPA list to see whether chemicals are toxic or hazardous.

The EPA list is a prime factor in dealing with chemical substances in the workplace. If a chemical is found to be on the EPA list, the prudent manager will get rid of it and find a safe substitute. If there is no suitable substitute, a procedure must be introduced to control the chemical in the factory.

The classic example of a hazardous material in recent years has been asbestos. Before the late 1960s, this heat-shielding, insulating material was not known to be a health hazard; therefore, its use was considerable. Once the hazard was identified, manufacturing engineering organizations began to eliminate use of asbestos where possible and restrict it where that was not feasible. An example of each would be (1) to eliminate asbestos blankets for heat treating and for welding preheat use and (2) to seal asbestos pipe insulation so that powder from decaying insulation would not become airborne. Asbestos is a familiar hazard. Unfortunately, there are thousands more that must be dealt with, and manufacturing engineering resources must be allocated for the task.

## Safety Glasses Usage Enforcement

The planned action activity is to ensure that all personnel requiring safety glasses (impact-resistant and/or side shield types) are provided with them. This is done by requiring all employees to receive safety glasses from the designated dispenser and to wear them on the job. This often involves contracting for an optician to provide prescription services.

## Accident and Environmental Incident Investigation

This is the same activity as the OSHA/EPA incident evaluation except that it covers non-OSHA/EPA areas of concern.

## Noise Control

This is a vast subject that is covered adequately in the large numbers of textbooks in the field of acoustics and noise engineering. We will concern ourselves here with the management aspects.

Noise levels have been designated by OSHA for various industries. The standard is set at 85 dBA for an 8-hour exposure, and some wish to lower that to 80 dBA. Manufacturing engineering must ensure that its factory is in compliance. To do this, the manager of manufacturing engineering must determine the level of noise in the factory and then take any necessary corrective action.

The preferred way to determine the ongoing or base-level noise is to conduct a noise contour mapping of the factory and the office areas. This is done by setting instruments in place to gather data, and it provides the basis for measuring future engineering solutions. To be acceptable to the government agencies, noise contour mapping must be done by competent personnel using recognized instruments. Therefore, management must either give its personnel proper formal training or rely on competent outside services. Either alternative is costly. The training approach is preferable because with trained in-house personnel it is possible to deal with noise problems as they occur. When we hire consultants we have a solution to a particular problem, but gain little knowledge to assist in solving future problems.

Once noise problems are identified they must be solved. The management approach has been to assign advanced manufacturing engineering to achieve the engineering solutions required by government regulations. This means that the operators must be isolated from the noise or the noise eliminated. Wearing protective devices can be a temporary solution, but management must continue to show progress in achieving an engineering solution, otherwise the company is liable for fines. I have found that by working with company research laboratories and universities it is possible to meet the intent of the law. The experience with the portable grinder tool described earlier in this chapter is an example of an engineering solution that successfully satisfied the requirement.

## Manufacturing Engineering Project Environmental Control and Safety Evaluations

This is essentially the same activity as the job hazard analysis except that it deals with capital equipment projects to ensure that they will be safe to use. To facilitate this, the manufacturing engineer responsible for safety and environmental control is often asked to review appropriation requests during the formulation stage to ensure that there are no safety or environmental problems.

## Pollution Control Monitoring

Once hazardous chemicals have been identified, monitoring procedures must be established to ensure that they remain under control. Also, firms that discharge chemicals into the atmosphere, ground, or waterways as a by-product of their manufacturing processes must monitor these chemical discharges. Virtually all discharges are included in government regulations and levels of discharge are set. Discharging at levels higher than those set by regulations can result in serious civil and/or criminal penalties.

Manufacturing engineering is responsible for installing the necessary pollution control equipment as part of the capital equipment purchasing program. Once the equipment has been installed, manufacturing engineering must monitor it for proper use, report to the appropriate agencies, and make corrections and repairs when the equipment is not functioning properly. This means that manufacturing engineering must maintain audit logs and incident logs. The audit log shows when inspections have taken place, and the incident log records out-of-tolerance conditions and shows what was done to correct the situation.

## Grievance Consultation

Grievance consultation is a service manufacturing engineering provides to employee relations for employee grievances with a technical content. A significant number of grievances have to do with safety of the workplace, hence the inclusion of this item in the environmental control and safety program. Manufacturing engineering does not, as a rule, sit in on grievance sessions. It is only required to provide technical facts to employee relations. However, manufacturing engineering is responsible for ensuring that employee relations personnel understand the facts as presented.

## SUMMARY

The environmental control and safety responsibilities within manufacturing engineering constitute a very broad and expanding activity that cannot be taken lightly and must be properly managed. Certainly, they are as important as the more traditional responsibilities of manufacturing engineering.

The key to managing this activity is to divide it into awareness and planned action segments. The awareness portion develops a heightened sense of need for proper safety actions, while the planned action activities are specific steps taken to make sure the workplace is as safe as modern technology and management practices can make it. The key planned action activity is the job hazard analysis, a combined AME and MP&WM activity focused on making the workstation as safe as possible.

Throughout this chapter the key emphasis has been on understanding that a safe workplace environment can be a positive factor in enhancing productivity.

## REVIEW QUESTIONS

1. Explain why it is good business sense for a factory to have a strong environmental control and safety program.

2. What is meant by direct and indirect causes with respect to environmental and safety programs?

3. Why are accidents called productivity deflators?

4. What is the basic theory on which effective factory safety programs are built?

5. Under what set of circumstances would management of the environmental control and safety program not be assigned to manufacturing engineering?

6. A manufacturing engineer is assigned responsibility for writing the specifications for acquiring a new CNC horizontal boring mill. Make a list of questions the engineer might ask to ensure that the specification complies with the following codes: fire, structures above the floor, hazardous and toxic substances, noise abatement, and safety apparel.

7. Describe how a safety awareness program can be measured for effectiveness.

8. Describe how a safety planned action activity program can be measured for effectiveness.

9. Why is a regularly scheduled safety meeting as important for engineers as it is for workstation operators?

10. What is the purpose of a job hazard analysis?

11. A new cutting fluid is to be implemented for all the lathes of a job shop. Explain how the manufacturing engineer would determine beyond a reasonable doubt that the cutting fluid is not toxic or hazardous.

12. What is a noise contour map and how is it used in a factory?

CHAPTER
# SEVENTEEN

## THE INTEGRATED PRODUCTIVITY IMPROVEMENT PROGRAM

The preceding chapters covered the many aspects of manufacturing engineering in a relatively isolated sense in order to bring out salient points. However, we all realize that the real world does not allow isolation of subsets of problems that are solved independently of the others. In reality, each problem subset imposes certain constraints on the solution of the others and the whole.

**Example:** A production-level problem of output below expectation may require another workstation for its solution, but this may be economically unfeasible because the funds had to be used to solve a materials problem. The materials problem may be caused by a design that is not producible, requiring more exotic and hence more expensive material. Thus the new facility cannot be purchased because funds are needed to buy material at a greater cost since a producibility engineering problem went unsolved.

This may be a roundabout way of coming to the point, but it shows that real problems are interdependent on many circumstances and events. This leads us to the subject matter for this chapter: the integrated productivity improvement program.

### NEED FOR AN INTERACTIVE SOLUTION

By integrated we mean that all the resources the company has will be brought to bear on the problem at hand. We are looking for a solution that transcends manufacturing engineering, and even the manufacturing function in total, requiring the resources of most if not all functions of the company. We wish to discover how to make quantum jumps in improving productivity and thus profitability. Manufacturing engineering skills will be important because they are involved in all technical aspects of the factory operations. But others will also be involved because productivity is a measure of total output divided by total costs to obtain a cost per product value. The productivity improvement problem is to reduce the cost per product value; the lower this becomes, the more profitable the company becomes. It is easy to see that much more than manufacturing costs are involved. Beyond the labor and materials costs we must consider the entire overhead cost structure of all segments of the company. Therefore, the impacts of design engineering, marketing, finance, and employee relations on cost must also be analyzed and reduced to solve the productivity problem.

This overall approach to improving productivity is the only rational one. It makes no sense to have manufacturing drive down factory floor operating expenses if design engineering is adding cost to the product. If we are fortunate, the net effect is close to zero. If both functions are working to reduce costs—manufacturing in operating costs, design engineering in costs of tolerances required and materials selected—then the reductions are additive and the results are significantly greater than zero and beneficial.

Traditionally, each function has looked after its realm of responsibility in relative isolation from the other functions. This method is never optimal, because we are not sure we are spending company resources where we obtain the best results. Each function is competing against the other functions for limited resources and the resources can unknowingly be misapplied.

**Example:** Assume we have a business that costs $60 million annually to run. In attempting to reduce costs, we should be directing our activities where they will have the greatest effect. The major categories of costs are

Direct labor: Cost of hands-on labor to make the products.
Overhead: Costs of all other labor and support activities such as engineers' salaries, heating the building, and so on.
Materials: Cost of supplies directly used to make the products.

The $60 million operating costs can be divided into these three categories. The relative percentages vary from company to company and from product to product. A heavy-apparatus job shop would typically divide the costs as follows:

| | |
|---|---|
| Direct labor | $10 million |
| Overhead | $20 million |
| Materials | $30 million |
| Total | $60 million |

If we have limited resources (i.e., people to effect positive changes in the company's profitability), the most attractive area for reduction would obviously be the materials area. A goal to reduce the cost of materials by 5% would result in a $1.5 million reduction; while the same goal in direct labor productivity would result in only a $500,000 improvement. Clearly, most of the resources should be aimed at reducing materials costs, but the productivity improvement programs set up by most companies give only a small probability of the real target being addressed.

All too often, each function is left to establish a cost reduction plan relatively independently. The result is that the overall picture of opportunity is never seen and we have tactical rather than strategic plans for improvement. For example, within manufacturing a great deal of effort may be expended to reduce direct labor costs by achieving greater workstation effectiveness rather than to reduce the cost of materials, because direct labor productivity is under the control of the manufacturing function. Manufacturing has shop operations to monitor and control work attention time and adherence to methods, manufacturing engineering to design efficient workstations and methods to reduce the time to do the required work, quality assurance to monitor manufacturing losses and feed back corrective action requests, and even materials to batch stock to make effective production runs. Therefore, manufacturing can make an all-out assault on direct labor costs, that is, productivity.

Let us look at the materials situation. Here manufacturing can only affect the purchase price negotiations and perhaps argue successfully for reduced tolerance rigidity so that less scrap is created. But the real saving is in substitutions for materials—using cheaper grades of steel, for example. These substitutions are the domain of design engineering, not manufacturing. Hence the big reduction is not sought as it should be because it lies in the sphere of responsibility of design engineering and not manufacturing. Why does design engineering not respond to this challenge? There are a few good reasons. First, if left in isolation, design engineering would consider productivity improvement to be the ability to produce designs faster and with fewer people. This affects the design engineering budget in a positive way. Second, design engineers tend to design Cadillacs when stripped-down Chevrolets would be adequate. This is referred to as protecting the design margins. Third, producibility engineering is a manufacturing activity; therefore, there is little reason to expect design engineering to search effectively for the cheapest material if using it requires a design compromise. The result of these productivity improvements in isolation is an extreme underestablishment of the materials cost improvement goal.

The solution is to have an integrated attack on the most lucrative area of the cost reduction potential. In the example above, this would mean that manufacturing and design engineering should pool their resources to cut materials costs. This would be done even at the expense of other projects, such as improving workstation performance in manufacturing and reducing drafting time per drawing in design engineering. In an integrated program the producibility and advanced manufacturing engineers would look for ways to make the products with cheaper materials, while the design engineers would evaluate the changes in design needed to allow such materials to be used. In addition, manufacturing engineering would work on methods with the primary objective of saving material and the secondary objective of saving worker time. Purchasing would focus its negotiated price activities on the lower-cost materials that design engineering is striving to use. In this integrated approach the different functions are working toward the same goal. When the functions worked independently and only looked at their portion of the business, this goal was secondary in importance. Thus when the functions are taken out of isolation and given the opportunity to see the area where the greatest overall gain can be made, we can have a quantum jump improvement in productivity.

This is the rationale for the integrated productivity improvement program. Now let us see how an integrated productivity improvement program is established, starting with an explanation of productivity.

## TYPES OF PRODUCTIVITY

The basic definition of productivity measurement is output divided by input. To give this useful meaning we divide our productivity measurements among three broad categories selected to show the contributions of the entire organization against established goals. These three categories are (1) productivity, direct labor; (2) productivity, all other; and (3) quality costs.

The first two are positive measurements, while quality costs are expressed in a negative sense. In the first two measurements the higher the absolute value of the ratio of output to input, the better the performance of the organization. The opposite is true for quality costs. The cost to correct quality problems would be a negative factor for profit, while the input is the cost of total work—both direct and all other required to produce both good and bad products.

The productivity measurement must be divided into three segments, otherwise it is too general to be useful. We cannot add direct labor to indirect labor and overhead and come up with a meaningful number. Even though we use dollars as the basic measurement, this does not help management. We would have no way to compare the specific factory with others because we would not know the ratio between productivity, direct labor and productivity, all others. Therefore, we separate them. We can have a direct comparison with all other factories in terms of direct labor productivity. We can also have a comparison of productivity for everything else by using the "all other" category. Because different companies may have different ways of doing the work to support the factory operator, the category productivity, all other is very broad. It measures the productivity of all the support functions and materials and does not try to segregate any component out. Segregating other components, even materials, would result in a very narrow measurement that would be difficult to compare with many other organizations.

The quality cost measurement is a mitigating factor. It is used to evaluate the purity of the other two productivity measurements. If the two productivity measurements (direct labor and all other) are within targeted goals but the quality costs are high, quality is being sacrificed to meet the output goals. This can have detrimental long-range effects on the business by giving it a reputation for poor quality. Hence, the quality cost measurement is an important modifier in interpreting the results of the other two measurements.

An important fact about productivity measurements is that they are not only for the manufacturing section. Only one of them—productivity, direct labor—is a measurement solely of the manufacturing function, and it applies to a portion of manufacturing, the shop operations subfunction. All other measures of productivity encompass the entire organization.

Productivity, all other measures manufacturing engineering in terms of the effectiveness of utilizing funds to develop tooling and workstations via the cost of equipment purchases and the salaries of the engineers involved. The measurement includes the cost of materials purchased by the materials subfunction, which is affected by the success of negotiations to obtain lower prices. Design engineering is included in the cost of salaries and the cost of design as related to productivity and selection of materials. Finance and the employee relations functions are also part of the measurement via the payment of salaries to their personnel and the effectiveness with which they work. Thus, productivity, all other, measures the efficiency of all the functions in all things except direct labor applied to the product.

Since the "all other" category is usually the largest cost of doing business, companies show great interest in improving this aspect of productivity. In fact, it should be of more concern than the category of productivity, direct labor. The value companies place on this measurement is usually shown by the control of people hired for various functions. This is the "head count" control placed on operating components by senior management. The theory is that controlling the number of people will limit salary costs and offer fewer opportunities for those on the payroll to spend money for unnecessary things. By controlling the head count, only the high-priority items will be done; the lower-priority items that contribute less to profitability will not be done. Unfortunately, the proper head count is not always easy to assess. Therefore, this approach to improve productivity, all other is not a simple one to use and can damage a business by allowing necessary activities to be eliminated because of a lack of people to do them.

Quality costs measure all functions because all functions can participate in the creation of defective products. Scrap or repair costs can be due to design engineering specifying defective material, or manufacturing engineering devising ineffective methods that allow operators to make wrong choices in producing a part, or an operator simply making a mistake.

Quality costs can also be attributed to the functions with less direct inputs into producing the products: marketing, finance, and employee relations. These would be classified as indirect causes of quality costs. They would consist of the types of actions that limit the margin of error manufacturing and design engineering should have in the course of their work.

Marketing is responsible for establishing the contracts to which the company works on specific projects. The content of these contracts determines the magnitude of exposure to quality costs.

> **Example:** If an agreement is entered into with a customer to warranty a product for 2 years instead of the normal 1 year, there is an additional risk to the company. The margin of error in producing the product is somewhat reduced because it must perform correctly for an additional year. The risk of the customer mistreating the equipment and causing a failure extends over a longer period. In order not to incur the quality cost, the manufacturer will have to prove that the customer did not follow the correct operating procedures, or allowed some other circumstance to cause the failure. Even if the manufacturer is not at fault, there will still be additional costs. The manufacturer will have to use design engineering and manufacturing engineering resources to defend the company. Therefore, there is a buildup of warranty expenses even if the company is not at fault. Warranty costs are typically considered together with scrap and repair costs. Hence, by agreeing to a longer warranty period, marketing can have a detrimental effect on the quality cost measure of productivity.

The impact of employee relations on quality costs may be even more distant, but it exists. An example concerns placement of operators on jobs. If employee relations allows easy displacements of people from one job to another during an increase or decrease of the work force, a detrimental effect on quality costs can result. Employee relations is responsible for negotiating work rules related to seniority, employee "bumping" rights, grouped vacation periods, and so forth. Although in theory employee relations only represents management in total during these negotiations with employee groups or unions, in practice they play a considerable role in setting policy, since they tell management what they believe to be obtainable in negotiations and what they believe management will have to concede to achieve or maintain harmony in the labor force. Senior management tends not to vary considerably from the recommendations of its experts in employee relations. Therefore, employee relations is a powerful indirect force in increasing or decreasing quality costs through its establishment of work rules and seniority rules.

Bumping means that a senior worker can displace a worker with fewer years of service on a job assignment when the work force is being reduced. Similarly, when the work force is being expanded, the most senior members of the hourly ranks have the first opportunity to bid on new jobs as they become available. In either case the displacement causes a chain of personnel moves, thus disrupting the workplace. The more people being moved, the more unfamiliar the work becomes to the work force. This leads to less than optimum performance, in both a direct labor sense and a quality sense. The quality sense is the more problematic since quality errors found outside the factory usually cost 10 times as much to fix as errors found during manufacture, whereas the loss of direct labor productivity never even approaches twice the cost under optimum conditions. The amount of exposure to excess costs due to bumping is up to employee relations in their establishment of the rules. If the provision for bumping is very generous—that is, many people are allowed to displace others because job descriptions are broad and require minimal qualifications—then the exposure to extra costs is great. On the other hand, if employee relations narrowly defines jobs and maximizes qualification requirements, bumping becomes a smaller problem. Narrowly defining jobs and strictly interpreting requirements means that only a qualified senior can displace a junior. This is what management strives for. In practice, it means that a senior employee can only move into a job he has held before and performed satisfactorily. If employee relations must deviate from this, the amount of deviation is the amount of increased quality cost risk to the business.

The finance function's contribution to quality costs is similar to that of employee relations. It is indirect but can have a sizable impact. There are two major ways in which finance can affect quality costs: in the purchase of equipment phase and in the collection of funds phase.

The purchase of equipment phase involves the longest range. Financial analysts exert an influence in evaluating plant appropriation requests. Manufacturing engineering must rely on finance to make the final decision on the validity of the financial data submitted in support of a capital equipment project. Unfortunately, most capital equipment project justifications require assumptions concerning the savings to be achieved. If finance takes a very strict approach to the justification, it may not be possible to acquire the desired equipment, and the company may be forced to make do with lesser equipment or no new equipment at all.

**Example:** A lesser vertical boring mill may have a more difficult time holding required machining tolerances because it is less rigid than the more costly preferred machine. Both machines may be capable of holding the required tolerance, but the cheaper one may require closer attention on the part of the operator, resulting in a higher probability of undesirable quality costs.

In this case, manufacturing engineering may have bought the lesser machine because it was the only one the finance function would endorse. There are many actors and many circumstances involved in capital equipment purchases. However, this is a valid example of the influence of finance causing a greater probability of incurring higher quality costs.

The second way in which finance may cause quality costs is more immediate and direct. One area of quality costs is due to defective material or subcontract work. Theoretically, no costs should be absorbed as a result of these categories, but in practice many such costs are unnoticed or uncollectable. The uncollectable costs are those on which finance has some positive or negative influence. It is the responsibility of the manufacturing function to initiate charges to vendors for defective work or material. This is done through the quality control report process, whereby defects are reported and costs incurred to repair or make the product over again are recorded. If finance has an aggressive policy of tabulating such costs and encouraging the purchasing unit to recover them, the company's exposure to such quality costs is usually minimized. However, if the finance function has poor records of cost because its data collection system is not detailed enough, the costs associated with vendor-caused defects are difficult to collect. Therefore finance has a responsibility for seeing that extra costs due to poor vendor quality are recoverable. Finance can thus have an effect on quality costs.

## PRODUCTIVITY MEASUREMENT EQUATIONS

Productivity measurement equations are algebraic notations that describe the existing state of productivity. They describe mathematically what would otherwise take a paragraph or more to communicate.

The productivity equations yield dimensionless numbers that show whether a profit is being made. But it takes a chronological series of these numbers to determine whether productivity is improving, staying the same, or diminishing. Let us look at the equations and review their components.

$$\text{Total Productivity Measurement (TPM)} = \frac{\text{output}}{\text{input}} \qquad (1)$$

This is the overall measurement used to determine whether an operation is profitable. It is also used to detect productivity trends. The output is usually sales dollars and other payments and the input is total cost of operations. A few examples illustrate how the calculated ratio can show the state of the business.

**Example (a):** Sales for company A for 1993 were $180.6 million while costs of operations were $182.4 million. These costs included direct labor, indirect labor, material, and all overhead costs. Therefore, the values input into Eq. (1) are:

$$\text{TPM} = \frac{\$180.6 \text{ million}}{\$182.4 \text{ million}} = 0.99$$

The result, 0.99, is a dimensionless number less than unity. By inspection we can see that company A lost money in 1993. If a loss is suffered the total productivity measurement value will always be less than unity. This is the first important observation concerning productivity formulas; a loss is always less than 1.00, breakeven always equals 1.00, and a profit is always greater than 1.00.

**Example (b):** In 1994 and 1995 company A had sales of $124.3 million and $117.1 million, with respective operating costs of $123.8 million and $114.4

million. We now have data to evaluate the trend of productivity improvement. Including 1993 data, we have

$$\text{1993:} \quad \text{TPM} = \frac{\$180.6 \text{ million}}{\$182.4 \text{ million}} = 0.99$$

$$\text{1994:} \quad \text{TPM} = \frac{\$124.3 \text{ million}}{\$123.8 \text{ million}} = 1.004$$

$$\text{1995:} \quad \text{TPM} = \frac{\$117.1 \text{ million}}{\$114.4 \text{ million}} = 1.024$$

These results show a positive or improving productivity trend. We have gone from a value less than unity, a loss; to a value slightly above breakeven; to a greater than unity value, a profit, in a 3-year period. We can also determine the percent productivity improvement from year to year. This is the second important observation concerning use of the TPM. The percent productivity improvement would be

$$\text{1994/1993:} \quad \frac{1.004}{0.99} = 1.014; \quad 1.4\%$$

$$\text{1995/1994:} \quad \frac{1.024}{1.004} = 1.0199; \quad 1.99\%$$

These calculations show that company A is not improving very rapidly. In fact, we would wonder whether any progress is being made at all. This leads us to example (c).

**Example (c):** Company A, not satisfied with progress shown since its loss year of 1993, decides to initiate a bold integrated productivity improvement program. Management has decided that total productivity must improve by 10% in 1996 over 1995. The question is how much they can afford to spend on all operations if the sales forecast is $122.0 million. The TPM equation becomes:

$$\text{TPM} = \frac{\$122.0 \text{ million}}{x} = 1.10$$

and $x$, the total operating costs, equals $110.9 million. This is the third important observation. The TPM permits quick evaluation of the cost reductions that must be achieved to reach desired goals. By neglecting to perform this simple calculation, many managers overestimate goals and consequently fail. Productivity improvements are an essential part of any realistic business plan. Let us look at the goal set by company A. In order to achieve the stated goal, costs must be reduced from $114.4 million to $110.9 million—a 3.15% reduction. This appears to be achievable; therefore, the plan is practical. If the reduction was over 10% we would have a stretched goal. If it was over 15% we would have to say that the goal was not practical.

$$\text{Productivity, direct labor} = \frac{\text{sales}}{\text{direct labor costs}} \qquad (2)$$

Here sales are identical to the output part of the TPM equation. Direct labor costs refer to the salaries and benefits paid to the factory work force responsible for actually making the product.

This subset of the TPM deals only with the major shop operations portion of the productivity program. The analysis of the numerical values calculated is the same as that for Eq. (1), but there is a much narrower interpretation base. For the productivity, direct labor measurement we are concerned with the practicality of lowering the costs. Of all factory measurements, this is the most analyzed. We calculate the hours required to produce the products and then measure the efficiency of producing the products against the planned time. Productivity, direct labor is capable of being very precisely measured and proposed improvement plans can be evaluated well before they are approved.

**Example:** A program is proposed to reduce overtime by improving the attendance rate of the direct labor work force. By knowing the percentage attendance improvement targeted for, we can calculate the number of additional hours made available to produce products, hence the number of overtime hours that can be reduced. This reduces direct labor costs because overtime labor hours usually cost 50% more than normal labor hours.

$$\text{Productivity, all other} = \frac{\text{sales}}{\text{indirect labor} + \text{material cost} + \text{overhead}} \qquad (3)$$

This measures everything but direct labor. Direct labor is very well documented and measured, while the other cost factors are not. The determination of exact times to perform direct labor is a well-established engineering discipline; therefore, the measurement of performance (e.g., productivity, direct labor) is more accurate and specific than the other productivity measurements. We do not want to combine all other labor because only the direct labor portion based on scientific time standards would be accurate, and we would be in jeopardy of skewing the measurement because of the lack of preciseness of the indirect labor portion.

Because direct labor is much more precisely measurable than all other cost factors, we choose to measure it separately. However, because the other cost factors are less precisely controlled, it makes little sense to measure the improvement in performance of these areas separately. Therefore, we choose to combine all the generally measured cost factors into one measurement.

There is another good reason for lumping together all cost factors except direct labor—the advantage of averaging. If we can combine enough general measurement categories, we should, according to statistical theory, be able to say that the positive and negative biases of the individual measurements balance each other. The resulting productivity measurement should then fairly represent the actual situation. This is the case with the productivity, all other measurement, where we combine the costs of indirect labor, materials, and all other overhead factors. If we inadvertently classified as indirect labor a cost that should be charged as a service fee associated with the cost of materials, it would not matter because the value of the denominator would still be the same. This is also the case for the overhead charges. It does not matter whether we are talking about fixed overhead such as the cost of heating the plant with steam, or variable overhead such as the cost of using steam as an energy source for a process. For the purpose of productivity measurement, we do not need to know exactly the account into which the steam costs are placed.

By grouping all other expenses in one productivity measurement category, we can measure the productivity of all the support functions. We can determine whether we have the level of support required for the particular manufacturing entity. The percentage improvements of the factors productivity, direct labor and productivity, all other should be approximately the same. If they are not, there is an imbalance that must be corrected. The imbalance analysis is not precise, but we can use the following guidelines.

1. If productivity, direct labor is growing faster than productivity, all other, and the rate of productivity, direct labor is greater than 5%, then the support functions are probably too ponderous. There is probably excess staff and expenses and lack of standardization. This is indicated by the fact that the direct labor operation is doing well but the staff functions are slowing down. Direct labor performance is outstripping its support, probably because the support is incapable of reacting faster, and reaction time is usually a function of the number of people that must be stopped and pointed in another direction.

2. If productivity, direct labor is growing more slowly than productivity, all other, and the rate of productivity, direct labor is greater than 5%, then the support functions are probably too lean. The support staff appears to be reacting to change very well but does not have the ability to implement change in the way the direct labor work force is instructed. An erroneous decision often made in this situation is to increase the number of foremen. Usually this will not work because the new foremen will be in the same situation as the old foremen, who need new manufacturing instructions. When this is the case, it is essential to have a sufficient manufacturing engineering work force to fully translate the productivity improvements into the direct labor factor. An example would be the development of a new weld technique that reduces the cost of materials and has the potential to shorten the time cycle, but is not translated into lower labor costs because there is insufficient support staff to do so.

3. If productivity, direct labor and productivity, all other show less than 5% improvement, an analysis is of little value. An improvement of 5% or less is in the scatter band of accuracy of the measurements. Rather than evaluating whether

the support functions are too large or too small, it is important to ensure that all programs are investigated to achieve a greater than 5% productivity improvement factor.

The techniques of using Eq. (3) are the same as those for Eq. (1).

$$\text{Quality costs} = \frac{\text{scrap} + \text{repair costs} + \text{warranty costs}}{\text{total costs}} \quad (4)$$

This measurement, as discussed previously, is the modifier of all the other measurements. If this measurement is numerically high, we have significant problems. We have processes that are out of control because we are experiencing the high cost of fixing poorly made items, or making them again, or, worse yet, fixing the products in the customer's factory and possibly causing the customer loss of income.

The goal, of course, is to have the ratio in Eq. (4) equal to zero. A good quality cost measurement depends on the type of business we are engaged in. In general, an excellent rating would be in the range of less than 0.5%. Fair to good would be in the range of 0.5 to 2.0%. Anything above 2.0% would be considered poor. I must emphasize that these are guidelines and not rules. Each manager must determine what the ratio value should be for a specific operation.

Scrap and repair costs are the two components of the cost factor called manufacturing losses. Scrap refers to the disposal of parts that are defective and either cannot be repaired or are uneconomical to repair. Repair costs are the extra costs incurred to fix defective parts.

It is customary to set an upper-bound budget for manufacturing losses. This is the largest amount of money the company has decided it can spend to fix and remake products. It is determined by past experience and a judgment of what quality improvement programs can do to reduce the incidence of scrap and rework. Therefore, a factor in improving the quality cost ratio is the effectiveness of the manufacturing loss reduction program.

Warranty costs are those associated with fixing deficiencies or replacing whole units after the product's ownership is transferred to the customer. They are the most onerous costs a producer must contend with. Here we are admitting that we were incapable of fixing all problems before the product left the factory.

I have always considered warranty costs to be a measure of the effectiveness of manufacturing engineering. A case can be made that if manufacturing engineering was doing its job correctly, products of poor quality would never have been shipped. Manufacturing engineering is responsible for designing the factory system, training operators, and creating workable designs from proposed designs via producibility engineering. Therefore, having contributions in virtually all phases of the manufacturing process, manufacturing engineering must take a large part of the responsibility for warranty costs.

Manufacturing engineering cannot force operators to work conscientiously, but through process control and MP&WM it has adequate information on the state of operator performance. Thus it should not be surprised if less than adequate quality is being produced. Hence manufacturing engineering must in a sense be the conscience of the organization, seeing to it that shop operations management is cognizant of the current quality status, and taking corrective action as required.

The other major contributing factor in warranty costs is design. No remedial actions within the manufacturing function can correct for deficient design. Therefore, when warranty costs are incurred it is critical to determine whether the cause is design or manufacturing. If we erroneously assume that the cause is a manufacturing inadequacy, many corrective action programs will be initiated that will not fix the problem. This leads to considerable problems beyond money and time wasted. It can create distrust between organizations, which can ultimately result in management paralysis and a business out of control.

The utilization of the quality cost equation is similar to that of the other productivity measurements. We can either work from the desired ratio number and back to the numerator—that is, knowing the total budgeted costs, determine repair cost, scrap cost, and warranty costs. Or we can set the repair, scrap, and warranty budgets and then work out the ratio. In the former case we are taking what is considered an acceptable goal and working out how much repair, scrap, and warranty cost can be tolerated. In the latter case we are comparing current and recent results and determining what can be achieved in the next reporting period. Of course, the latter result may be too conservative and the former too optimistic. Therefore, setting the parameters must involve a subjective blend of both.

## IMPLEMENTING THE INTEGRATED PRODUCTIVITY IMPROVEMENT PROGRAM

We will now look at aspects of putting together an integrated program that will produce productivity savings. The process of operating an integrated program is another way of creating an operating plan. Once the plan is put together and approved, it is necessary to implement it and check progress. The basic techniques for creating plans, implementing them, and checking progress have been discussed throughout this text. Our purpose here is to show how these efforts are coordinated and made into an effective integrated productivity improvement program.

### Objectives and Goals Input

Chapter 2 is dedicated to this very important management concept. The objective and goals end product is a set of action plans that allow achievement of projects, and hence of their supportive goals.

For an integrated cross-functional plan, we must revisit the objectives and goals exercise to make sure that manufacturing engineering is not working at cross purposes to other functions and subfunctions. We must also be sure that the manufacturing engineering objectives and goals are not so narrowly defined that they miss the intent of an integrated program. It would be a serious error for senior management to allow objectives and goals statements to stand as presented until a review of overall problems is conducted. The objectives and goals statements for each function and subfunction must be reviewed against the perceived needs of the entire organization to ensure that vital concerns of the company are not falling into cracks between the boundaries of responsibility of the respective functions and subfunctions. By doing this we can take the necessary steps to modify various objectives and goals statements so that overall needs take priority over narrower needs of the issuing agency.

> **Example:** Manufacturing engineering objectives and goals may include an additional drill press, while the overall needs of the business may dictate that improvements in the appearance of products, even though only cosmetic, is more important in achieving the desired sales volume. The objectives and goals statement of manufacturing engineering may have to be changed to emphasize projects dealing with product appearance rather than a product production capital equipment program.

Here we have an example of a business need taking precedence over a purely technical need. The outcome should be improving the business results and ultimately allowing more funds to be available for technical needs.

## The Productivity Seminar

A productivity seminar is an optional step in forming an operating plan, but it is an excellent tool for determining whether the objectives and goals statements have found all the action plans to be pursued. It is also an effective means of creating a cross-functional dialog, and a method by which senior management can review the adequacy of all the objectives and goals statements. Productivity seminars are a technique for obtaining a wide variety of ideas.

A seminar is a call to action to solve a specific set of problems or to teach a particular set of skills. The productivity seminar is an example of the former. It is dedicated to identifying a set of integrated productivity enhancement activities to allow the company to reach a stated goal.

Five broad steps are involved in a successful productivity seminar. These are: (1) announcement of the productivity seminar, (2) selection of attendees, (3) the productivity seminar procedure, (4) incorporation of results into the operating plan, and (5) reporting back of results and plans to attendees.

**Announcement of the productivity seminar.** This step informs the organization that a productivity seminar is to be held. It states the purpose of the seminar, the goal to be attained, and how the results will be implemented. Usually we point out the need for having such an activity and state why it is important. For example,

> We have come a long way in 1993 in terms of improving productivity and sales performance of our business. However, we are still not at the point of generating acceptable profit levels

and there are new challenges and opportunities facing us in 1994. Therefore, we are scheduling a productivity seminar to take place on December 1, 1993 in order to develop ideas for meeting these challenges.

The statement of purpose tells everyone that a greater effort will be required to meet the generic challenge. Now it is necessary to state what the challenge is. We do this by introducing the goal of the productivity seminar. The goal is defined in the same way as in the objectives and goals systems. It must be a measurable statement of intent bounded by a time period. A productivity seminar goal might be formulated as follows:

Our segment of the company has been issued a mandate to achieve a $4 million net profit in 1994. To do this we must reduce operating expenses by $1.5 million and all support function costs by $500,000 for this year's operation. We must therefore develop integrated ideas to save $2 million.

Once the purpose and goal have been published, it is time to tell the attendees what will happen to their ideas. This can be done in the following manner:

In this meeting we will identify specific opportunities available to help us meet our commitments. The ideas generated by consensus in the meeting will be included in our 1994 operating plan.

The targeted audience has been informed about the purpose, goal, and use of the results. We have set the stage for a successful productivity seminar.

**Selection of attendees.** The announcement of attendees is usually made simultaneously with the seminar announcement. The procedure is to make the announcement in a letter addressed to those who should attend.

The selection of attendees depends on the specific goals of the productivity seminar. For an integrated productivity improvement program, particularly one intended to solve cross-functional productivity problems, the attendees must represent all phases of the operation. Therefore the first rule is: (1) Select a cross-sectional population representing all constituencies.

We want people who can generate ideas for actions to meet the goal. These will be individuals who are primarily knowledgeable in their areas and also have a good understanding of how the business is conducted and how its products are used. Since the latter type of people are not always available, we must be content with the former. The second rule is: (2) Select people who are knowledgeable in their areas of responsibility.

Another criterion is that of group interaction. We want people who express themselves well and who are creative. Therefore the third rule is: (3) Select the natural leaders—those who get things done and whom others will follow.

Finally, we want the key managers to participate because they are the persons with responsibility for conducting the operating plan. Therefore, the fourth rule is: (4) Select the key managers as participants.

By following these four selection rules we ensure that the participation is cross-functional and will include the people who are the key contributors to the ongoing success of the business.

**The productivity seminar procedure.** The procedure is a controlled brainstorming activity. It is made up of four parts: (1) identification of all relevant ideas, (2) evaluation of the ideas, (3) recording of the results of part 2, and (4) ranking the ideas in order of priority. For a productivity seminar we can divide the activities into three chronological sessions, with the first session covering part 1, the second covering parts 2 and 3, and the third covering part 4.

An example of this technique is as follows:

1. Divide the participants into approximately 10-person teams. Give each team the same task. Identify ideas for:
   a. Productivity improvements, direct labor.
   b. Quality improvements.
   c. Productivity improvements, all other.

   Specify:
   a. Expected savings for the current period.
   b. Major measurement milestones.
   c. Time and cost to implement the idea.
   d. Function or subfunction to lead the project.
   e. Category the project falls into (productivity improvement, direct labor; quality improvement; or productivity improvement, all other). Only one category per idea is allowed.

   I have found that six teams is the upper limit of practicality. For each team, select a chairman and a recorder.

2. For period two, divide and shuffle again. This time have only three teams. Assign each team to analyze only one set of ideas:
   a. Team 1: productivity improvement, direct labor.
   b. Team 2: quality improvements.
   c. Team 3: productivity improvement, all other.

   For its assigned area, each team will have the following tasks.
   a. Rank all ideas by importance.
   b. Evaluate feasibility of achievement.
   c. Establish priorities for recommended project ideas.
   d. List ideas deferred with a reason for the deferment.

   As with the first session, select a chairman and recorder, but do not select people who have already served in either of these capacities. This is done to maximize leadership participation in the activity and have the participants develop a sense of personal responsibility for reaching a successful conclusion.

3. Period three is the reporting and final results portion. Here the period two chairpersons report the results of their team's efforts to all the seminar attendees. This allows additional inputs to be made and a consensus to be reached on the results for the three categories.

Such a seminar may take 1 or 2 days, depending on how much time is available. It is important that no distractions interfere with this process. The participants must focus on the task at hand and not be distracted by other responsibilities. Therefore, if a 2-day program is planned, it should take place on 2 consecutive days.

There should also be a sense of urgency—that is, the participants should have a time deadline. Remember that we are looking for ideas, not fully detailed plans. Therefore, we put a time limit on the process in order to minimize the tendency to drift into detailed project planning. Developing of the plans to implement the ideas will take place later.

Finally, someone must be in charge of the productivity seminar and must make sure that the program flows smoothly. This seminar leader must ensure that each team at each session is actually doing the assigned tasks. This is done by briefly visiting each team and observing its performance. The leader participates if asked to do so, but not as a leader. The seminar leader also makes sure that each team is aware of its deadline, encourages teams where necessary, and deftly steers teams away from irrelevant tangents. He must continually challenge the participants to produce and at the same time maintain the schedule. The pressure of the schedule ensures that ideas, and not position papers, are generated.

**Incorporation of results into the operating plan.** The ideas generated by the productivity seminar are of no value unless they are acted upon. The mechanism for doing this is the operating plan, which will be discussed later in this chapter. The operating plan is the summary of all ideas and plans that management is supporting for action in the current time period.

The good ideas should be included in the operating plan. But first the productivity seminar ideas must be reviewed to make sure there are no duplications of planned activities already in the objectives and goals and in the present form of the operating plan. It is likely that duplication will exist. The participants in the seminar were probably the same people who had made inputs in the objectives and goals exercise. This would be especially true for the manufacturing engineers and other manufacturing subfunction representatives, because the objectives and goals discipline is very strong within the manufacturing function.

When duplication is found, we must first compare costs and results between the productivity seminar idea and the goal from the objectives and goals statement. This is an opportunity to review the older input (the goal) against possibly fresher or more detailed inputs. If nothing else, the productivity seminar provides an opportunity to clarify the operating plan by updating or corroborating the data. This alone would be a useful result of the seminar.

Once all duplications are culled out we begin the validation and planning cycle for the remaining ideas. First, we must review the ideas to see whether they are still valid. Then we must determine whether the benefits contemplated are accurate. This is done by planning the activity in the same way as the planning described for goals and projects in Chapter 2. We then have a plan and an ob-jectively determined savings. Finally, we must evaluate the productivity seminar idea for cost to implement. If the cost is beyond what the company can afford, the idea must be dropped, possibly to be implemented in future years.

After the productivity seminar idea has passed all these hurdles, it is intro-duced into the operating plan. This means that the idea has been translated into the objectives and goals system for tracking and implementation.

**Reporting back of results and plans to attendees.** This is done to keep up interest in productivity improvement and to show the people who attended that their ideas are indeed shaping the direction the company is taking. It can be done in several different ways. A good approach is to have each subfunction manager report via his operating plan and point out the plan segments that trace their in-ception or at least their verification or updating to the productivity seminar.

Using the operating plans to present results also gives management the op-portunity to show the key contributors exactly what the organization is attempting to do and how it will be done. It is important for management to communicate with their people and for everyone to know what they and their company are trying to achieve. With this approach we can show an integrated plan whereby many functions and subfunctions are working on the same problem.

## The Operating Plan

An operating plan is a summarized statement of the total objectives and goals statements and other independently obtained ideas. It shows in matrix form all the approved objectives and goals and ideas for the current time period. It rep-resents a further sifting of plans at a higher management level which ensures that the activities finally selected by each function and its subfunctions are truly in-tegrated to meet the needs of the particular business. By putting the objectives and goals in the matrix, management has a clearer picture of what planned ac-tivities will have the greatest beneficial impact on the business. This overview gives management the ability to set priorities and to add to, or subtract from, activities in order to continually optimize the chances of achieving the overall productivity goal. The senior managers can advance, hold, or delay implemen-tation of activities as the current year evolves. The headings to be found on the operating plan matrix are: goal, measurement, resources, costs, and risks and contingency plans.

**Goal.** This is the goal derived either from the objectives and goals statement or from another source of ideas such as a productivity seminar. In the objectives and goals format we had a statement and a measurement in one or two sentences. In the operating plan we combine several statements, from several different objectives and goals inputs, but do not continue with the measurement portion, which is found under the next heading.

**Measurement.** This is the second half of the goals statement, as indicated above. It represents a summarized condensation of all objectives and goals statements and other ideas that can be combined. The measurements, as explained in Chapter 2, must be bounded by time and be quantifiable.

**Resources.** Here we have a summary of the manpower and other resources that will be brought to bear on the particular program to achieve the goal. It is interesting to note that by adding the man-years required to achieve all the goals on the operating plan we can validate its chances for success. If we require more man-years than are available, the operating plan goals will probably not be totally achieved.

**Costs.** This is a list of the funds necessary to achieve the goals. Here again, we can use this column to evaluate the probability of achieving all the goals. If the sum of money required to achieve all the goals is more than the sum available, we come to the conclusion that not all the goals will be achieved.

**Risks and contingency plans.** In this last column the manager lists the important risks in achieving the goals and then determines what contingency plan should be followed if the risks cannot be overcome. This is a vital portion of the operating plan. It is the pragmatic input that brings overly optimistic plans back to reality and practicality. It is important to give personnel the opportunity to strive for the goals, but it is equally important to have a backup plan that can be quickly implemented if the goals are missed.

In reviewing a section of the operating plan matrix, shown in Fig. 17.1, one very important feature should stand out. All of the goals are "doable." There are no plans to do things that would revolutionize the way the business is conducted. Also note that we are involving cross-functional responsibilities within the matrix. We have employee relations responsible for certain aspects of the plan, and manufacturing engineering and shop operations are also involved. This is truly an integrated approach. We wish to achieve a productivity improvement of 10% by having three different organizations carry out segments of a plan. The segments add up to a total, and only the total plan can achieve the goal.

## SUMMARY

The operating plan matrix shows the need for integrated planning. Without an integrated approach none of the functions or subfunctions would be interested in pursuing such an activity. Employee relations would not on its own aim for a 10% productivity improvement of direct labor. Manufacturing engineering would not try to improve planning to allow shop operations to schedule at short intervals

| Goal | Measurement | Resources | Costs | Risks and contingency plans |
|---|---|---|---|---|
| 8. Improve direct labor 10% by reducing workstation variance 8%<br><br>(a) reducing workstation variance<br><br>(b) Reducing overtime to 7.5% of work-in-process labor | (a) Achieve reduction in measured workstation variance from 125% '93 average to 115% '94 average. Save $250K in DL cost<br><br>(b) Achieve reduction in WIP overtime from 10% '93 average to 7.5% '94 average. Save $150K in DL cost. | (a) (b) Implement multicode for operators; employee relations spec. 1/2 man-year<br><br>Improve planning to allow short interval scheduling Mfg. Eng.—2 man-years<br><br>Implement attendance and absenteeism follow-up programs; foremen ongoing | Improved planning will require additional disk storage for the computer—$65K<br><br>Short interval scheduling will cost $100K to develop software | *Risks:* funds not available to add disk storage or software development<br><br>*Contingency:* Investigate and implement group incentive pay plan to improve productivity; probably will not achieve goal for one additional year |

**Figure 17.1** Section of operating plan matrix.

in order to gain greater control unless shop operations was committed to a co-operative effort. Similarly, shop operations would not undertake such an effort without the support of manufacturing engineering and employee relations.

If we did not integrate the approach through a tool such as the operating plan, we would end up with each function, and indeed the subfunctions within it, working only on items over which it has complete control. Such a confederation of semiautonomous functions cannot operate as a business team. To achieve their functioning as a business team is the goal of the integrated productivity improvement program. It is done through the use of the operating plan, a thorough understanding of the meaning of the productivity measurements, and continual education of all the key contributors in the organization for the overall needs and goals of the company.

With an integrated program to improve productivity we can use the productivity measurements to set and measure progress toward achieving goals. We can be assured that programs requiring cooperative efforts of the various functions will be coordinated, minimizing the likelihood that the organization will be splintered and the functions will strive toward unrelated goals. Manufacturing engineering plays a very important role in the integrated approach because it is the technical arm of manufacturing and the liaison with marketing, finance, and design engineering. For this reason managers within the manufacturing engineering subfunction must be well versed in the techniques of operating plan implementation and productivity measurement.

## REVIEW QUESTIONS

1. Discuss the reasons why the productivity improvement program cannot be solved by manufacturing engineering independently but requires an integrated approach.

2. Describe the three basic categories of operating costs and explain how they relate to a manufacturing company.

3. Explain why a functional approach as compared to an integrated approach to productivity improvement often fails to achieve the desired results.

4. Discuss the reasons for dividing the productivity measurement into productivity, direct labor; productivity, all other; and quality costs.

5. Why is quality costs considered a productivity measurement?

6. What is the rationale behind head count control? How does it affect productivity improvement capabilities?

7. How does bumping affect productivity improvement capabilities?

8. How should manufacturing engineering mitigate possible adverse affects of the finance function on productivity improvements.

9. Construct a productivity trend chart for a company that has recorded the following data. Calculate the year-to-year changes in productivity levels. Dollars are shown in millions.

| Year | Sales | Cost |
|------|-------|------|
| 1985 | 51.6 | 51.9 |
| 1986 | 50.3 | 52.8 |
| 1987 | 57.8 | 53.2 |
| 1988 | 80.6 | 73.7 |
| 1989 | 76.6 | 71.4 |
| 1990 | 62.1 | 63.0 |
| 1991 | 64.2 | 63.8 |
| 1992 | 42.8 | 40.1 |

10. Why is productivity, direct labor segregated out while all other productivity measurements are essentially group measurements?

11. In a manufacturing company productivity, direct labor is growing at a 7.3% annual rate, while productivity, all other is growing at a 4.0% rate. What does this imply about the support staff of the company?

12. Why are warranty costs a direct measure of the effectiveness of manufacturing engineering?

13. What is the purpose of a productivity seminar?

14. Considering an operating plan as only a productivity improvement tool is incorrect. Explain why.

15. If the objectives and goals statements clearly show the company's wishes for the budget year, why is an operating plan required?

16. Explain why it is necessary to include risks and contingency plans in an operating plan.

# GLOSSARY

Accountability, theory of—the basis for operator certification for quality assurance or quality control. People will make sure a job is done correctly if they must identify themselves as the performer of the work.

Action plans—with respect to work planning, very specific, detailed items to be accomplished by a specific individual by a certain date.

AME—advanced manufacturing engineering, a unit within the manufacturing engineering function.

Archival data collection document—a history document for equipment.

Area planning—an activity within advanced manufacturing engineering.

Artificial intelligence—a set of computer software that can add data to its data base, and give the appearance of learning.

Batch—a grouping of parts or orders together for group manufacturing.

Bid—submittal of costs and performance characteristics by vendors offering to sell equipment.

Bill of materials—BOM, a listing of materials in sequence order of manufacture of a product.

Bottleneck—a choke point in the manufacturing process as a result of line imbalances.

Budget classification—in reference to capital equipment programs, the categories established for identifying project types.

CAD/CAM—computer-aided design/computer-aided manufacturing, a generic term meaning extensive coordinated use of computers in industry.

CAD/CAM triad—the three basic aspects of CIM: machine/process control, design and planning control, and production and measurements control.

CAM—computer-aided manufacturing, use of computer technology to assist in the management of manufacturing activities.

Capability—the physical limitations of what can be produced in a factory, including size and weight of parts, processes available, and materials that can be worked on.

Capability matrix—a format for tabulating the ranges and permutations of a factory's ability to produce an array of products.

Capacity—the number of items that the factory has the capability to produce in a given period of time.

Capacity requirements planning—the long-range scheduling program of MRP II.

Capital equipment—facilities, machines, etc., that can be depreciated for income tax purposes.

CAPP—computer-aided process planning, the use of computers for creating, storing, cataloging, and retrieving plans.

Cartesian coordinate system—the three-coordinate $(x, y, z)$ system for locating a point in space, used to define positions in space for CNC machines.

Cellular layout—a factory layout based on group technology principles, in which a variety of types of equipment are located adjacent to each other to manufacture a family of parts.

CIM—computer-integrated manufacturing, a generic term meaning achieving communications excellence via computer synergism throughout the business entity.

Classification and coding—a technique for cataloging parts in accordance with geometric, manufacturing, and material specifications.

CNC—computer numerical control, an advanced type of numerical control.

Common data base—electronically stored product data useful to and accessible by multiple functions.

Communications excellence—the ultimate purpose of CIM, the key to enhanced productivity.

Concept design—the rationalization of a design in terms of science.

Concurrent engineering—the process of team approach to do design and manufacturing engineering activities in parallel instead of series format.

Control chart—a format for recording results of statistical sampling measurements.

Cookbook quality plan method—a method of writing quality plans in which steps are outlined in chronological sequence.

CPM—critical path method, a technique for scheduling and monitoring the accomplishment of a complex project.

Daily accounts controls—a technique for controlling specific expense categories within established budgets.

Daily operating status report—a machine tool and equipment status document pertaining to availability for use.

Data collection—the pulling together of information in a systematic way, usually referring to a computer interface.

Day work system—a work measurement system based on pay for hours present at the workplace.

Defect analysis report—a technique for analyzing quality control reports for trends.

Design for manufacturability—see Concurrent engineering.

Design review—a review of designs by producibility engineering to check for compliance to producible design criteria.

Direct labor—effort applied at workstations to transform raw materials into finished products.

Down equipment—facilities that are in a state of disrepair that precludes their use for production.

Dynamic scheduling—master scheduling capable of being altered almost at will via CIM techniques.

Empirical decision-making—the simplest and most restrictive technique for finding families of parts.

Environmental control—a body of knowledge dedicated to identifying and controlling hazardous and toxic substances so that the probability of harm is minimized.

Error detection program—a subprogram of computer data collection systems used to verify the correctness of entries.

Estimated time standard—a time standard based on estimated cycle times.

Expert system—a type of Artificial Intelligence computer program.

Facilities engineering—a subunit of maintenance engineering dealing with machine tools and process equipment.

Facilities list—a list of equipment used for the facilities program.

Facilities program—a capital equipment purchase and upgrade plan linked to the objectives and goals.

Family of parts—parts that are similar in geometry or in the manufacturing procedures used to make them; determined by group technology classification coding, parts flow analysis, or empirical decision methods.

Family-of-parts programming—a generic numerical control program for categories of parts where there are limited geometric variables.

Farm out—have production work done in factories other than the company's factory.

Fire codes—safety codes pertaining to flammable materials and fire prevention.

Fire fighting—a technique for solving short-term quality problems.

Firm savings—calculable savings associated with plant appropriation requests.

Flexible automated factory—a factory in which CNC machines and robots directed by central and process computers are used to produce manufactured parts.

Flexible machining center—a complex CNC machine capable of performing many different types of machining operations on varied workpieces and positions.

Gantt chart—a bar chart used to monitor the progress of a project.

Goal—measurable statement of specific intent bounded by a specific period of time.

Group technology—a philosophy of manufacturing that exploits the principle of sameness, in which parts are grouped according to similarities in geometry and/or manufacturing process in order to approach flow-type manufacturing in job shop environments.

Group technology classification and coding—see Classification and coding.

Hazardous substances—chemicals that can cause biological damage after long-term exposure.

HBM—horizontal boring mill, a classification of machine tools.

Ideal capacity—the number of products a factory is able to produce based on theoretical feeds and speeds that can be achieved with the equipment.

Incentive system—a work measurement system based on pay per number of parts or operations completed per time period.

Indirect labor—work done to support the direct labor activities but does not add value to the material being worked on.

Inference engine—a part of an expert systems data base.

Interactive graphics—a computer-controlled device used to create engineering drawings on a video display tube and then perform engineering calculations and evaluations. Sometimes known as an "engineering workstation."

Interference time—nonproductive time during which an operator is available to work but cannot work due to outside circumstances.

Job deletion routine—a subprogram of the computer data collection system that reports completion of work to the master schedule.

Job effectiveness report—the basic efficiency measurement system for maintenance work performance.

Job hazard analysis—an evaluation of methods for possible safety hazards.

Job log—a format for entering and controlling maintenance work orders.

Job rate evaluation—method of comparing the work of one job skill with another in order to set pay scales.

Job shop—batch, intermittent manufacturing.

Job shop layout—a pre-group technology layout in which like machines and equipment are located together.

Just in case—an inventory strategy stressing emergency buffer stock at strategic locations.

Just in time—JIT, a popularized version of good engineering practice resulting in elimination of waste in the manufacturing process.

Knowledge base—the storage of data in an expert system data base.

Layout—an engineering procedure for placing equipment on the factory floor to optimize product flow.

Level loading—a technique for balancing production throughput over time.

Limping equipment—facilities that cannot be used to their full potential because of partial malfunctions.

Loss—a negative value of selling price minus the cost to produce.

LRF—long-range forecast, a document spelling out the goals of a business over a period of time, usually 3 to 5 years.

Machine tool capability and replacement log—chronological capability evaluations of equipment used for shop scheduling and replacement.

Maintenance engineering—a unit within the manufacturing engineering function.

Manufacturing facilities design—the specific design for jigs, fixtures, and processes necessary to implement the producibility design.

Manufacturing losses analysis—a systematic way to determine the causes of processing and manufacturing errors.

Manuscript—a sequential listing of steps to be followed by a CNC machine; part of the programming process.

Master schedule—overall sequenced schedule of multiple orders through a job shop.

MCU—machine control unit, the microprocessor or minicomputer that controls a CNC machine.

Mean—a measurement of central tendency, an average of many data points.

Methods—specific procedures for performing work on a workstation or series of workstations.

Methods analysis—a technique for developing a method.

Methods sheet—format for specifying a method.

Monthly quality report—a typical format for documenting product quality status.

MP&WM—methods, planning, and work measurements; a unit within the manufacturing engineering function.

MRP—materials requirements planning, the coordinative computer program for dynamic scheduling of purchasing and inventory control activities.

MRP II—manufacturing resources planning, the addition of labor and facilities scheduling to MRP to form the coordinative CIM computer program for dynamic production scheduling.

N/C—numerical control, a generic term for computer control of machine tools.*

N/C parts programming—the concept of writing a manuscript for input into a machine control unit to instruct an N/C machine in how to produce a part.*

N/C tool path—the connected points in space an N/C machine will process through as instructed by the machine control unit.*

NDT—nondestructive test; an evaluation procedure using x-ray, liquid dye penetrant, ultrasonics, and so forth to examine structures to ensure that flaws, cracks, and other abnormalities are not present above design allowances.

*Note. The term N/C has generally been superceded by CNC to indicate an advanced control logic in the MCU.

Networking chart—a technique for showing relations between steps of a complex project and planning strategies for successful completion. The critical path method format is an example of a networking chart.

NOI—notification of intent; a preliminary indication of a need to purchase capital equipment.

Noise control—programs and techniques for keeping factory noise levels below allowed levels.

Normal distribution—a statistical distribution described by the mean and its three sigma standard deviations. A useful tool for quality control analysis to determine whether a manufacturing process is in control.


Objectives—broad-based generalized statements of intent.

Objectives and goals concept—a management technique for controlling a results-oriented organization. Based on the concept of focusing from broad-based statements of policy to measurable statements of intent bounded by time.

Operating plan—an integrated productivity improvement program and business strategy including an objectives and goals statement.

Operations sequencing—the short-range scheduling program of MRP II.

Operator certification—a technique of self-inspection for quality control.

Order entry system—a method of managing maintenance work.

Orders release—the medium-range scheduling program of MRP II.

Overhead—operating costs exclusive of direct labor.


PAR—plant appropriation request; a document outlining the reasons and justifications for spending funds for capital equipment.

Participatory programs—programs to enable all employees to have inputs into factory operation and problem solving.

Parts flow analysis—a technique for finding families of parts for cellular manufacturing.

Payback period—a criterion for evaluating the worth of plant appropriation requests based on dividing the gross costs of the project by the annualized savings. This yields a time period in years.

PC form—process control form, a document for recording in-process or finalized dimensions and other data; a quality control form.

Performance requirements—with reference to purchase of capital equipment, the detailed specifications stating the parameters of acceptability.

Periodic review of operations—a technique for evaluating the success or failure of operations as compared to the short-range forecast.

Physical capacity—ability to produce to design specification by the due date.

Practical capacity—capacity values downgraded for unavoidable delays.

Preliminary layout—a first attempt to place equipment in the factory to optimize product flow.

Present worth method—a financial analysis procedure using the time value of money concept to evaluate the worth of plant appropriation requests.

Preventive maintenance—work done on equipment to minimize future failures and downtime.

Principle of sameness—the basic theoretical concept of group technology. Parts are grouped on the basis of similar geometry or processing and are produced on the same equipment in a batch mode.

Principles of motion economy—an industrial engineering body of knowledge based on the physics of human body motions related to energy expenditure.

Process control—a unit within the manufacturing engineering function.

Producibility design—the process of customizing the design for the production source.

Producibility engineering—a unit within the manufacturing engineering function.

Product flow—the sequence of making a part in a factory, from raw material to finished product.

Production control system—a methodology of organizing to produce products in a factory and monitoring the results.

Productivity—in general, a measurement of output of products and goods divided by costs for design, labor, and material; a trend measurement of an organization's effectiveness.

Productivity, all other—the measurement of all cost contributors except direct labor.

Productivity chart—a chart used to plot productivity measurements.

Productivity, direct labor—the measurement of direct labor as a cost contributor.

Productivity rate measurement—a specific productivity measurement of quantity per time period.

Productivity seminar—a technique for developing an operating plan.

Profilometer—a device for measuring surface finish or smoothness.

Profit—a positive value of selling price minus cost to produce.

Project—a specific plan with measurable steps that leads to accomplishment of a goal.

QCR—quality control report; a format for recording out-of-tolerance conditions and subsequent dispositions.

Quality assurance—a marketing-oriented document verification procedure used to ensure product compliance with specifications.

Quality circles—a type of participatory program whereby ideas for improving productivity are elicited from production workers in formal meetings and thoroughly investigated for implementation by management.

Quality control—a body of knowledge dedicated to ensuring that products manufactured are in accordance with design requirements.

Quality costs measurement—a measurement of negative productivity.

Rapid emergency repair—the expeditious repair of down and limping equipment to minimize production disruptions.

Real-time measurement—the measurement and reporting of manufacturing activities as they occur.

Robot—a mechanical device capable of being programmed to move or manipulate in three-dimensional space and capable of doing work in that space imitating, in a general way, motions that can be accomplished by a human being.

ROI—return on investment; a financial calculation procedure for evaluating the worth of a plant appropriation request.

Routing—a sequence in order of manufacture of workstations used to make a product.

Sequence of operations—the chronological listing of items to be accomplished in producing a product.

Setup—the procedure of placing and adjusting a workpiece on a machine tool or equipment for the purpose of doing work.

Schedule control chart—a format for monitoring the progress of action plans.

Scientific method—the classical method of observation, development and testing of hypotheses, repeated observation, and iteration to a conclusion.

Scientific time standard—a time standard based on the principle of motion economy and other precise data related to machine and human movement.

Shop maintenance information system—a system for cataloging all events pertinent to the functional performance of equipment.

Short-range forecast—a financial forecast of expenses and credits, usually for up to 3 months.

Standard deviation—a statistical measurement of the dispersion of data from the mean, used in statistical tolerancing and quality control analysis.

Static scheduling—master scheduling of a complex nature that is very difficult to change.

Subjective savings—intangible savings associated with plant appropriation requests.

Systems countil—a temporary cross-functional organization for the implementation of CIM in a factory.

Team production—a type of participatory program whereby productivity is improved by allowing groups of workers to decide how they will divide and perform tasks to produce a finished product.

Technical capacity—ability to conceive and design products.

Time standard—a calculation of the time it should take to perform a specified task.

Throughput time—the portal-to-portal time to complete a job in the factory.

Tolerance—allowable deviations from a design dimension.

Tollgate inspection—a quality control status review conducted on semifinished or finished parts to ensure that all specifications for design and manufacturing have been complied with.

Toolroom—a special-purpose manufacturing area used to make jigs and fixtures, parts for repairs, and prototype products.

Total productivity measurement—an overall measurement of a company's productivity.

Ultrasonic test—a form of nondestructive testing utilizing ultrasound energy.

Variance control programs—activities designed to minimize the difference between planned times to accomplish tasks and actual times.
VBM—vertical boring mill, a classification of machine tools.
Vendor—a supplier of raw materials and components.

Warner diagram—a statistical diagram used in conjunction with setting tolerances, typically by comparing load characteristics with material strength characteristics.
Weekly problem sheet—raw quality data disseminated for resolution.
Work measurements—the broad concept of developing and measuring compensation systems and production effectiveness.
Work planning—a technique for establishing individual action plans, schedules, and measurements of accomplishments.
Workable design—a design that achieves the design engineer's goals while being economical to produce.
Working memory—the memoirs of an expert used to develop a data base for an expert system.
Workplace—the specific location where manufacturing is accomplished; also denotes a factory environment. A workplace can contain many workstations.
Workplace layout—a graphical description of a workplace.
Workstation—a place where contributed value is added to materials in the process of becoming a finished part (or product).

# SELECTED RELATED READINGS

Bajaria, Hans J. (ed.), *Quality Assurance: Methods, Management and Motivation,* Society of Manufacturing Engineers, Dearborn, Mich., 1981.

Barish, Norman N., and Kaplan, Seymour, *Economic Analysis for Engineering and Managerial Decision Making,* McGraw-Hill, New York, 1978.

Bolz, Roger W., *Production Processes, The Productivity Handbook,* Conquest, Winston-Salem, N.C., 1977.

Boothroyd, Geoffrey, *Fundamentals of Metal Machining and Machine Tools,* Scripta, Washington, D.C., 1975.

Brigham, Eugene F., *Fundamentals of Financial Management,* Dryden, Hinsdale, Ill., 1978.

Buffa, Elwood S., *Modern Production Management,* 4th ed., Wiley, New York, 1973.

Childs, James J., *Principles of Numerical Control,* 3d ed., Industrial Press, New York, 1982.

Cokonis, T. J. (ed.), *Computers in Engineering 1982,* vol. 1, American Society of Mechanical Engineers, New York, 1982.

DeVries, W. R. (ed.), *Computer Application in Manufacturing Systems,* American Society of Mechanical Engineers, New York, 1980.

Diehl, George M., *Machinery Acoustics,* Wiley, New York, 1973.

Feigenbaum, A. V., *Total Quality Control,* 3d ed., McGraw-Hill, New York, 1991.

Grant, Eugene L., and Grant, Ireson W., *Principles of Engineering Economy,* 5th ed., Ronald, New York, 1970.

Gross, Erwin E., Jr., and Peterson, Arnold P. G., *Handbook of Noise Measurement,* 7th ed., Gen Rad, Inc., Concord, Mass., 1974.

Grove, Andrew S., *High Output Management,* Random House, New York, 1983.

Hajek, Victor G., *Management of Engineering Projects,* McGraw-Hill, New York, 1977.

Hall, Robert W., *Zero Inventory,* Dow Jones–Irwin, Homewood, Ill., 1983.

Ham, I., Hitomi, K., and Yoshida, T., *Group Technology Applications to Production Management,* Kluwer-Nijhoff, Boston, 1985.

Harmon, Paul, and King, David, *Expert Systems,* Wiley, New York, 1985.

Hicks, Philip E., *Introduction to Industrial Engineering and Management Science,* McGraw-Hill, New York, 1977.

Hine, Charles R., *Machine Tools and Processes for Engineers,* McGraw-Hill, New York, 1971.

Kapoor, S. G., and Martinez, M. R. (eds.), *Statistics in Manufacturing,* American Society of Mechanical Engineers, New York, 1983.

Kepner, Charles H., and Tregoe, Benjamin B., *The Rational Manager,* McGraw-Hill, New York, 1965.

Kirkpatrick, Elwood G., *Quality Control for Managers and Engineers,* Wiley, New York, 1970.

Koenig, Daniel T., *Computer Integrated Manufacturing, Theory and Practice,* Hemisphere, a division of the Taylor & Francis Group, New York, 1990.

Koenig, Daniel T., Interrelationships Between Methods Engineering and Productivity Engineering, ASME Paper 81-DE-3, presented at the Design Engineering Conference and Show, April 27–30, 1981.

Koenig, Daniel T., The Pragmatic Application of CAD/CAM in Non Aerospace Job Shops, ASME Paper 82-DE-19, presented at the Design Engineering Conference and Show, March 29–April 1, 1982; reprinted as How CAD/CAM Can Improve Job Shop Productivity, in *Design Engineering,* Nov. 1982, Maclean Hunter Ltd., Toronto, Ontario.

Koenig, Daniel T., Gongaware, Terry, and Ham, Inyong, Application of Group Technology Concept for Plant Layout and Management of a Miscellaneous Parts Shop, in *Ninth North American Manufacturing Research Conference Proceedings, May 19–22, 1981,* Society of Manufacturing Engineers, Dearborn, Mich., 1981.

Kops, L. (ed.), *Toward the Factory of the Future,* American Society of Mechanical Engineers, New York, 1980.

Koren, Yoram, *Computer Control of Manufacturing Systems,* McGraw-Hill, New York, 1983.

Lazarus, Harold, and Tomeshi, Edward A., *People-Oriented Computer Systems, the Computer Crisis,* Van Nostrand Reinhold, New York, 1975.

Leibried, K. H. J., and McNair, C. J., *Benchmarking, a Tool for Continuous Improvement,* Harper Business, New York, 1992.

Len, M. C., and Martinez, Miguel R. (eds.), *Computer Integrated Manufacturing,* American Society of Mechanical Engineers, New York, 1983.

Lubben, Richard T., *Just In Time, an Aggressive Manufacturing Strategy,* McGraw-Hill, New York, 1988.

Martin, Charles C., *Project Management: How to Make It Work,* Amacom, New York, 1976.

Mayer, Raymond R., *Production and Operations Management,* 3d ed., McGraw-Hill, New York, 1975.

Moder, Joseph J., and Phillips, Cecil R., *Project Management with CPM and Pert,* 2d ed., Van Nostrand Reinhold, New York, 1970.

Moore, Franklin, G., *Manufacturing Management,* Irwin, Homewood, Ill., 1955.

Orlicky, Joseph, *Material Requirements Planning,* McGraw-Hill, New York, 1975.

Ostwald, Phillip F. (ed.), *Manufacturing Cost Estimating,* Society of Manufacturing Engineers, Dearborn, Mich., 1980.

Ott, Ellis R., *Process Quality Control, Troubleshooting and Interpretation of Data,* McGraw-Hill, New York, 1975.

Park, William R., *Cost Engineering Analysis,* Wiley, New York, 1973.

Pollack, Herman W., *Tool Design,* Reston Publishing Co., Reston, Va., 1976.

Radford, J. D., and Richardson, D. B., *Production Engineering Technology,* 2d ed., Crane, Russak, New York, 1974.

Raiffa, Howard, *The Art and Science of Negotiation,* Belknap Press of Harvard University Press, Cambridge, Mass., 1982.

Reintjes, J. Francis, *Numerical Control, Making a New Technology,* Oxford University Press, New York, 1991.

Schonberger, Richard J., *Japanese Manufacturing Techniques,* Free Press, New York, 1982.

Shamblin, James E., and Stevens, G. T., Jr., *Operations Research, A Fundamental Approach,* McGraw-Hill, New York, 1974.

Vough, Clair F., with Asbell, Bernard, *Tapping the Human Resource, A Strategy for Productivity,* Amacom, New York, 1975.

Wight, Oliver W., *Production and Inventory Management in the Computer Age,* CBI Publishing Co., Boston, Mass., 1974.

# INDEX